Chi-Keong Goh and Kay Chen Tan

Evolutionary Multi-objective Optimization in Uncertain Environments

Studies in Computational Intelligence, Volume 186

Editor-in-Chief
Prof. Janusz Kacprzyk
Systems Research Institute
Polish Academy of Sciences
ul. Newelska 6
01-447 Warsaw
Poland
E-mail: kacprzyk@ibspan.waw.pl

Further volumes of this series can be found on our homepage: springer.com

Vol. 164. Nadia Nedjah, Luiza de Macedo Mourelle, Janusz Kacprzyk, Felipe M.G. França and Alberto Ferreira de Souza (Eds.)
Intelligent Text Categorization and Clustering, 2009
ISBN 978-3-540-85643-6

Vol. 165. Djamel A. Zighed, Shusaku Tsumoto, Zbigniew W. Ras and Hakim Hacid (Eds.)
Mining Complex Data, 2009
ISBN 978-3-540-88066-0

Vol. 166. Constantinos Koutsojannis and Spiros Sirmakessis (Eds.)
Tools and Applications with Artificial Intelligence, 2009
ISBN 978-3-540-88068-4

Vol. 167. Ngoc Thanh Nguyen and Lakhmi C. Jain (Eds.)
Intelligent Agents in the Evolution of Web and Applications, 2009
ISBN 978-3-540-88070-7

Vol. 168. Andreas Tolk and Lakhmi C. Jain (Eds.)
Complex Systems in Knowledge-based Environments: Theory, Models and Applications, 2009
ISBN 978-3-540-88074-5

Vol. 169. Nadia Nedjah, Luiza de Macedo Mourelle and Janusz Kacprzyk (Eds.)
Innovative Applications in Data Mining, 2009
ISBN 978-3-540-88044-8

Vol. 170. Lakhmi C. Jain and Ngoc Thanh Nguyen (Eds.)
Knowledge Processing and Decision Making in Agent-Based Systems, 2009
ISBN 978-3-540-88048-6

Vol. 171. Chi-Keong Goh, Yew-Soon Ong and Kay Chen Tan (Eds.)
Multi-Objective Memetic Algorithms, 2009
ISBN 978-3-540-88050-9

Vol. 172. I-Hsien Ting and Hui-Ju Wu (Eds.)
Web Mining Applications in E-Commerce and E-Services, 2009
ISBN 978-3-540-88080-6

Vol. 173. Tobias Grosche
Computational Intelligence in Integrated Airline Scheduling, 2009
ISBN 978-3-540-89886-3

Vol. 174. Ajith Abraham, Rafael Falcón and Rafael Bello (Eds.)
Rough Set Theory: A True Landmark in Data Analysis, 2009
ISBN 978-3-540-89886-3

Vol. 175. Godfrey C. Onwubolu and Donald Davendra (Eds.)
Differential Evolution: A Handbook for Global Permutation-Based Combinatorial Optimization, 2009
ISBN 978-3-540-92150-9

Vol. 176. Beniamino Murgante, Giuseppe Borruso and Alessandra Lapucci (Eds.)
Geocomputation and Urban Planning, 2009
ISBN 978-3-540-89929-7

Vol. 177. Dikai Liu, Lingfeng Wang and Kay Chen Tan (Eds.)
Design and Control of Intelligent Robotic Systems, 2009
ISBN 978-3-540-89932-7

Vol. 178. Swagatam Das, Ajith Abraham and Amit Konar
Metaheuristic Clustering, 2009
ISBN 978-3-540-92172-1

Vol. 179. Mircea Gh. Negoita and Sorin Hintea
Bio-Inspired Technologies for the Hardware of Adaptive Systems, 2009
ISBN 978-3-540-76994-1

Vol. 180. Wojciech Mitkowski and Janusz Kacprzyk (Eds.)
Modelling Dynamics in Processes and Systems, 2009
ISBN 978-3-540-92202-5

Vol. 181. Georgios Miaoulis and Dimitri Plemenos (Eds.)
Intelligent Scene Modelling Information Systems, 2009
ISBN 978-3-540-92901-7

Vol. 182. Andrzej Bargiela and Witold Pedrycz (Eds.)
Human-Centric Information Processing Through Granular Modelling, 2009
ISBN 978-3-540-92915-4

Vol. 183. Marco A.C. Pacheco and Marley M.B.R. Vellasco (Eds.)
Intelligent Systems in Oil Field Development under Uncertainty, 2009
ISBN 978-3-540-92999-4

Vol. 184. Ljupco Kocarev, Zbigniew Galias and Shiguo Lian (Eds.)
Intelligent Computing Based on Chaos, 2009
ISBN 978-3-540-95971-7

Vol. 185. Anthony Brabazon and Michael O'Neill (Eds.)
Natural Computing in Computational Finance, 2009
ISBN 978-3-540-95973-1

Vol. 186. Chi-Keong Goh and Kay Chen Tan
Evolutionary Multi-objective Optimization in Uncertain Environments, 2009
ISBN 978-3-540-95975-5

Chi-Keong Goh
Kay Chen Tan

Evolutionary Multi-objective Optimization in Uncertain Environments

Issues and Algorithms

Springer

Dr. Chi-Keong Goh
Data Storage Institute,
Agency for Science, Technology and Research
5 Engineering Drive 1
Singapore 117608
Email: Goh_Chi_Keong@dsi.a-star.edu.sg

Dr. Kay Chen Tan
Department of Electrical and Computer Engineering
National University of Singapore
4 Engineering Drive 3
Singapore 117576
E-mail: eletankc@nus.edu.sg

ISBN 978-3-540-95975-5 e-ISBN 978-3-540-95976-2

DOI 10.1007/978-3-540-95976-2

Studies in Computational Intelligence ISSN 1860949X

Library of Congress Control Number: 2008942381

Typeset & Cover Design: Scientific Publishing Services Pvt. Ltd., Chennai, India.

Printed in acid-free paper

9 8 7 6 5 4 3 2 1

springer.com

Preface

Many real-world problems involve the simultaneous optimization of several competing objectives and constraints that are difficult, if not impossible, to solve without the aid of powerful optimization algorithms. What makes multi-objective optimization so challenging is that, in the presence of conflicting specifications, no one solution is optimal to all objectives and optimization algorithms must be capable of finding a number of alternative solutions representing the tradeoffs. However, multi-objectivity is just one facet of real-world applications. Most optimization problems are also characterized by various forms of uncertainties stemming from factors such as data incompleteness and uncertainties, environmental conditions uncertainties, and solutions that cannot be implemented exactly.

Evolutionary algorithms are a class of stochastic search methods that have been found to be very efficient and effective in solving sophisticated multi-objective problems where conventional optimization tools fail to work well. Evolutionary algorithms' advantage can be attributed to it's capability of sampling multiple candidate solutions simultaneously, a task that most classical multi-objective optimization techniques are found to be wanting. Much work has been done to the development of these algorithms in the past decade and it is finding increasingly application to the fields of bioinformatics, logical circuit design, control engineering and resource allocation. Interestingly, many researchers in the field of evolutionary multi-objective optimization assume that the optimization problems are deterministic, and uncertainties are rarely examined. While multi-objective evolutionary algorithms draw its inspiration from nature where uncertainty is a common phenomenon, it cannot be taken for granted that these algorithms will hence be inherently robust to uncertainties without any further investigation.

The primary motivation of this work is to provide a comprehensive treatment on the design and application of multi-objective evolutionary algorithms for multi-objective optimization in the presence of uncertainties. Chapter 1 provides the necessary background information required to appreciate this work, covering key concepts and definitions of multi-objective optimization as well as a survey of the state-of-the-arts which highlights the major design issues of multi-objective evolutionary techniques.

The rest of this work is divided into three parts, which each part considering a different form of uncertainties: 1) noisy fitness functions, 2) dynamic fitness functions, and 3) robust optimization. The first part comprises of Chapters 2-4 and addresses the issues of noisy fitness functions. Chapter 2 investigates the effect of noise on multi-objective evolutionary algorithms and Chapter 3 provides a comprehensive survey of noisy evolutionary multi-objective optimization literature and presents a comparative study between existing algorithms for noisy multi-objective optimization. As a specific instance of a noisy multi-objective problem, Chapter 4 presents a hybrid multi-objective evolutionary algorithm for the evolution of artificial neural network classifiers.

Part II is concerned with dynamic multi-objective optimization and comprises of Chapters 5 and 6. Chapter 5 provides a survey of dynamic evolutionary multi-objective optimization literature as well as a discussion on the different types of dynamic multi-objective test functions and performance indicators. Chapter 6 extends the notion of coevolution to track the Pareto front in a dynamic environment. Since problem characteristics may change with time, it is not possible to determine one best approach to problem decomposition. Therefore, this chapter introduces a new coevolutionary paradigm that incorporates both competitive and cooperative mechanisms observed in nature to facilitate the adaptation and emergence of the decomposition process with time.

The final part of this work addresses the issues of robust multi-objective optimization where the optimality of the solutions is sensitive to parameter variations. Analyzing the existing benchmarks applied in the literature reveals that the current corpus has severe limitations. Therefore, Chapter 7 presents a robust multi-objective test suite with noise-induced solution space, fitness landscape and decision space variation. In addition, the vehicle routing problem with stochastic demand (VRPSD) is presented a practical example of robust combinatorial multi-objective optimization problems. A survey of existing robust multi-objective evolutionary techniques are presented in Chapter 8 and simulations are conducted to solve the test suite suggested in Chapter 7. In Chapter 9, a hybrid MOEA is developed to optimize robust route schedules for the VRPSD problem.

Data Storage Institute,
National University of Singapore,

Chi-Keong Goh
Kay Chen Tan

Contents

Part II: Tracking Dynamic Multi-objective Landscapes

Chapter 1
Introduction

Optimization may be considered as a decision-making process to get the most out of available resources for the best attainable results. Simple examples include everyday decisions, such as the type of transport to take, which clothes to wear and what groceries to buy. For these routine tasks, the decision to be made for, say, the cheapest transport can be exceedingly clear. Consider now, the situation where we are running late for a meeting due to some unforeseen circumstances. Since the need for expedition is conflicting to the first consideration of minimizing cost, the selection of the right form of transportation is no longer as straight-forward as before and the final solution will represent a compromise between the different objectives. This type of problems, which involves the simultaneous consideration of multiple objectives, are commonly termed as multi-objective problems.

Many real-world problems naturally involve the simultaneous optimization of several competing objectives. Unfortunately, these problems are characterized by objectives that are much more complex as compared to routine tasks and the decision space is often so large that it is often difficult, if not impossible, to be solved without advanced and efficient optimization techniques. In addition, as reflected by the element of uncertainty in the example given above, the magnitude of this task is exacerbated by uncertainties such as the presence of noise and time-varying components that are inherent to real-world problems. Multi-objective optimization in the presence of uncertainties is of great importance in practice, where the slight difference in environmental conditions or implementation variations can be crucial to overall operational success or failure.

1.1 Multi-objective Optimization

Real-world optimization tasks are typically represented by its mathematical model and the specification of multi-objective criteria captures more information about the modeled problem as several problem characteristics are taken into consideration. For instance, consider the design of a system controller

C.-K. Goh and K.C. Tan: Evolutionary Multi-objective Optimization, SCI 186, pp. 1–40.
springerlink.com

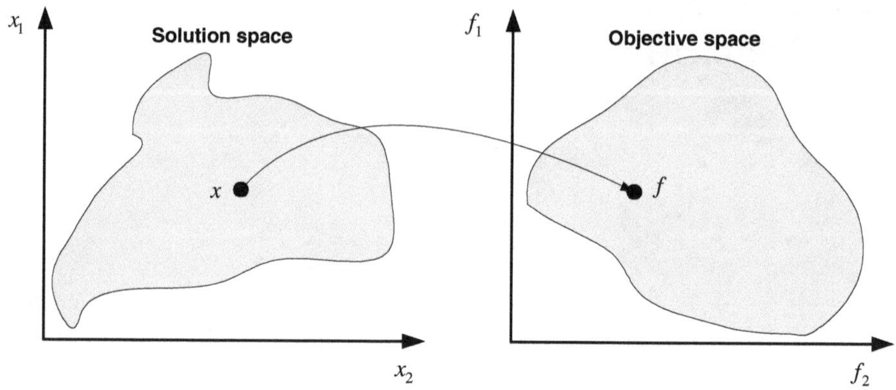

Fig. 1.1 Illustration of the mapping between the solution space and the objective space

that can be found in process plants, automated vehicles and, household appliances. Apart from obvious tradeoffs between cost and performance, the performance criteria required by some applications, such as fast response time, small overshoot, and good robustness, are also conflicting in nature [47, 83, 186, 265].

Without any loss of generality, a minimization problem is considered here and the multi-objective problem can be formally defined as

$$\min_{\mathbf{x}\in\mathbf{X}^{n_x}} \mathbf{f}(\mathbf{x}) = \{f_1(\mathbf{x}), f_2(\mathbf{x}), ..., f_M(\mathbf{x})\} \tag{1.1}$$

$$\text{s.t. } \mathbf{g}(\mathbf{x}) \geq 0, \mathbf{h}(\mathbf{x}) = 0$$

where $\mathbf{x}$ is the vector of decision variables bounded by the decision space, $\mathbf{X}^{n_x}$ and $\mathbf{f}$ is the set of objectives to be minimized. The terms "solution space" and "search space" are often used to denote the decision space and will be used interchangeably throughout this book. The functions $\mathbf{g}$ and $\mathbf{h}$ represent the sets of inequality and equality constraints that define the feasible region of the n_x-dimensional continuous or discrete feasible solution space. The relationships between the decision variables and the objectives are governed by the objective function $\mathbf{f} : \mathbf{X}^{n_x} \longmapsto \mathbf{F}^M$. Figure. 1.1 illustrates the mapping between the two spaces. Depending on the actual objective function and constraints of the particular multi-objective problem, this mapping is not unique and may be one-to-many or many-to-one.

1.1.1 *Totally Conflicting, Non-conflicting, and Partially Conflicting Multi-objective Problems*

One of the key differences between single-objective and multi-objective optimization is that multi-objective problems constitute a multi-dimensional objective space, $\mathbf{F}^M$. This leads to three possible instances of multi-objective problems, depending on whether the objectives are totally conflicting, non-conflicting, or partially conflicting. For multi-objective problems of the first category, the conflicting nature of the objectives are such that no improvement can be made without violating any constraint. This result in an interesting situation where all feasible solutions are also optimal. Therefore, totally conflicting multi-objective problems are perhaps the simplest of the three since no optimization is required. On the other extreme, a multi-objective problem is non-conflicting if the various objectives are correlated and the optimization of any arbitrary objective leads to the subsequent improvement of the other objectives. This class of multi-objective problems can be treated as a single-objective problem by optimizing the problem along an arbitrarily selected objective or by aggregating the different objectives into a scalar function. Intuitively, a single optimal solution exists for such a multi-objective problem.

More often than not, real-world problems are instantiations of the third type of multi-objective problems and this is the class of multi-objective problems that we are interested in. One serious implication is that a set of solutions representing the tradeoffs between the different objectives is now sought rather than an unique optimal solution. Consider again the example of cost vs performance of a controller. Assuming that the two objectives are indeed conflicting, this presents at least two possible extreme solutions, each representing the best achievable situation for one objective at the expense of the other. The other solutions, if any, making up this optimal set of solutions represent the varying degrees of optimality with respect to the two different objectives. Certainly, our conventional notion of optimality gets thrown out of the window and a new definition of optimality is required for multi-objective problems.

1.1.2 Pareto Dominance and Optimality

The concepts of Pareto dominance and Pareto optimality are fundamental in multi-objective optimization, with Pareto dominance forming the basis of solution quality. Unlike single-objective optimization, where there is a complete order (i.e, $f_1 \leq f_2$ or $f_1 \geq f_2$), $\mathbf{X}^{n_x}$ is partially-ordered when multiple objectives are involved. In fact, there are three possible relationships between the solutions, which are defined by Pareto dominance.

Definition 1.1: Weak Dominance:$\mathbf{f}_1 \in \mathbf{F}^M$ weakly dominates $\mathbf{f}_2 \in \mathbf{F}^M$, denoted by $\mathbf{f}_1 \preceq \mathbf{f}_2$ iff $x_{1,i} \leq x_{2,i}$ $\forall i \in \{1, 2, ..., M\}$

Definition 1.2: Strong Dominance: $\mathbf{f}_1 \in \mathbf{F}^M$ strongly dominates $\mathbf{f}_2 \in \mathbf{F}^M$, denoted by $\mathbf{f}_1 \prec \mathbf{f}_2$ iff $x_{1,i} \leq x_{2,i}$ $\forall i \in \{1, 2, ..., M\}$ and $x_{1,j} < x_{2,j}$ $\exists j \in \{1, 2, ..., M\}$

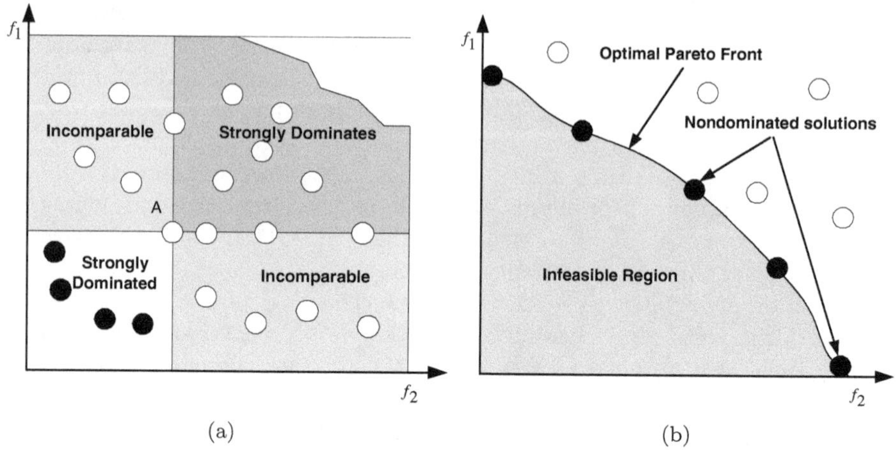

Fig. 1.2 Illustration of the (a) Pareto Dominance relationship between candidate solutions relative to solution A and (b) the relationship between the Approximation Set, PF^A and the true Pareto front, PF^*

Definition 1.3: Incomparable: $\mathbf{f_1} \in \mathbf{F}^M$ is incomparable with $\mathbf{f}_2 \in \mathbf{F}^M$, denoted by $\mathbf{f}_1 \sim \mathbf{f}_2 \ iff \ x_{1,i} > x_{2,i} \ \exists i \in \{1, 2, ..., M\}$ and $x_{1,j} < x_{2,j} \ \exists j \in \{1, 2, ..., M\}$

With solution A as our point of reference, the regions highlighted in different shades of grey in Figure. 1.2(a) illustrate the three different dominance relations. Solutions located in the dark grey region are dominated by solution A because A is better in *both* objectives. For the same reason, solutions located in the white region dominates solution A. Although A has a smaller objective value as compared to the solutions located at the boundaries between the dark and light grey regions, it only weakly dominates these solutions by virtue of the fact that they share a similar objective value along either one dimension. Solutions located in the light grey regions are incomparable to solution A because it is not possible to establish any superiority of one solution over the other: solutions in the left light grey region are better only in the second objective while solutions in the right grey region are better only in the first objective. It can be easily noted that there is a natural ordering of these relations: $\mathbf{f}_1 \prec \mathbf{f}_1 \ \Rightarrow \ \mathbf{f}_1 \preceq \mathbf{f}_1 \ \Rightarrow \ \mathbf{f}_1 \sim \mathbf{f}_2$.

With the definition of Pareto dominance, we are now in the position to consider the set of solutions desirable for multi-objective optimization.

Definition 1.4: Pareto Optimal Front: The Pareto optimal front, denoted as PF^*, is the set of non-dominated solutions with respect to the objective space such that $PF^* = \{\mathbf{f}_i^* | \nexists \mathbf{f}_j \prec \mathbf{f}_\mathbf{i}^*, \mathbf{f}_j \in \mathbf{F}^M\}$

Definition 1.5: Pareto Optimal Set: The Pareto optimal set, denoted as PS^*, is the set of solutions that are non-dominated in the objective space such that $PS^* = \{\mathbf{x}_i^* | \nexists \mathbf{F}(\mathbf{x}_j) \prec \mathbf{F}(\mathbf{x}_i^*), \mathbf{F}(\mathbf{x}_j) \in \mathbf{F}^M\}$

The set of tradeoff solutions is known as the Pareto optimal set and these solutions are also termed "non-inferior", "admissible" , or "efficient" solutions. The corresponding objective vectors of these solutions are termed "non-dominated" and each objective component of any non-dominated solution in the Pareto optimal set can only be improved by degrading at least one of its other objective components [243].

1.1.3 Multi-objective Optimization Goals

An example of the PF^* is illustrated in Figure. 1.2(b). Most often, information regarding the PF^* and its tradeoffs are either limited or not known *a priori*. It is also not easy to find a nice closed analytic expression for the tradeoff surface because real-world multi-objective problems usually have complex objective functions and constraints. Therefore, in the absence of any clear preference on the part of the decision-maker, the ultimate goal of multi-objective optimization is to discover the entire tradeoff. However, by definition, this set of objective vectors is possibly an infinite set as in the case of numerical optimization and it is simply not achievable.

On a more practical note, the presence of too many alternatives could very well overwhelm the decision-making capabilities of the decision-maker. In this light, it would be more practical to settle for the discovery of as many non-dominated solutions as possible within our limited computational resources. More precisely, we are interested in finding a good approximation of the PF^* and this approximate set, PF^A, should satisfy the following optimization goals.

- Minimize the distance between the PF^A and PF^*.
- Obtain a good distribution of generated solutions along the PF^A.
- Maximize the spread of the discovered solutions.

An example of such an approximation is illustrated by the set of non-dominated solutions denoted by the filled circles residing along the PF^* in Figure. 1.2(b). While the first optimization goal of convergence is the primary consideration of all optimization problems, the second and third optimization goals of maximizing diversity are entirely unique to multi-objective optimization. The rationale of finding a diverse and uniformly distributed PF^A is to provide the decision maker with sufficient information about the tradeoffs between the different solutions before the final decision is made. It should also be noted that the optimization goals of convergence and diversity are somewhat conflicting in nature, which explains why multi-objective optimization is much more difficult than single-objective optimization.

1.2 Evolutionary Multi-objective Optimization

Traditional operational research approaches to multi-objective optimization typically entail the transformation of the original problem into a single-objective problem and employ point-by-point algorithms, such as branch-and-bound, to iteratively obtain a better solution. Such approaches have several limitations, including the requirement of the multi-objective problem to be well-behaved, i.e. differentiability or satisfying the Kuhn-Tucker conditions, sensitivity to the shape of the Pareto-front, and the generation of only one solution for each simulation run. On the other hand, metaheuristical approaches that are inspired by biological or physics phenomena, such as evolutionary algorithms and simulated annealing, have been gaining increasing acceptance as a much more flexible and effective alternative to complex optimization problems in the recent years. This is certainly a stark contrast to just two decades ago, as Reeves remarked in [220] that an eminent person in operational research circles suggested that using a heuristic was an admission of defeat!

Among these metaheuristics, MOEA is one of the more popular stochastic search methodologies to solve multi-objective problems. Emulating the DarwinianWallace principle of "survival-of-the-fittest" in natural selection and adaptation, MOEAs have the distinct advantage of being able to sample multiple solutions simultaneously. Such a feature provides the MOEA with a global perspective of the multi-objective problem as well as the capability to find a set of Pareto-optimal solutions in a single run. Applying genetic operators, such as the selection process and crossover operator, allows the MOEA to intelligently sieve through the large amount of information embedded within each individual representing a candidate solution and exchange information between them to increase the overall quality of the individuals in the population. In this section, state-of-the-art MOEAs, multi-objective test problems, and performance indicators that are used for algorithmic performance evaluation in this book are discussed.

1.2.1 MOEA Framework

Many different evolutionary techniques for multi-objective optimization have been proposed since the pioneering effort of Schaffer in [231], with the aim of fulfilling the three optimization goals described previously. Most of these MOEAs are largely based on the computational models of genetic algorithms (GAs) [121], evolutionary programming (EP) [80], and evolutionary strategies (ES) [219]. Interestingly, ES is the only paradigm developed for the purpose of optimization; GAs are designed as a general adaptive system, while EP are developed as a learning process to create artificial intelligence.

Recent years have also seen the emergence of other biologically inspired models, such as particle swarm optimization (PSO), differential evolution (DE), cultural algorithms (CA), and artificial immune systems (AIS) for multi-objective optimization. While all these algorithms are different in methodology, particularly in the generation of new candidate solutions, the distinctions between them have become increasingly vague as researchers sought to exploit the advantages offered by the different algorithms in a

Fig. 1.3 Framework of MOEA

```
P←Population Initialization
A←Create External population or Archive
While (Stopping criteria not satisfied)
    P←Eval(P, A)
    P←Diversity(P, A)
    A←Update(P, A)
    S←Selection(P, A)
    P←Variation(S)
End While
```

common platform. Moreover, multi-objective optimization requires researchers to address many similar issues that are unique to multi-objective problems, regardless of the computational model applied. Therefore, no distinction will be made between the different evolutionary computation models and all these techniques developed for multi-objective optimization are referred to as MOEA.

One distinct feature that characterizes state-of-the-art MOEAs such as non-dominated sorting genetic algorithm II (NSGAII) [61], Pareto archived evolution strategy (PAES) [166], Pareto envelope based selection algorithm (PESA) [45], incrementing multi-objective evolutionary algorithm (IMOEA) [258] and strength Pareto evolutionary algorithm 2 (SPEA2) [298] from early research efforts is the incorporation of elitism. Elitism involves two closely related process, 1) the preservation of good solutions and 2) the reinsertion of these solutions into the evolving population. While the general motivations may be similar, these algorithms can be distinguished by the way in which the mechanisms of elitism and diversity preservation are implemented.

The general MOEA framework can be represented in the pseudocode shown in Fig. 1.3 and it can be shown that most MOEAs fit into this framework. There seem to be many similarities between single-objective evolutionary algorithms (SOEAs) and MOEAs with both techniques involving an iterative adaptation of a set of solutions until a pre-specified optimization goal/stopping criterion is met. What sets these two techniques apart is the manner in which solution assessment and elitism are performed. This is actually a consequence of the three optimization goals described in Section 1.1.3. In particular, solution assessment must exert a pressure to drive the solutions toward the global trade-offs as well as to diversify the individuals uniformly along the discovered PF^A. The archive updating and selection process must also take diversity into consideration to encourage and maintain a diverse solution set.

The optimization process starts with the initialization of the population. This is followed by evaluation (Eval) and density asssessment (Diversity) of candidate solutions. After which, good solutions are updated into an external population or archive (Update). MOEAs perform the archiving process differently, some of which maintains a fixed sized archive while others store only non-dominated solutions. Nonetheless, in most cases, a truncation process will be conducted based on some density assessment to restrict the number of archived solutions. Both NSGAII and SPEA2 maintains a fixed sized archive which includes both dominated and non-dominated solutions while PAES and PESA stores only non-dominated solutions. For the truncation process, PAES and PESA employ a hyper-grid measure while SPEA, NSGAII and IMOEA employ Euclidean-based measures.

The selection process typically involves the set of non-dominated solutions updated in the previous stage. For NSGAII, SPEA2 and PESA, tournament selection is conducted directly on the archive. In [252], the archive of non-dominated solutions and evolving population is combined before tournament selection is performed. Bosman and Thierens [24] noted that diversity usually serves only as a secondary selection criteria to the optimization goal of convergence. As a specific instance, NSGAII applies the crowded comparison operator only to break any tie in rank occurred during the tournament selection. On the other hand, the selection process in PESA is based on the degree of crowding or the squeeze factor only.

After the selection process, variation operators are applied to explore and exploit the selected individuals to generate a new population of solutions. Different methods of generating individuals can be found in the literature. Uniform crossover and bit-flip mutation have been used for NSGAII and SPEA2. In AIS-inspired MOEAs [42, 189], cloning and hypermutation are applied while EDA-based MOEAs [25, 202] enforce sampling from leant probabilistic models. Variation operators associated with the various paradigms have been applied across the different computational model resulting in very similar implementations, a point mentioned earlier. Some recent examples include the introduction of recombination into the AIS-inspired MOEAs in [143, 248] and the hybridization of clonal selection and hypermutation with PSO-inspired MOEAs [292].

1.2.2 Basic MOEA Components

The framework presented in the previous section serves to highlight the primary components of the MOEA, elements without which the algorithm is unable to fulfill its basic function of finding PF*satisfactorily. More elaborate frameworks with different concerns exist in the literature. For instance, Bosman and Thierens [24] presented a framework that considers how MOEAs can be constructed to control the emphasis on the exploration and exploitation of diversity or proximity. In another work, Laumanns *et al* [176] focused on the design and incorporation of elitism into MOEAs.

Fitness Assignment

As illustrated in Figure 1.4, solution assessment in MOEA should be designed in such a way that a pressure $\overleftarrow{P}_n$ is exerted to promote the solutions in a direction normal to the tradeoffs region and at the same time, another

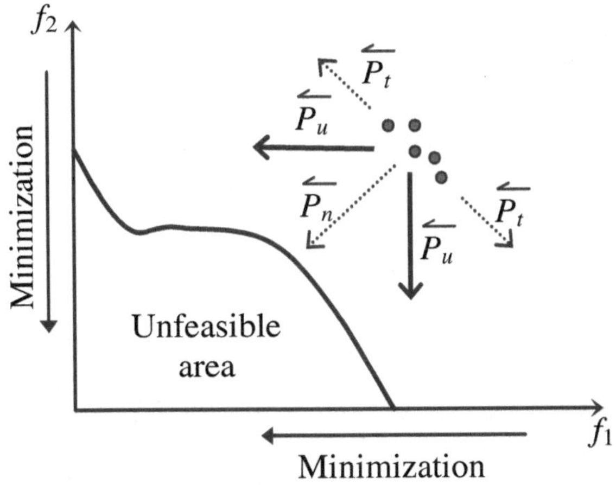

Fig. 1.4 Illustration of selection pressure required to drive evolved solutions towards PF*

pressure $\overleftarrow{P}_t$ to promote the solutions in a direction tangentially to that region. These two orthogonal pressures result in the unified pressure $\overleftarrow{P}_u$ to direct the evolutionary search in the multi-objective optimization context. Based on the literature, it is possible to identify three different classes of fitness assigment: 1) Pareto-based assignment, 2) aggregation-based assignment and 3) indicator based assignment.

Pareto-based Fitness Assignment: Pareto-based MOEAs have emerged as the most popular approach [257] since Fonseca and Fleming [84] put into practise the notion of dominance suggested in [100]. On its own, Pareto dominance is unable to induce $\overleftarrow{P}_t$ and the solutions will converge to arbitrary portions of the PF^A, instead of covering the whole surface. Thus Pareto-based fitness assignments are usually applied in conjuction with density measures, which can be incorporated in two ways. The first approach which is commonly known as fitness sharing aggregates the Pareto-based fitness and some form of density measure to form a scalar fitness. In this case, the aggregation function must be carefully constructed to maintain a balance between $\overleftarrow{P}_t$ and $\overleftarrow{P}_n$. This approach has been applied by sucessfully in works such as [82, 188, 298]. The second approach adopts a two stage process where comparison between solutions is conducted based on Pareto-fitness before density measure is used. Note that this indirectly assigns higher priority levels to proximity. Another interesting consequence is that $\overleftarrow{P}_n$ will be higher in the initial stages of the evolution. On the other hand, when the solutions begin to converge to the PF^*, the influence of $\overleftarrow{P}_t$ becomes more dominant as most of the solutions are equally fit. This approach is used in algorithms such as PAES, NSGAII and IMOEA.

However, Fonseca and Fleming [84] highlighted that Pareto-based assignment may not be able to produce sufficient selection pressure in high-dimensional objective and it also has been shown empirically that performances of Pareto-based MOEAs do not scale well with respect to the number of objectives in [125, 153]. To understand this phenomenon, let us consider a M-objective problem where M>>2. Under the definition of Pareto dominance, as long as a solution has one objective value that is better than another solution, never mind the degree of superiority, it is still considered to be non-dominated when if it is grossly inferior in the other M-1 objectives. Intuitively, the number of non-dominated solutions in the evolving population grows with the number of objectives resulting in the lost of selection pressure.

To this end, some researchers have sought to relax the definition of Pareto-optimality. Ikeda *et al* [132] proposed the α-dominance scheme which considers the contribution of all the weighted difference between the respective objectives of any two solutions under comparison to preventing the above situation from occuring. Laumanns *et al* [173] suggested an ϵ-dominance scheme which has the interesting property of ensuring convergence and diversity. In this scheme, an individual dominates another individual only if it offers an improvement in all aspects of the problem by a pre-defined factor of ϵ. A significant difference between α-dominance and ϵ-dominance is that a solution that strongly dominates another solution also α-dominates that solution while this relationship is not always valid for the latter scheme. Another interesting alternative in the form of fuzzy Pareto-optimality is presented by Farina and Amato [74] to take into account the number and size of improved objective values.

Aggregation-based Fitness Assignment: Aggregation of the objectives into a single scalar is perhaps the simplest approach to generate PF^A. Interestingly, unlike the Pareto-based approach, aggregation-based fitness induces $\overleftarrow{P}_u$ directly. However, aggregation is usually associated with several limitations such as its sensitivity to PF* shape and the lack of control of the direction of $\overleftarrow{P}_u$ resulting in the contrasting lack of interest paid by EMOO researcher as compared to Pareto-based techniques. Ironically, the failure of Pareto-based MOEAs in high-dimensional objective space may well turn the attention towards the use of aggregation-based fitness assignment in MOEAs.

The multi-objective genetic local search (MOGLS) [138, 139, 140, 141] is a well-known instance of aggregation-based MOEA that has been demonstrated to be capable of evolving uniformly distributed and diverse PF^A. Different search trajectories are generated during the evolution through the use of random weights in [138, 139] while Jaszkiewicz [140, 141] applied different instances of predefined utility functions. Jin *et al* investigated two different approaches in [147]. In the first method, each individual is assigned its own weights that will be regenerated every generation while the second method periodically change the weights along the evolutionary process. The most significant result of this work is that both methods are able to converge on concave PF* empirically, which is against conventional wisdom on the limitations of aggregation. According to [148], this is because the aggregation-based MOEA will transverse the entire Pareto front regardless of PF* shape and the archive plays a significant role in retaining the non-dominated solutions found.

Instead of performing the aggregation of objective values, Hughes [126, 127] suggested the aggregation of individual performance with respect to a set of predetermined target vectors. In this approach, individuals are ranked according to their relative performance in an ascending order for each target. These ranks are then sorted and stored in a matrix such that is may be used to rank the population, with the most fit being the solution that achieves the best scores most often. It has been shown to outperform non-dominated sorting applied in NSGAII in high-dimensional multi-objective problems [126].

At this point of time, it seems that Pareto-based fitness are more effective in low-dimensional multi-objective problems while aggregation-based fitness has an edge with increasing number of objectives. Naturally, some researchers have attempted to marry both methods together. For example, Turkcan and Akturk [269] proposed an hybrid multi-objective fitness assigment method which assigns a non-dominated rank that is normalized by niche count and an aggregation of weighted objective values. On the other hand, Pareto-based fitness and aggregation-based fitness are used independently in various stages of the evolutionary process in [75, 198].

Indicator-based Fitness Assignment: Since the performance of MOEAs are assessed and compared using performance indicators, it is therefore desirable to maximize algorithmic performance according to these measures. Fleischer [79] is probably the first to suggest that multi-objective performance indicators can be used to guide the evolutionary process and recasted the multi-objective problem as a single-objective problem that maximizes the hypervolume metric of the discovered PF^A. In [69], hypervolume is used as the selection criteria to remove the worst individuals in the worst-ranked PF* after non-dominated sorting to maintain a fixed population size. Zitzler and Kunzli [296] took a step further and applied binary indicators directly to determine the relative fitness of the evolving individuals. The utility of indicator-based fitness has also been investigated in [16]. While no clear guidelines on the choice of metrics exist at this time, it is clear that the selected measure must be able to provide an indication of solution quality in the aspects of diversity and convergence in order to exert the $\overleftarrow{P}_u$.

Diversity Preservation

Density Assessment: A basic component of diversity preservation strategies is density assessment. Density assessment evaluates the density at different sub-divisions in a feature space, which may be in the parameter or objective domain, before any action is taken to influence the survival rate of the solution points [154]. Depending on the manner in which solution density is measured, the different density assessment techniques can be broadly categorized under 1) Distance-based, 2) Grid-based, and 3) Distribution-based. One of the basic issues to be examined is whether density assessment should be computed in the decision space or objective space. Horn and Nafpliotis [123] stated that density assessment should be conducted in the feature space where the decision-maker is most concerned about its distribution. Since we are interested in obtaining a well-distributed and diverse PF^A, most works reported in the EMOO literature applied density assessment in the objective space. There are also researchers who performed density assessment in the decision space [243] as well as in both objective and decision spaces simultaneously [64, 119, 223]. In fact, there may be little correlation between diversity in the two feature spaces. Tan *et al* [252] pointed out that it essentially depends on what is desired for the underlying problem.

Distance-based assessments is based on the relative distance between individuals in the feature space. Examples include niche sharing [84, 123, 243], crowding [61], clustering [50, 300], lateral interference [155], Pareto potential regions [113] and *k-th* nearest neighbor [2, 298]. Niche sharing or simply niching is by far the most popular distance-based approach.

Niching is originally proposed by Goldberg [101] to promote population distribution to prevent genetic drift as well as to search for possible multiple peaks in single-objective optimization. The main limitation of this method is that its performance is sensitive to the setting of niche radius. Fonseca and Fleming [84] gave some bounding guidelines of appropriate niche radius values for multi-objective problems when the number of individuals in the population and the minimum/maximum values in each objective dimension are given. However, such information are often not known *a prior* in many real-world problems. Tan *et al* [255] presented a dynamic sharing scheme where the niche radius is computed online based on the evolved tradeoffs.

The *k-th* nearest neighbor is another approach which requires the specification of an external parameter. Zitzler *et al* [298] adopted k as the square root of the total population size based in some rule-of-the-thumb used in statistical density estimation. In [2, 228], average Euclidean distance between the two nearest solutions is used as the measure of density. Like niching, this approach is sensitive to the setting of the external parameter, which in this case is k. The *k-th* nearest neighbor can also be misleading in situations where all the nearest neighbors are located in a similar region of the feature space. In certain sense, the nearest neighbor is similar to the method of crowding. However, crowding do not have such bias since it is based on the average distance of the two points on either side of current point along each dimension of the feature space.

Crowding, clustering and lateral interference are instances of distance-based assessments that are not influenced any external parameters. Nonetheless, distance-based assessment schemes are susceptible to scaling issues and their effectiveness are limited by the presence of non-commensurable objectives.

Grid-based assessment is probably the most popular approach after niching and it can be found in [42, 43, 45, 166, 188]. In this approach, the feature space is divided into a predetermined number of cells along each dimension and distribution density within a particular cell has direct relation to the number of individuals residing within that cell location. Contrary to distance-based assessments methods which includes methods that are very different, both conceptually and in implementation, the main difference among the various implementation of this approach, if any, lies in the way in which the cells and individuals are located and referenced. For example, the cell location of an individual in PAES and PESA is found using recursive subdivision. However, in [188], the location of each cell center is stored and the cell associated with an individual is found by searching for the nearest cell center. The primary advantage of grid-based assessment is that it is not affected by the presence of non-commensurable objectives. However, this technique depends heavily on the number of cells in the feature space containing the population and it works well if knowledge of the geometry of the PF^* is known *a priori*. Furthermore, it's computational requirements are considerably more than distance-based assessments.

Distribution-based assessment is rather different from distance-based and grid-based methods because distribution density is based on the probability density of the individuals. The probability density is used directly in [25] to identify least crowded regions of the PF^A. It has also been used to compute the entropy as a means to quantify the information contributed by each individual to the PF^A in [49, 158, 248]. Like grid-based methods, it is not affected by non-commensurable objectives. The tradeoff is that it can be computationally more intensive compared to distance-based measures because it involves probability density estimation functions such as Parzen window in estimating the distribution density of the individuals. On the other hand, the computational effort is a linear function of population size which is advantageous for large population sizes. While some distribution-based methods require external parameter setting such as the window width in Parzen window estimation [248], there exist an abundance of guidelines in the literature.

Finally, an empirical investigation is conducted in [154] on the effectiveness of the various density assessment methods in dealing with convex, non-convex and line distributions. In general, the study shows that all techniques under investigation are able to improve distribution quality in the sense of uniformity. But the findings also suggest that it is not possible to ascertain which method is better for which type of problem distribution because of the interactions between density assessment and genetic selection.

Encouraging Density Growth: Apart from inducing appropriate $\overleftarrow{P}_t$ and $\overleftarrow{P}_u$ to generate new and diverse solutions, other means of encouraging diversity growth can also be found in the literature. For instance, in [266], Toffolo and Benini applied diversity as an objective to be optimized. Specifically, the multi-objective problem is transformed into a two-objective problem with genetic diversity as one of the objectives and the other objective being the ranks with respect to the objectives of the original multi-objective problem.

Mating restriction is another alternative approach and it is extended from SOEA where it is originally intended to promote diversity in the population. Mating restriction has been applied in [84, 112, 137] and it works by preventing similar Parents from participating in the recombination process together in order to avoid the formation of lethal individuals. However, contrary results on the effectiveness of mating restriction in promoting diversity has been reported in [136]. In particular, Ishibuchi and Shibata [136] noted that mating restriction improves convergence at the expense of solution set diversity.

Diversity can also be encouraged through the simultaneous evolution of multiple isolation subpopulations. In [60, 198, 230], each subpopulation is guided towards a particular region of the evolved PF*. Okuda *et al* [203] assigned one subpopulation for each objective and used an additional subpopulation as a normal MOEA solving for the multi-objective problem. The best individuals from the SOEA subpopulations are migrated to the MOEA subpopulation.

Elitism

The use of the elitist strategy is conceptualized by De Jong in [62] to preserve the best individuals found to prevent the lost of good individuals due to the stochastic nature of the evolutionary process in SOEA. When appropriate individuals are reinserted or retained in the evolving population, elitism can improve convergence greatly, although it maybe achieved at the risk of premature convergence. Zitzler [301] is probably the first to introduce elitism into MOEAs, sparking off the design trend of a new generation of MOEAs [41]. Elitism can be considered as an indispenable component of MOEA, having being shown to be a theoretical necessity for MOEA convergence [162, 225, 226].

Archiving: The first issue to be considered in the incorporation of elitism is the storage or archiving of elitist solutions. Archiving usually involves an external population or archive as the repository and this process is much more complex than in SOEAs since we are now contenting with a set of Pareto-optimal solutions instead of a single solution. However, the PF* is an infinite set which raises the natural question of *what should be maintained?*. Without any restriction on the archive size, the number of non-dominated solutions can grow exceedingly large. Therefore, in the face of limited computing and memory resources in implementation, it is sometimes unwise to store all the non-dominated or elitist solutions found.

Most works enforce a bounded set of elitist solutions which requires a truncation process when the size of the elitist solutions exceeds a predetermined bound. This leads to the interesting question of *which* solution should be kept? Some works [61, 249, 298] maintain a fixed-size archive which updates dominated solutions as long as space is available, while others store strictly non-dominated solutions only [45, 166, 251, 252]. In either case, it is only natural to truncate the archive based on some form of density assessment discussed earlier when the number of elitist solutions exceeds the upper bound. However, other measures, such as hypervolume [161] and relaxed forms of Pareto dominance, have been applied as well [63, 206].

For bounded archiving, two implementations of truncation can be found in the literature, i.e. batch and recurrence mode. The truncation criteria will be based on the density assessment process described earlier. In the batch mode, all solutions in the archive will undergo density assessment and the worst individuals are removed in a batch. On the other hand, in the recurrence mode, an iterative process of density assessment and truncation is repeated to the least promising solution from the archive until the desired size is achieved. While the recurrence-mode of truncation has higher capability to avoid the extinction of local individuals, which somehow leads to the discontinuity of the discovered Pareto front, compared to the batch-mode truncation, the recurrence-mode truncation often requires more computational effort.

The restriction on the number of archive solutions leads to two phenomena [76] which have a detrimental effect on the search process. The first is the shrinking PF^A phenomenon which results from the removal of extremal solutions and the subsequent failure to rediscover them. In the second phenomenon, non-dominated solutions in the archive are replaced by least crowded individuals. In the subsequent generations, new individuals that would have been dominated by the removed solutions are updated into the archive only to be replaced by solutions dominating them. Repeated cycles of this process is known as the oscillating PF^A. The alternative and simplest approach is, of course, to store all the non-dominated solutions found [72, 77, 199, 209]. One potential problem is the computational cost involved with the pairwise comparison between a new individual and each of the archived solutions. To this end, more efficient data structures have been proposed in [76].

Reinsertion: The next issue to be considered is the introduction of elitist solutions into the evolving population. Empirical investigations are also conducted in [175, 209] and the results demonstrate the advantages of elitism in improving convergence.

One problem faced is the “exploration-exploitation” dilemma; a higher degree of exploitation attained through the reintroduction of elitist solutions leads to the lost of diversity, while too much exploration leads to slow convergence rates. The consequence of the lack of necessary diversity to fuel the evolutionary process is a PF^A that fails to span the entire PF^* uniformly and, in the worst case, premature convergence to local optimal solutions.

To this end, elitist schemes that sought to balance the tradeoffs between exploration and exploitation have been proposed recently. Bosman and Thierens [24] highlighted the importance of improving diversity through elitism and presented a general framework for MOEAs which allows designers to control the balance between diversity and proximity exploration. Designing an elitist scheme along the same lines, Tan *et al* [252] proposed an enhanced

Table 1.1 Definition of ZDT Test Functions

Problem	Definition
ZDT1	$f_1(x_1) = x_1$, $\frac{g(x_2,...x_m)=1+9\sum_{i=2}^{m}(x_i}{(m-1))}$, $h(f_1, g) = 1 - \sqrt{\frac{f_1}{g}}$ where $m = 30$, $x_i \in [0, 1]$
ZDT2	$f_1(x_1) = x_1$, $\frac{g(x_2,...x_m)=1+9\sum_{i=2}^{m}(x_i}{(m-1))}$, $h(f_1, g) = 1 - \frac{f_1}{g}^2$ where $m = 30$, $x_i \in [0, 1]$
ZDT3	$f_1(x_1) = x_1$, $\frac{g(x_2,...x_m)=1+9\sum_{i=2}^{m}(x_i}{(m-1))}$, $h(f_1, g) = 1 - \sqrt{\frac{f_1}{g}} - \frac{f_1}{g}\sin(10\pi f_1)$ where $m = 30$, $x_i \in [0, 1]$
ZDT4	$f_1(x_1) = x_1$, $g(x_2, ...x_m) = 1 + 10(m-1) + \sum_{i=2}^{m}(x_i^2 - 10\cos(4\pi x_i))$, $h(f_1, g) = 1 - \sqrt{\frac{f_1}{g}}$ where $m = 10$, $x_1 \in [0, 1]$, $-1 \leq x_i < 1$, $\forall i = 2, ..., 10$
ZDT6	$f_1(x_1) = 1 - exp(-4x_1) \cdot \sin^6(6\pi x_1)$, $g(x_2, ...x_m) = 1 + 9 \cdot (\frac{\sum_{i=2}^{m} x_i}{m-1})^{0.25}$, $h(f_1, g) = 1 - (\frac{f_1}{g})^2$ where $m = 10$, $x_i \in [0, 1]$

exploration strategy in which the ratio of solutions selected based on ranking and diversity is adapted based on an online performance measure. Solutions selected on the basis of rank are subjected to normal genetic operators, while those selected based on niche count undergo local search to improve solution distribution. In [57], controlled elitism is explored in NSGAII where the number of individuals from each non-dominated front is fixed by a user-defined parameter. Furthermore, each front is allowed to have an exponentially reducing number of solutions.

1.2.3 Benchmark Problems

Benchmark problems are used to reveal the capabilities, important characteristics, and possible pitfalls of the algorithm under validation. In the context of multi-objective optimization, these test functions must pose sufficient difficulties to impede MOEAs' search for Pareto optimal solutions. Deb [54] has identified several characteristics that may challenge MOEAs' ability to converge and maintain population diversity. Multi-modality is one of the characteristics that hinders convergence in MOEA. Multi-modality essentially refers to the presence of multiple local Pareto fronts. Convexity, discontinuity, and non-uniformity of the Pareto front may prevent the MOEA from finding a diverse set of solutions.

The problems of ZDT1, ZDT2, ZDT3, ZDT4, ZDT6, TLK, DTLZ2, DTLZ3, FON, and KUR are selected to validate the effectiveness of multi-objective optimization techniques in converging and maintaining a diverse Pareto solution set in this book. This set of test problems is characterized by the different features mentioned above and should be a good test suite for a fair comparison of different multi-objective algorithms. Many researchers, such as [45, 61, 258, 275, 299], have used these problems in the validation of their algorithms under noiseless conditions. The optimal Pareto fronts for the dual-objective problems are shown in Figure 2.

The test problems of ZDT1 through ZDT6 are constructed by Zitzler *et al* [299] based on the guideline described by Deb [54]. Formally, the ZDT test problems have the following functional structure.

$$\begin{aligned} \min f_1(\mathbf{x_{d1}}) &= x_1 \\ \min f_2(\mathbf{x_{d2}}) &= g(\mathbf{x_{d2}}) \cdot h(f_1, g) \end{aligned} \qquad (1.2)$$

where $\mathbf{x_{d1}}, \mathbf{x_{d2}} \in \mathbf{x}$, and the g and h functions control the problem difficulty and the shape of the Pareto front, respectively. By having independent functions relating to convergence and diversity, this framework facilitates the incorporation of various problem features to construct different test problems. Table 1.1 summaries the definition and features of the various ZDT test functions.

DTLZ2 and DTLZ3 belong to the DTLZ test suite proposed by Deb *et al* in [61], which is different from most existing multi-objective test problems in the sense that these test problems are scalable in the number of objectives. In the light of recent studies [125, 153] reporting on MOEA' apparent inability to scale up its performance with high dimensional space, both DTLZ2 and DTLZ3 will undoubtably be useful in the investigation of MOEA capability to handle high dimensional objective spaces. DTLZ3 is also characterized by the presence of multiple local fronts. The definitions of DTLZ2 and DTLZ3 are given below,

$$\begin{aligned} \min f_1(\mathbf{x}) &= \big(1 + g(\mathbf{x}_M)\big) \cdot \cos(0.5\pi x_1) \cdots \cos(0.5\pi x_{M-1}) \\ \min f_2(\mathbf{x}) &= \big(1 + g(\mathbf{x}_M)\big) \cdot \cos(0.5\pi x_1) \cdots \sin(0.5\pi x_{M-1}) \\ &\vdots \\ \min f_M(\mathbf{x}) &= \big(1 + g(\mathbf{x}_M)\big) \cdot \sin(0.5\pi x_1) \\ \min g(\mathbf{x}_M) &= \textstyle\sum_{x_i \in \mathbf{x}_M} (x_i - 0.5)^2 \end{aligned} \qquad (1.3)$$

where $M = 5$, $\mathbf{x}_M = \{x_M, ..., x_{M+9}\}$, $x_i \in [0, 1]$

$$\begin{aligned}
\min f_1(\mathbf{x}) &= \big(1+g(\mathbf{x}_M)\big)\cdot\cos(0.5\pi x_1)\cdots\cos(0.5\pi x_{M-1})\\
\min f_2(\mathbf{x}) &= \big(1+g(\mathbf{x}_M)\big)\cdot\cos(0.5\pi x_1)\cdots\sin(0.5\pi x_{M-1})\\
&\vdots\\
\min f_M(\mathbf{x}) &= \big(1+g(\mathbf{x}_M)\big)\cdot\sin(0.5\pi x_1)\\
\min g(\mathbf{x}_M) &= 100\Big\{|\mathbf{x}_M + \textstyle\sum_{x_i\in\mathbf{x}_M}(x_i-0.5)^2-\cos\big(20\pi(x_i-0.5)\big)\Big\}
\end{aligned}\tag{1.4}$$

where $M=5$, $\mathbf{x}_M=\{x_M,...,x_{M+9}\}$, $x_i\in[0,1]$

FON [82] is a two-objective minimization test problem that has been widely used in the literature. Besides having a non-convex Pareto front, there are high interactions between decision variables and this problem has a large and non-linear tradeoff curve that is suitable for challenging an algorithm's ability to find and maintain the entire Pareto front uniformly.

$$\begin{aligned}
f_1(x_1,...,x_8) &= 1-\exp\big[-\textstyle\sum_{i=1}^{8}(x_i-\frac{1}{\sqrt{8}})^2\big],\\
f_2(x_1,...,x_8) &= 1+\exp\big[-\textstyle\sum_{i=1}^{8}(x_i-\frac{1}{\sqrt{8}})^2\big],
\end{aligned}\tag{1.5}$$

where $-2\le x_i<2$, $\forall i=1,2,...,8$

KUR [165] is characterized by an optimal Pareto front that is non-convex and disconnected, i.e. it contains three distinct disconnected regions on the final tradeoff. The decision variables correspond to the global tradeoff for KUR are difficult to be discovered since they are disconnected in the decision variable space. Like FON, there are high interactions between the decision variables which will pose problems to the MOEAs.

$$\begin{aligned}
f_1(x_1,x_2) &= \textstyle\sum_{i=1}^{2}\big[-10\exp(-0.2\sqrt{x_i^2+x_{i+1}^2})\big],\\
f_2(x_1,x_3) &= \textstyle\sum_{i=1}^{3}\big[|x_i|^{0.8}+5\cdot\sin(x_i^3)\big],
\end{aligned}\tag{1.6}$$

where $x_i\in[-5,5]$.

1.2.4 Performance Metrics

Performance analysis of different MOEAs essentially boils down to the evaluation of the approximate Pareto front obtained within some computational budget. Then performance metrics or indicators play an important role as functions that return a scalar quantity, reflecting the quality of the scrutinized solution set with respect to some measure. In single-objective optimization, this quality comes in the form of the objective function. For multi-objective optimization, however, quality can be defined in a variety of ways, for example, the uniformity of solutions, the dominance relationship between two solution sets, and the closeness to the Pareto-optimal front.

There have been increasing concerns on the choice of performance metrics. To this end, Knowles and Corne [163] and Zitzler *et al* [297] have discussed at length, the suitability and limitations of various performance metrics. Comparative studies performed by researchers, such as Jaszkiewicz [141], Deb *et al* [61], Tan *et al* [252], Veldhuizen and Lamont [275], made use of a suite of unary performance metrics pertinent to the optimization goals of proximity, diversity, and distribution. The metrics used in this book are described below. Appropriate performance indicators for measuring uncertainties will be discussed in the relevant chapters.

Proximity Indicator: The metric of generational distance (GD) gives a good indication of the gap between the PF* and the evolved PF^A. Mathematically, the metric is a function of individual distance given as,

$$\mathrm{GD} = \frac{1}{n_{PF}} \cdot \left(n_{PF} \sum_{i=1}^{n_{PF}} d_i^2 \right)^{\frac{1}{2}} \tag{1.7}$$

where $n_{PF} = |\mathrm{PF}^A|$, d_i is the Euclidean distance (in objective space) between the i-th member of PF^A and the nearest member of PF^*. Intuitively, a low value of GD is desirable, which reflects a small deviation between the evolved and the true Pareto front. However, this metric gives no indication of diversity achieved by the various algorithms. In order to evaluate the true ability of the algorithm, GD has to be complemented by diversity indicators.

Diversity Indicator: One of the primary concerns regarding the use of unary diversity indicator is that a good measure of diversity is meaningless if the approximate Pareto front is far away from the ideal tradeoffs. Taking into account these concerns, we propose a simple modification of maximum spread (MS) to measure how well the true Pareto front is covered by the discovered Pareto front

$$\mathrm{MS'} = \left\{ \frac{1}{M} \sum_{i=1}^{M} \left(\frac{\min[\overline{f}_i^A, \overline{f}_i^*] - \max[\underline{f}_i^A, \underline{f}_i^*]}{\overline{f}_i^* - \underline{f}_i^*} \right)^2 \right\}^{\frac{1}{2}} \tag{1.8}$$

where $\overline{f}_i^A$ and $\underline{f}_i^A$ are the maximum and minimum of the i-th objective in PF^A, respectively; $\overline{f}_i^*$ and $\underline{f}_i^*$ are the maximum and minimum of the i-th objective in PF^*, respectively. The greater the MS' is, the larger the area of PF^* is covered by the PF^A. The modified metric is illustrated in Figure. 1.5.

Distribution Indicator: Further, the uniformity among the distributed points or individuals is also an important issue in order to ensure consistent transition among the solution points when searching for the most suitable solution from the best possible compromise. The metric of spacing [233] gives an indication of how evenly the solutions are distributed along the discovered front. It is defined as,

$$\mathrm{S} = \frac{1}{\bar{d}'} \cdot \left(\frac{1}{n_{PF}} \cdot \sum_{i=1}^{n_{PF}} (d'_i - \bar{d}')^2 \right)^{\frac{1}{2}} \tag{1.9}$$

$$\bar{d}' = \frac{1}{n_{PF}} \sum_{i=1}^{n_{PF}} d'_i$$

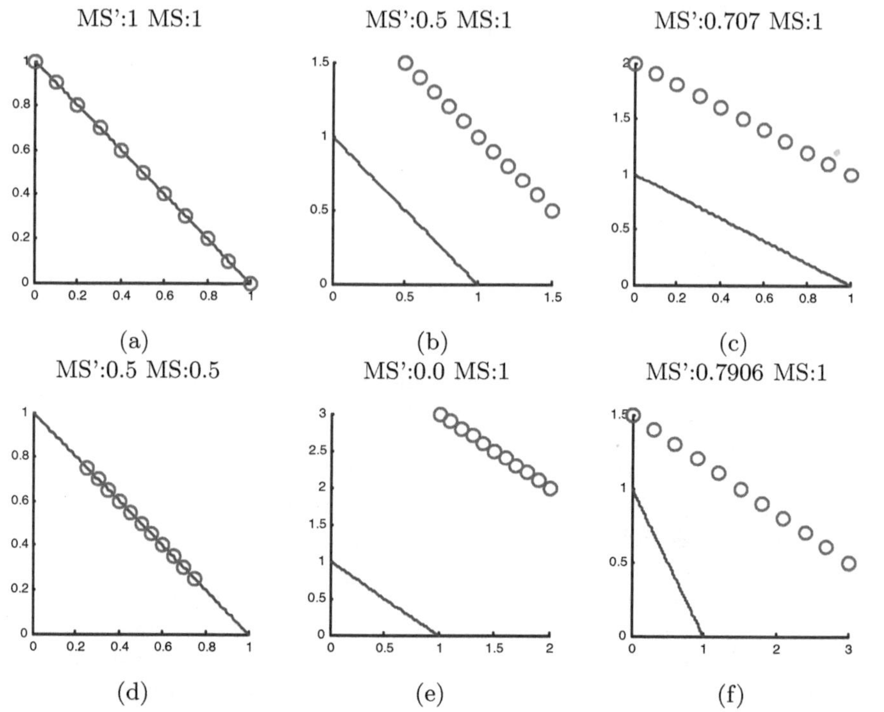

Fig. 1.5 Different Characteristics exhibited by MS' and MS. MS' takes into account the proximity to the ideal front as well.

where $n_{PF} = |\mathrm{PF}|$, d'_i is the Euclidean distance (in the objective domain) between the i-th member and its nearest member in PF^A.

General Quality Indicator: By taking into account performance in diversity and proximity, the metric of hypervolume (HV) provides a general quality measure of the solution set. In order to calculate a normalized value and eliminate bias, Veldhuizen and Lamont [275] expressed the metric of HV as a ratio between the HV of PF^A and PF^*,

$$\mathrm{HVR} = \frac{\mathrm{HVR}(\mathrm{PF}^A)}{\mathrm{HVR}(\mathrm{PF}^*)} \tag{1.10}$$

$$\mathrm{HV} = \mathrm{volume} \bigcup_{i=1}^{n_{PF}} v_i \tag{1.11}$$

where $n_{PF} = |\mathrm{PF}^A|$. Mathematically, for each member f_i^A in the nondominated set, a hypercube v_i is constructed with a reference point and the member f_i^A as the diagonal corners of the hypercube. The reference point can be found by constructing a vector of the worst objective function values.

Pareto Dominance Indicator: In [297], Zitzler *et al* showed that no combination of unary performance metrics can provide a clear indication of whether an evolved set is better than another in the Pareto dominance sense. Therefore, an n-ary Pareto dominance indicator is proposed in this book as a complement to the above metrics. Considering the different PFs, $A_1, A_2, ..., A_n$, evolved by n algorithms, this metric measures the ratio of non-dominated solutions that are contributed by a particular solution set A_i to the non-dominated solution set provided by all solution sets. Mathematically, the non-dominance ratio (NR) is given by,

$$\mathrm{NR}(A_1, A_2, ..., A_n) = \frac{|B \cap A_1|}{|B|}$$
$$B = \{b_i | \ \forall\ b_i \nexists a_j \in (A_1 \cup A_2 ... \cup A_n) \prec b_i\} \tag{1.12}$$

where A_1 is the solution set under evaluation.

1.3 Empirical Analysis and Performance Assessment Adequacy for EMO Techniques

As researchers continue to design more advanced multi-objective evolutionary techniques and explore other EC paradigms, the issue of assessing the weaknesses and strengths of the various methods is becoming increasingly important. This section provides a holistic perspective towards the empirical investigation of MOEAs and presents a conceptual framework in which researchers should consider in the design and implementation of MOEA experimental study. This framework comprises of a structural design plan and a general theory of adequacy in the context of MOEA. The former considers the practical aspects of experimental study and involves the delineation of essential components, description, and discussion of design and implementation issues. On the other hand, the latter considers the theoretical aspects of experimental study and a set of axioms for adequacy are formulated based on various insights gained from the state-of-the-arts. With this framework in place, we hope to motivate discussions in this area from different perspectives to improve empirical study techniques.

1.3.1 Preliminary Discussions

Despite almost two decades of research, performance assessment remains one of the most difficult challenges to the researchers of MOEAs. As with all stochastic methods, the evaluation of EA performance is not a trivial task. One implication of stochasticity is that the capability of the algorithm cannot be precisely determined before its actual application. It gets even more complicated in the context of multi-objective optimization where the conflicting goals of multi-objective optimization have a profound impact on MOEA performance assessment. Zitzler *et al* [297] states that *it is not clear what quality means with respect to approximations of the Pareto-optimal set: closeness to the optimal solutions, or other properties?* Bosman and Thierens [24] further noted that state-of-the-art MOEAs have similar or incomparable performances because of the conflicting optimization goals. Performance assessment can be conducted either empirically or theoretically. At present, there are several limitations to the theoretical approach which lead to the adoption of experimental studies as the *de facto* approach for the evolutionary computation community. On the other hand, empirical assessment has its fair share of limitations, some of which are inherent.

Performance Assessment Issues

Contemporary views held that theoretical investigation is an essential fixture of technique evaluation for black box optimization techniques such as EA. The main detractors of evolutionary optimization assert that theoretical validation is the only way to have any assurance of algorithmic reliability. The situation is best described by Goldberg:

> "...why in the design, testing and utilization of a material machine are we quite content to adopt normal design or engineering method and when we turn to what I'll call a conceptual machine [EAs] do the rules change in the direction of rigor and proof? [99]'

Theoretical analysis contributes to the understanding of evolutionary dynamics and will be crucial to the design of competent evolutionary optimization techniques. However, the theoretical approach lacks the flexibility and practicality of experimental investigation. In the literature, theoretical treatment of algorithmic performance includes verification of convergence and stability for particular algorithms as well as run-time analysis for comparative purposes. Either way, due to the stochastic nature of EAs and its complex relationship with the optimization problem, it is difficult, if not impossible, for any formal mathematical treatment of algorithmic performance. Researchers can either get lost in the mire of complexity or resort to substantial simplifications before any analysis can be done. Markov chains and, in Goldberg's words, a variety of patch-quilt models have featured prominently in these studies.

Secondly, existing theoretical studies [225, 226] are focused on obtaining proofs of convergence under certain general conditions. Moreover, these assumptions are often weak and Powell (1998) noted that there is hardly any correlation between the algorithms that are in regular use for practical applications and the algorithms that enjoy guaranteed convergence in theory. It is worth mentioning that a proper framework for theoretic means of comparing algorithmic performance is formulated by He and Yao [167] only recently. Furthermore, the only known theoretic-comparative study for multi-objective optimization [174] is valid for the solitary class of simple discrete problems. As a consequence of the lack of fidelity and generality, these studies are mere simulacrum of the original quest for rigor. In order for any theoretical study to stand up against any scrutiny, there must be a complete axiomatization of particular EA and the various problems to be solved. It also follows that the formulation of a systematic proofing of EA performance is required. Unfortunately, these requirements are far beyond our knowledge.

Empirical study is attractive because it can be used to assess algorithmic performance once the basic algorithm is in place. In particular, experiments can easily be conducted to determine the conformance of algorithmic behavior to design specifications or the fine-tuning of algorithm parameters during any stage of the design process. When evaluating the performance of MOEA empirically, the particular algorithm/s will be applied to optimize a set of pre-selected test functions and the evolved approximate sets will be taken as an indication of algorithmic performance. However, it is clear that most empirical assessments in the literature do not provide any insight to algorithmic behavior apart from the fact that it is either working or not. In other words, there is a distinct lack of any discussion on why and how the algorithm fails or succeeds. Thus, in order to enhance our understanding of the MOEA mechanism, it is necessary to conduct empirical analysis of sufficient depth.

While guidelines for the design of experimentation involving heuristics go as far back as the 1980s [15, 48, 103, 106], similar works pertinent to evolutionary computation are only reported recently. Eiben and Jelasity considered the generality of the analysis drawn from experimental results reported in the evolutionary computation (EC) literature and identified some of the limitations regarding to the choice of test problems and performance metrics in [68]. In the domain of evolutionary multi-objective optimization, studies have focused on the development of performance metrics and test functions, resulting in great strides in these areas. Initial empirical studies are usually based on simple extension of single-objective problems, which reveal little or no characteristic of the algorithm under observation. To this end, Veldhuizen and Lamont [276] are probably the first to formalize a set of standard test problems for MOEA, while Deb [54] identified the different difficulties that may pose problems for MOEA. In order to quantify the evolved tradeoffs, many different metrics have been proposed over the years. For instance, the metric of generational distance measures proximity, the metric of maximum spread measures diversity, while the metric of spacing measures distribution. The fact that any interpretation is largely dependent on the accuracy of performance indicators has initiated much research in the recent years [116, 163, 297]. In particular, Zitzler *et al* [297] proved that many popular metrics are not compliant with respect to non-dominance relationship. Although much work has been done to improve the reliability of empirical studies, there are little or no discussions at all on how it should be conducted with adequate substantiality on their statements made on the performance and behavior of the evaluated algorithm. In particular, there is a lack of discussion on the following issues:

- the development and implementation of empirical studies,
- the appropriate depth of empirical analysis,

- the adequacy of empirical studies for claims made on algorithmic performance and behavior.

1.3.1.1 Easy and Hard Problems in Empirical Assessment

Empirical assessment involves the consideration of several issues, some of which are easy to resolve while the rest may take considerable more time before any consensus may be formed. In this chapter, we will simply call these issues as the easy and hard issues of performance assessment. Since the early days of development, EA researchers have taken a rather pragmatic or engineering approach towards performance assessment; it is decomposed into simpler sub-problems and investigated separately, while alternatives are sought for issues that are deemed too hard to solve. Intuitively, many of the easy problems of empirical assessment are associated with these smaller components. Conversely, the hard issues not only include issues that researchers avoid but also issues that got lost in the decomposition.

The easy problems of empirical assessment include the observation of algorithmic performance, the construction of benchmarks, the design of performance indicators, and the identification of deficiencies in existing works. Readers should not be misled into thinking that these easy problems can be easily addressed - the construction of benchmarks that can adequately challenge the capabilities of EAs, for instance, is the accumulation of two decades worth of research. These problems are so termed easy because the requirements can be expressed objectively, though not necessarily through theoretical or mathematical means. In the context of multi-objective optimization, it is clear that the evolved approximation sets are the subject of investigation. benchmarks should impede the discovery of near-optimal, diverse and well-distributed solution sets, indicators should measure approximation set quality in terms of multi-objective optimization goals, deficiency of indicators can be identified based on non-conformance to some dominance relationship, and run-time analysis should encompass both convergence and diversity.

Furthermore, there exist theories and principles from which guidance can be drawn. Initial efforts in the design of multi-objective benchmarks are influenced by the work established in the domain of single-objective optimization and operational research. Therefore, most research efforts on performance assessment are focused on these areas. Since the scope is properly defined, the results can be likewise presented in a straightforward manner. All of the easy problems are associated with the direct testing or examination of performance. For example, algorithms are tested upon benchmarks and their performances are measured, in turn, by performance indicators. If observations and testing were all there was to performance assessment, then performance assessment would not be much of a problem. Although we still lack a complete understanding of these issues, we have a clear idea of how we should go about developing them.

On the other hand, the hard problems of performance assessment are those that are susceptible to subjectivity and these problems evolve about the notion of *adequacy*. Most researchers agree that the successful application of MOEA on a test suite is no guarantee of success on practical problems [276]. A typical test suite employed in recent works [42, 58, 251, 252] consists of a set of four to twelve test functions characterized by different problem difficulties, which have been identified by Deb [54]. While various multi-objective test problems have been employed in the literature, these test suites are generally designed based on the works of [54, 55, 276, 300]. This naturally leads to the question of whether exhaustive testing, defined in terms of all possible combinations of problem difficulties, is a viable option for algorithm evaluation. The answer is an emphatic no for two reasons

- It is an impractical attempt to achieve an adequate experimental study.
- The validity of the concept of problem difficulty hinges on our perceived notion of how a real-world problem behaves.

By comparing the two types of problems, one observation that can be made is that the hard problems are supersets of the easy problems. In other words, these are the big picture considerations that were lost in the decomposition of these issues. Based on the state-of-the-arts, empirical investigation is decomposed into the aspects of benchmark generation and performance indicator design. Even though, these easy issues have been individually addressed relative to the said requirements, taken together, do they really represent an adequate empirical assessment?

These problems are termed hard because of the degree of ambiguity involved. As a first step to answering these problems, the next sections will present an experimental framework for MOEA assessment as well as a conceptualization of test adequacies.

1.3.2 Systematic Design for Empirical Assessment

MOEA design involves an iterative process of designer intuition and validation, where performance assessment is carried out to compare, examine, and improve algorithmic design. In addition, it plays a crucial role in improving our understanding of EA dynamics and the interplay between the different components. The knowledge gained will greatly aid the development of more powerful algorithms. The relationship between algorithm design and the empirical study is illustrated in Figure. 1.6. The design of the experimental study involves the consideration of several issues, which can be classified into three distinct phases, 1) specification, 2) design and 3) execution. Other experimentation design process have also been described in [15]. This design plan allows researchers to investigate the different aspects of empirical study as well as the interplay between the different issues. In the subsequent sections, the various design and implementation issues will be discussed.

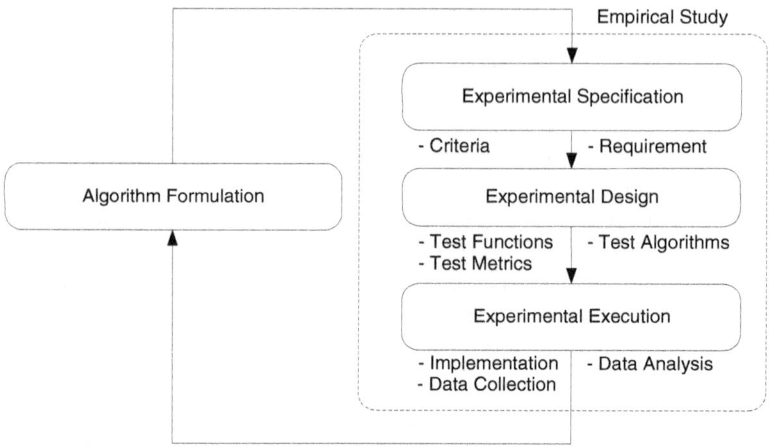

Fig. 1.6 Conceptual framework for the design and implementation of MOEA experimental study

Specification

The design plan begins with the specification phase where the motivations of the work and criteria of the empirical study are laid out. The motivations can range from the validation of one's design intuition to the understanding of some theory to the solving of real-world problems. On the other hand, the criteria determine the extent and depth at which the empirical study is carried out. Since the criteria will dictate the requirements of the empirical study, it should be defined before the design and implementation of the actual experimental study. It is worth noting that although the criteria can be laid down specifically by the researcher, it is usually inferred implicitly from the goal of the study as well as the venue of publication; journal publications definitely warrant a more in-depth empirical study compared to a conference publication.

Table 1.2 describes the different levels of empirical study, highlighting the different types of criteria and the corresponding experimental requirements. Type I and II assessments represent the most preliminary level of empirical study and will examine the algorithm's ability to achieve the goals of proximity, diversity, and distribution. However, the two types of empirical studies are different in the sense that the latter involves a comparative assessment which evaluates the significance of the MOEA with respect to the state-of-the-arts. Type III assessment takes a closer scrutiny of algorithmic performance at the component level and typically involves a parameter sensitivity test. This allows a better understanding of the various operators constituting the algorithm and their relationships. Type IV will analyze the dynamics of the evolutionary process and unveil insights to the algorithmic behavior and

Table 1.2 Different categories of empirical assessment

Type	Criteria	Experimental Requirements	Significance
I	Assess algorithmic performance	Test algorithm capability to converge and maintain diversity	Analyzes whether algorithm is working
II	Assess relative algorithmic performance	Test and compare algorithm capability to converge and maintain diversity	Analyzes how well algorithm is working
III	Evaluate robustness of algorithmic performance	Test and analyze effects of algorithmic components and parameter variation on algorithmic performance	Analyzes what is working
IV	Verify algorithm correctness	Characterize population or individual dynamics	Analyzes why and how algorithm is working

characteristics. Essentially, different criteria demand different experimental requirements and the final decision will depend on the designer and the motivation in algorithm formulation. Note that an empirical study may consist of more than one type of assessment and each type will answer different questions specific to EMOO as noted in Table 1.2.

Design

Decisions made in the design phase should be reflective of the motivations and criteria specified in the prior phase. This is crucial since it determines the adequacy of the empirical observations to support any statement made by the MOEA researcher about algorithmic behavior or performances. Issues considered in this phase usually include 1) test problems, 2) test algorithms, 3) performance indicators, and 4) experimental data.

Test functions: Many guidelines for the construction of test functions and test suites have been suggested in the literature. At present, more than 50 multi-objective test functions with different features that can pose difficulties to the algorithm converging to the Pareto front and maintaining a diverse solution set have been applied in the literature. Most of these test functions are either continuous or discrete in nature, while mixed-types are rarely considered. Although it is not known how well these test functions reflect real world optimization problems, the community has mostly agreed with Deb

[54] and Zitzler *et al* [299] on the type of problem difficulties to be utilized in experimental studies. In this regard, researchers optimizing real world applications should report on how particular test suites can reflect the problem in hand.

Although many good guidelines have been presented, the choice of test functions should be dependent on the problem and issues considered by the MOEA designer. One approach is to consider the MOEA functionalities challenged by the features inherent to the test problems. Problem features are properties that collectively identify the difficulties posed to the MOEA [168] and they can be classified into two broad categories of primary and secondary features according to the functionalities challenged

- *Primary:* Bias, PF* and PS* geometry, multi-modality, deception, parameter interaction, high-dimensionality.
- *Secondary:* Robustness, dynamic landscape, noisy landscape, goal/preferences, constraints, real-world.

The latter category provides challenges beyond the difficulties posed by the former category to the basic ability of MOEA in discovering a near-optimal and diverse Pareto-front. The researcher should consider incorporating relevant secondary test functions only if it is the motivation of the experimental study to do so, for example, to assess the constraint handling capability of a particular proposed feature. On the other hand, regardless of the secondary MOEA features considered, primary test problems must be included to test the basic capability of MOEA.

Performance metrics: Although it is commonly agreed that performance indicators should be independent of the particular experiment [48], we believe that it is important to consider the motivation at hand. A researcher with the motivation of assessing the capability of a particular noise-handling feature may consider the issue of robustness as another performance measure. In certain cases, researchers considering specific domains may need to specify their own metrics. Furthermore, it is also important to consider how the different metrics can complement one another. While we are aware of other classification schemes, it is more useful to classify the different metrics into the following categories

- *Specific indicators:* Performance metrics that quantify the approximate solution set in one aspect of the three multi-objective optimization goals,
- *General indicators:* Performance metrics that provide a general indication of solution set quality by incorporating both convergence and diversity.
- *Pareto indicators:* Performance metrics that quantify the solution set based on dominance criterion.

The need of complementary indicators from the different categories is illustrated in Figure. 1.7. Considering the two approximate sets PF^{A1} and PF^{A2}, we will find that the metric of coverage will indicate that $\mathrm{PF}^{A2} \prec \mathrm{PF}^{A1}$, while

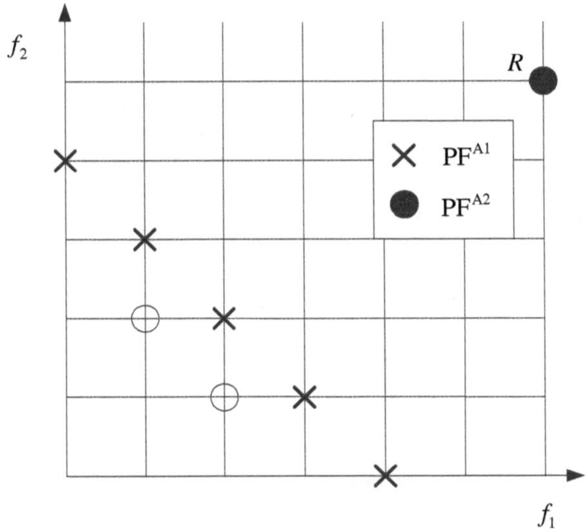

Fig. 1.7 Illustration of the necessity of complementary Performance indicators. The metric of coverage will indicate that approximate sets $\mathrm{PF}^{A2} \prec \mathrm{PF}^{A1}$ while the metric of HVR will with respect to R indicate the same performance.

the metric of HVR with respect to R will prefer PF^{A2}. In this specific instance, it is necessary to apply specific indicators, such as GD and MS, to understand this phenomenon.

Test algorithms: Apart from the selection of test problems and performance metrics, it is also necessary to have an appropriate set of test algorithms for any comparative study. Many comparative studies demonstrating the superiority of current state-of-the-art over earlier attempts and conventional approaches can be found in the literature. Depending on the issues and problem considered, this set of test algorithms may include deterministic methods, heuristics, MOEAs, and/or hybrid MOEAs. Common examples of test MOEAs that have been employed in recent studies [95, 120, 188] include NSGAII, SPEA2, and PAES. The selection criteria are that these algorithms must be truly reflective of the state-of-the-arts and of the issues considered. Otherwise, the experimental results that form the basis of analysis will not be reliable.

A further point to note is that the type of test algorithms also depends largely on the motivation of the study. If the proposed optimizing technique represents a novel computational paradigm, the set of test algorithms should be representative of the different evolutionary computational models in the literature, including MOPSO, EA, and simulated annealing. However, if the proposed technique represents an improvement over some existing approach, it is necessary to use a set of test MOEAs that represents the state-of-the-arts in that particular field. On the other hand, if improvements are done at the operator level, not only is it necessary to gauge the improvements brought about by the proposed operators in a baseline algorithm, the extendibility of the proposed operators should also be assessed.

Experimental Data: Appropriate experimental data must be collected during the experimentation process to facilitate analysis in the later phase. Depending on the defined criteria, the required data may include evolutionary trends, population distribution, and the final evolved approximate Pareto fronts. The type of data gathered is largely dependent on the type of analysis warranted. In the most general case, the coordinates of the final set of Pareto optimum solutions found in the objective and search space will suffice for the plotting of the Pareto front approximated and the calculation of performance metrics. It is highly likely that these data may require further processing before they could be interpreted and analyzed. Lastly, due to high computational cost, careful considerations must be made to avoid repeating the experiments again.

Execution

In the final phase of the experimental study, 1) implementation of the experimentation design is carried out and 2) analysis of the results is performed. Practicality and fairness of experimentation is of significant concern in this part of the experimental study.

Implementation: Actual realization of test algorithms, test functions, and data collection schemes selected in the design phase involve hardware and software considerations. Practical constraints include processor speed, memory requirement, single or distributed processing, etc. For instance, once the appropriate data is defined, plans must be made on how it can be collected and processed. The decision to perform online computation of evolutionary trends will ease memory requirements to store raw data at the expense of computational effort.

Relevant software consideration includes the operating system and programming language used to implement the algorithms. One particular issue in this stage is the implementation of test algorithms. Factors affecting software implementation includes the MOEA researcher programming skills and the availability of test algorithms for download. While it may be more practical to use existing codes, it is well-known that different implementations can have severe impact on algorithmic performance. Therefore, steps must be taken to ensure fairness of all experiments conducted. The alternative is to conduct all experiments on a common platform. Of similar interest, Bleuler *et al* [23] suggested the use of PISA as a common platform to study the performance of stochastic algorithms. In cases where the MOEA researcher need to code the test algorithms from scratch, the performances of the test algorithms should at least be comparable to, if not better than, the performances reported in the original works. At the very least, researchers should ensure that the same initial population is used in the experiments since the evolutionary optimization process is sensitive to the bias introduced by the initial population. During the experimentation, justifications must be made for parameter settings for parameters such as crossover, mutation rates, etc. On the part of the researchers, they should ensure that implementation details are comprehensively furnished for implementation.

Data Analysis: In this stage, the experimental data obtained will be analyzed and evaluated with respect to the algorithmic motivation and the experimental specifications. For typical EMO empirical studies, statistical information of the various test metrics will be reported, i.e. the mean, median, standard deviation, maximum and minimum, to quantify the reliability and validity of the algorithmic performance. These statistics can either be summarized in tabular form or displayed graphically via box plots. While the former displays the precise value obtained for the various metrics, the use of the latter can highlight the relative differences between test algorithms, which can be useful in comparative assessment. Nevertheless, a mere difference in the average of the metrics cannot be blindly regarded as a performance difference between the various MOEAs. As such, statistical techniques, such as ANOVA, t-test, etc, should be included to quantify the significance of the differences amongst the optimizers.

Besides the statistical information, the Pareto fronts obtained are usually illustrated to reflect clearly the multi-objective nature of the optimization problem under investigation. Graphical visualization of the Pareto front allows a quick overview on the performance of the MOEA and cross validation between the statistical results earlier. Generally, authors will present the Pareto front of a randomly chosen run or select the run that can best complement the statistical information displayed earlier.

The experimental analysis should be sufficiently adequate to quantify the correctness and viability of the MOEA, test and verify any hypothesis and assumption made during the algorithm design and uncover pertinent parameter relationship and/or dynamic behavior within the algorithm. Naturally, the analysis should be synchronized to the underlying motivation during the algorithm formulation and the depth of the analysis will depend on the desired criteria.

1.3.3 Survey on Experimental Specifications

In this section, we will discuss on the different levels of experimental studies that have been conducted in the evolutionary multi-objective optimization literature. Note that this survey is restricted to articles with empirical studies found in the IEEE Transactions on Evolutionary Computation from the period between year 2000 and 2006. It is meant to illustrate the different experimental specifications listed in Table 1.2 and highlight possible trends and deficiencies in the current practices.

Table 1.3 shows the list of papers and the types of empirical analysis present. The label (√) in the table represents the presence of the particular depth of analysis (columnwise) for the particular article (row-wise). A further distinction is made according to the general motivation of the different works. Works that sought to solve real world problems are denoted as RW, while works with the motivations of developing new algorithms or features are represented by Alg and Mec, respectively.

As mentioned earlier, more than one type of analysis may be present in the empirical study. At the same time, we observed that some types of analysis occur at a higher frequency for works with certain motivations.

Table 1.3 Levels of Experimental Study Analysis in the Literature

Reference	Year	Motivations	Type 1	Type 2	Type 3	Type 4
Obayashi *et al* [201]	2000	RW	√			
Hughes and Leyland [130]	2000	RW	√	√		
Jaszkiewicz [140]	2002	RW		√		
Goulermas and Liatsis [105]	2003	RW	√	√		√
Atkinson-Abutridy *et al*[8]	2003	RW	√		√	
Guan and Zhang [109]	2003	RW	√	√		
Ishibuchi *et al* [139]	2003	RW		√	√	
Jaszkiewicz [141]	2003	RW		√	√	
Weicker *et al* [280]	2003	RW		√		
Ascia *et al* [7]	2004	RW	√		√	
Khoshgoftaar *et al* [156]	2004	RW	√		√	
Shin *et al* [239]	2005	RW	√	√		
Teo and Abbass [259]	2005	RW	√			√
Benedetti *et al* [17]	2006	RW	√	√		
Abido [1]	2006	RW	√	√	√	
Erbas *et al* [71]	2006	RW	√	√	√	
Everson and Fieldsend [72]	2006	RW	√			
Cvetkovic and Parmee [51]	2002	Mec			√	√
Jensen [142]	2003	Mec		√		
Fieldsend *et al* [76]	2003	Mec		√		
Tan *et al* [258]	2001	Alg		√	√	√
Tan *et al* [251]	2006	Alg		√	√	√
Deb *et al* [61]	2002	Alg		√	√	
Yen and Lu [287]	2003	Alg		√	√	
Lu and Yen [188]	2003	Alg		√		
Farina *et al* [73]	2004	RW Alg	√			
Coello Coello *et al* [43]	2001	Alg		√	√	
Ho *et al* [120]	2004	Alg	√			
Garca-Pedrajas *et al* [87]	2005	Alg	√	√		√
Knowles [160]	2006	Alg		√		
Emmerich *et al* [70]	2006	Alg		√		

RW: Type 1 analysis is more commonly employed in this class of works and it allows for either the identification of the tradeoffs between the various objectives or verification of PF^A with known results. There are two main approaches in performing Type 1 analysis, 1) comparing the evolved solutions with those currently employed in practice [201] and 2) validating through practical realization [17]. Type 2 analysis, i.e. comparative studies with existing approaches, are also conducted in works such as [1, 105]. The frequency of Type 3 analysis, which provides more depth to the analysis, seems to increase after 2003 as researchers [1, 8, 17, 71, 139, 141, 239] sought to investigate algorithm robustness under parameter variations.

Alg: Type 2 analysis is commonly employed in this class of publications. The comparative studies conducted in the papers sampled are rather comprehensive, with problem test suites that challenge the MOEA in various aspects, state-of-the-art test algorithms, and a good coverage of performance indicators to assess algorithmic performances. Type 3 analysis is also performed in some of these works to evaluate the contribution of the components within the proposed algorithm assessed at the parameter level [43] or at the component level [258].

Of course, the parameters should ideally be related to the complements that are newly introduced by the proposed MOEA and not the existing evolutionary operators [61]. While it can be argued that the viability of the conventional operators should be re-assessed due to the new proposed operators, such analysis should ideally be performed only after a deeper understanding of the new operators is achieved. It is common that evolutionary trends of certain performance indicators might be used sometimes for comparison purposes. However, it should be noted that a mere evolutionary trend should not be viewed as a type IV analysis [70]; instead, the trend analysis must show the behavior of the algorithm [251, 258].

Mec: Amongst the various papers surveyed, only three of them involve proposals for operator improvements in MOEA. They include the reduction of the run-time complexity of NSGA-II [142], unconstrained elite archives [76], and the incorporation of preferences in evolutionary multi-objective optimization [51]. As such, in these papers, a comprehensive theoretical treatment of the operators can be found, which is normally used to satisfy the type 3 and 4 specifications. As such, type 2 analysis will form the main bulk of the empirical study. Type 1 analysis will not be necessary as the focus in these papers are the operators and they will be evaluated in a suite of well-studied benchmark problems.

1.3.4 Conceptualizing Empirical Adequacy

Empirical study forms one of the most integral components in EMOO. Besides understanding the algorithmic development of MOEAs, it is important to determine whether the empirical study conducted is adequate to derive any conclusive statement about its performance. Despite its importance, there is alarmingly little research in this area. Due to the lack of proper theoretical foundation, rules of thumb serve as the guidelines for researchers in designing their evaluation study. As such, there is a pressing need to establish standard guidelines in assessing the adequacy of the empirical study. For this purpose, a set of criteria to evaluate the adequacy of an empirical study will be formally defined. Also, a general axiomatic theory of adequacy will be developed to make explicit some underlying assumptions in general empirical studies.

An Inadequacy Criterion

The concept of (experimental) adequacy is first coined by Goodenough and Gerhart [104] in the field of software engineering in order to formalize a framework for software testing. Likewise, this section aims to motivate the use of similar concepts to guide our investigation of algorithmic capability and behavior. The adequacy criterion [295] can be regarded as a predicate that defines what must be exercised to constitute a comprehensive test, i.e. one which is able to substantiate the conclusive remark of the corresponding analysis. As a simple extension, the adequacy criterion C can be formulated as a function that takes in the MOEA or the algorithm incorporating the proposed features under evaluation, denoted as A_T, and the corresponding set of specifications S and empirical test T.

Definition 1.6: Adequacy Criterion: The adequacy criterion C is the function

$$C{:}A_T x S x T \rightarrow \{True, False\} \tag{1.13}$$

where $C(A_T,S,T)$=True means that the evaluation test T is adequate for testing A_T with criteria S, else T is inadequate.

However, as mentioned earlier, the notion of adequacy is susceptible to subjectivity and it is difficult to define an appropriate adequacy criterion. Hence, an inadequacy criterion will be defined instead, i.e. criteria that define the inadequacy of the evaluation suite. For our purpose, the inadequacy criterion is based on the segregation of empirical studies into various components discussed earlier.

Definition 1.7: Inadequacy Criterion: The inadequacy criterion C is the function

$$C{:}A_T x S x T_{TA} x T_P x T_M x T_{Imp} x T_{Ana} \rightarrow \{True, False\} \tag{1.14}$$

where T=False if any of the components of the test (Test problems, T_P; Test metrics, T_M; Test algorithms, T_{TA}; Implementation, T_{Imp}; Data Analysis, T_{Ana}) is not adequate for testing algorithm A_T against specification S, otherwise T is adequate.

This definition provides a simple approach to evaluate the adequacy of an empirical test in general, where each of the different components of the empirical test could be evaluated for its adequacy based on the experimental specification. If each of the components cannot be proven to be inadequate with respect to the experimental specification, this will imply that the empirical test is adequate to substantiate conclusive statement on the algorithmic performance and behavior of the MOEA.

Axiomatization of Empirical Analysis Adequacy

The notion of axioms for testing can be found in the field of software engineering [279] as requirements for test adequacy criteria. On the other hand, the axiomatic approach presented in this chapter serves as a guide to the adequacy of empirical analysis. The basic idea is to highlight the fundamental properties of the experimental studies and use it as a checklist if these properties are satisfied.

Performance analysis is ultimately based on actual experimental performance. Therefore, the first and foremost property of empirical analysis adequacy is the existence of an empirical study.

Axiom I. Non-Existence inadequacy: The absence of any empirical study is not adequate for any performance validation.

That is to say choreographed scenarios of proposed evolutionary techniques/ mechanism, no matter how detailed, is simply not adequate. One might argue the case where the algorithm can be *proved* convergent or even superior to existing approaches. However, due to the limitations of theoretical analysis pointed out earlier, this axiom stands valid and performance assessment is inadequate unless empirical validation is provided.

Of course, the mere existence of experimentation does not imply adequacy of the empirical investigation. Central to our notion of adequacy is the need for the actual experiments and analysis to reflect the specifications required for the particular work, a point has been pointed out in the framework of experimental design and laid down formally in (1.14). As a specific instance, consider the scenario where a new density assessment scheme is proposed. Since the role of any density assessment scheme in MOEA is to promote diversity and improve PF^A distribution, the criteria of the empirical study involves demonstrating how the scheme performs its role. Clearly, the lack of any test problems that challenge the scheme's ability to maintain a diverse and uniformly distributed PF^A or the lack of any specific indicators that measure these qualities will render the empirical study inadequate. Therefore, the next axiom is given as,

Axiom II. Compliancy: Any component of the empirical assessment that is not reflective of the experimental criteria is not adequate.

Having considered the properties of the empirical studies in general, we will now regard how the very nature of MOEA affects the analytical process. The first thought that crosses our minds is, probably, the fact that MOEAs are stochastic and multiple simulation runs must be performed to attain statistical consistency. However, apart from stochasticity, what we really need to take note of is the fact that the canonical MOEA is composed of several interacting elements. While it may seem reasonable to generalize the success of the MOEA on a particular set of test problems to its base components, it is clear, based on reasonable intuitive deduction and known empirical algorithmic behavior, that such generalization is not true without any further empirical verification. Thus, an adequate empirical study for the algorithm is not adequate for analysis of its individual components.

Axiom III. Anti-Generalization: An adequate empirical investigation of A_T with components O_1, O_2 is not adequate for the analysis of its individual components.

By the same account, the complex relationship between the different components and the MOEA implies that the sum of all individual parts is by no means any indication of true algorithmic performance. Thus any conclusion on algorithmic performance drawn from the empirical evaluation of it's individual components is not adequate.

Axiom IV. Anti-Decomposition: Adequate empirical investigations of A_T components O_1, O_2 is not adequate for the analysis of the MOEA.

These two properties account for component interactions within the same algorithm. Given that MOEAs share common density assessment schemes, genetic operators, and etc, and many mechanisms are proposed with the intention for general implementation, it is necessary to consider the extendibility of the various components as well. Clearly, we should not expect, for instance, the crowding distance operator of NSGAII to function as well when implemented in PAES. Thus the final axiom presented in this chapter is given by,

Axiom V. Anti-Extensionality: An adequate empirical investigation of A_{T1} with components O_1, O_2 is not adequate for the analysis of A_{T1} with components O_1, O_2.

1.3.5 Case Studies

Adopting one of our previous works in [252] as a case study, the adequacy of a few experimental setups are examined through the proposed axiomatic approach in this section. In this work, the notion of adaptation is extended to the mechanisms of elitism and mutation operator with the motivations of improving the exploratory and distribution capabilities of MOEAs.

- Adaptive mutation operator (AMO): The mutation rate is adapted with time to balance the exploration and exploitation effects of AMO.
- Enhanced exploration strategy (EES): The ratio between solutions selected for normal genetic operators and archived solutions selected for local search to improve solution distribution is adapted.

For the ensuing discussions, we assume the following settings: The proposed features are incorporated into a basic MOEA, denoted as ALG, and validated upon the test problems of ZDT4, ZDT6, FON and KUR. Test algorithms employed include NSGAII, SPEA2 and PAES while the performance metrics of GD, MS, S and HVR are used. In order to provide a fair comparison, all the algorithms are implemented with the same binary coding scheme of 30-bit per decision variable, tournament selection, uniform crossover, and bit-flip mutation whenever applicable. Furthermore, all simulations are implemented in C++ on an Intel Pentium 4 2.8 GHz computer and thirty independent runs are performed for each of the test functions in order to obtain the statistical consistency.

Scenario 1: A type II empirical study is conducted in which the performance of ALG is compared against the test algorithms. Part of the analysis made is as follow.

> "ZDT4 proved to be the most difficult problem faced by the algorithms since no algorithm, except ALG due to the exploratory effects of AMO, is able to deal with multi-modality effectively. The challenge of FON is to find and maintain the entire Pareto front uniformly. With the exception of the ALG, the other algorithms found it difficult to find a good spread and distribution. This is due to the incorporation of EES which facilitates the distribution of solutions. "

This is a reasonable test suite with ZDT6, FON and KUR challenging the MOEA's ability to maintain a uniformly distributed PF^A and ZDT4 posing convergence problems. The test algorithms are representative of the state-of-the-arts while the performance metrics consist of a general quality indicator and three specific indicators to evaluate PF^A quality in terms of convergence, distribution and diversity. Thus the compliancy axiom holds. However, the statements claiming EES and AMO contribution to the distribution and convergence of the PF^A is a clear violation of the anti-generalization axiom. For this particular empirical analysis to be adequate, such statements should either be withdrawn or validated by type III performance assessment.

Scenario 2: Following a type II empirical study in which the performance of ALG is demonstrated to be competitive in all aspects, a type III analysis is conducted to examine the individual contributions of AMO and EES. In this particular setup, the performance of the basic MOEA with and without the individual components of the proposed mechanisms is measured on all the test problems but only the metric of HVR is retained. Part of the analysis made is as follow.

> "ALG with only AMO outperforms the basic MOEA and ALG with only EES for ZDT4, demonstrating the exploratory effects of AMO. On the other hand, ALG with only EES outperforms the other algorithm settings for FON and ZDT6. This clearly illustrates the ability of EES to improve the distribution of solutions along PF^A. Thus, it is clear that EES and AMO will improve the performance of the MOEA incorporating them"

This time round, the anti-generalization axiom holds as the contributions of AMO and EES to the basic MOEA are analyzed individually. On the other hand, the decision to apply HVR alone violates the compliancy property. As mentioned in section 1.3.2, the metric of HVR is unable to provide specific information of a solution set regarding convergence, distribution or diversity. For all we know, the superior performances exhibited by ALG with only AMO for ZDT4 can be attributed to better diversity or distribution rather than convergence. The extendibility also fails to hold since EES and AMO may not improve the performance of NSGAII or SPEA2. It should also be noted that the anti-Decomposition will fail to hold had not the comparative study preceeded this.

1.4 Overview of This Book

Multi-objective optimization is a challenging research topic not only because it involves the simultaneous optimization of several complex objectives in the Pareto optimal sense, it also requires researchers to address many issues that are unique to multi-objective problems. Advances made in the field of EMOO is the result of two decades worth of intense research examining topics such as fitness assignment [74, 188], diversity preservation [154], balance between exploration and exploitation [24], and elitism [176].

The primary motivation of this book is to provide a comprehensive treatment on the design and application of MOEAs for multi-objective optimization in the presence of uncertainties. As mentioned right at the start of this chapter, the difficulty of multiple criteria decision making (MCDM) is exacerbated by the fact that real world problems are not deterministic in nature. While it has been shown that MOEAs are powerful and efficient optimizers of static multi-objective problems, their performance are rarely examined in the presence of uncertainties and it is unlikely that the state-of-the-arts are capable of handling the demands that the task entails.

This book is organized into three parts, with each part addressing a different form of uncertainty. The first part comprising of Chapters 2-4 focuses on the optimization of noisy multi-objective problems. Unlike existing studies of multi-objective evolutionary algorithms (MOEAs) [257, 299], Chapter 2 examines the performance of MOEAs in noisy environments. Chapter 3 describes a number of existing MOEAs for noisy multi-objective problems and simulation studies are performed to compare the noise-handling features of various MOEAs. Chapter 4 considers the design of artificial neural networks as a specific instance of noisy problem.

The second part starts with a survey of existing works for dynamic multi-objective optimization. A formal categorization of dynamic multi-objective test functions and the requirements of performance indicators for assessment of dynamic MOEAs are also provided in Chapter 5. Chapter 6 introduces a new coevolutionary paradigm that incorporates both competitive and cooperative mechanisms observed in nature to solve multi-objective optimization problems and to track the Pareto front in a dynamic environment. The main idea of competitive-cooperation coevolution is to allow the decomposition process of the optimization problem to adapt and emerge rather than being hand-designed and fixed at the start of the evolutionary optimization process. In particular, each species subpopulation will compete to represent a particular subcomponent of the multi-objective problem, while the eventual winners will cooperate to evolve the better solutions. Through this iterative process of competition and cooperation, the various subcomponents are optimized by different species subpopulations based on the optimization requirements of that particular time instant, enabling the algorithm to handle both the static and dynamic multi-objective problems.

The third and final part concentrates on the issues of robust multi-objective optimization. In particular, the suitability of existing robust test problems for multi-objective optimization is examined and a set of guidelines for the construction of robust multi-objective test problems is presented. The fundamental component of the robust test problems is a Gaussian landscape generator that facilitates the specification of robust optimization-specific features, such as noise-induced solution space, fitness landscape, and decision space variation. This generator is developed with the purpose of generating noise-sensitive landscapes in conjuction with existing multi-objective test problems. Subsequently, a robust multi-objective test suite is built upon the ZDT framework. In addition, the vehicle routing problem with stochastic demand (VRPSD) is as presented a practical example of robust combinatorial multi-objective optimization problems. Chapter 8 provides an overview of existing robust multi-objective evolutionary techniques. This chapter also explores the concept of inheritance to reduce computational effort and simulations are conducted to analyze the characteristic of the continuous test suite. In Chapter 9, a hybrid MOEA employing a route simulation method is developed to evolve robust routing schedules for the VRPSD problem.

1.5 Conclusion

In this chapter, we have covered the necessary concepts and definitions of multi-objective optimization and uncertainties to appreciate this book. This chapter also presented an introduction to MOEAs, with a general framework which illustrates the basic design issues of the state-of-the-arts. Subsequently, a survey of the state-of-the-arts based on the basic MOEA components of fitness assignment, diversity maintenance, and elitism is presented to highlight the development trends of multi-objective evolutionary techniques. This chapter also addresses the issue of adequacy for the empirical performance assessment of MOEA. For this purpose, a set of adequacy criteria which evaluates the general adequacies of an empirical study based on its various components is presented. Details of the various components on how they relate to the overall algorithmic development process and their adequacy under different experimental specifications are provided. Furthermore, a set of axioms that make explicit the underlying assumptions of general empirical study was formulated. Finally, the overview of this book is presented.

Part I

Evolving Solution Sets in the Presence of Noise

Chapter 2
Noisy Evolutionary Multi-objective Optimization*

In the previous chapter, we have described the multi-objective optimization problem and the challenges that it entails. However, the formulation presented in equation (1.1) assumes that the objectives can be found deterministically, which is hardly the case in many real world problems. Noise stems from several sources, including sensor measurement errors, incomplete simulations of computational models, and stochastic simulations. Apart from these external sources, noise can also be intrinsic to the problem. A good example is the evolution of neural networks where the same network structure can give rise to different fitness values due to different weight instantiations [144].

In noisy single-objective optimization, a solution may appear either better or worse than its true fitness. Because noise changes the way in which we perceive a solution, the selected or adopted solution may not necessarily be the best available in the evolving population. Detrimental impacts of noise observed by Beyer [21] include the reduction of the convergence rate and premature convergence to sub-optimal solutions. In the case of multi-objective optimization, the solutions can be incomparable as well. An instance of how noise can mislead the optimization process is illustrated in Fig. 2. In a deterministic setting, it is clear that solution A dominates solution B. However, under the influence of noise, enhanced solution B (B') will be judged to be the better solution. In fact, a strongly-dominated solution can easily become incomparable in the presence of noise and admitted into the archive of non-dominated solutions. Furthermore, the shape of the Pareto front may appear differently, providing the decision makers with the wrong information regarding the tradeoffs between the different solutions.

Therefore, we can expect the way in which noise changes the dominance relationship between the different solutions in the evolving population to pose great challenges to MOEA. In the following sections, we will describe how noise is modeled in optimization problems and how noise affects the performance of the MOEA.

C.-K. Goh and K.C. Tan: Evolutionary Multi-objective Optimization, SCI 186, pp. 43–54.
springerlink.com

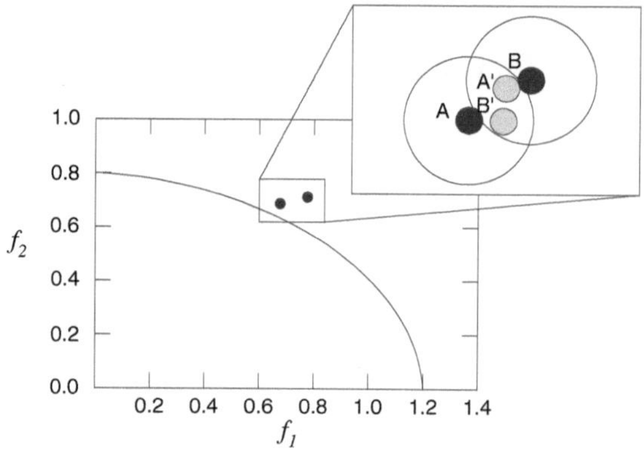

Fig. 2.1 Effects of noise on the perception of solution quality. The dark and grey closed circles denote the true and possible perceived quality of the solutions. The open circle surrounding each solution is the area of influence of noise.

2.1 Noisy Multi-objective Optimization Problems

As mentioned above, a distinctive feature of noisy fitness functions is that each evaluation of the same solution results in different objective values. Mathematically, for noisy multi-objective optimization, equation 1.1 can be rewritten as

$$\min_{\mathbf{x}\in\mathbf{X}^{n_x}} \mathbf{F}(\mathbf{x}) = \{f_1(\mathbf{x}) + \delta_1, f_2(\mathbf{x}) + \delta_2, ..., f_M(\mathbf{x}) + \delta_M\} \tag{2.1}$$

where δ_i is a scalar noise parameter added to the original objective function of f_i and $\mathbf{F}$ is the resultant objective vector.

The optimization of noisy problems is greatly influenced by the noise model adopted and the level of noise intensity. Several studies concerning evolutionary optimization in noisy environments, the vast majority of which conducted in the domain of SO optimization, have been reported [10, 13, 18, 21, 27, 99, 200, 217, 224, 227]. Most of these investigations are done on the basis of Gaussian noise. Notable exceptions include the investigation conducted by Arnold and Beyer [11], which revealed significant differences between the influence of Gaussian, Cauchy, and χ^2 distributed noise on the performance of $(\mu/\mu, \lambda)$-ES. In the context of multi-objective optimization, Teich [263] considered a uniform noise model, while Buche *et al* [32] incorporated the effects of outliers on the optimization process.

The common practice is to incorporate the selected noise model as an additive perturbation to the original test functions. Unlike the study of dynamic optimization problems (which we will cover later), there is no specific test problem or test suite for the analysis of the impact of noise on evolutionary optimization. The same guidelines for the selection of test problems are applicable to noisy multi-objective optimization as well. In fact, applying

Table 2.1 Summary of multi-objective test problems extended for noise analysis

Literature	Test Problems
Basseur and Zitzler [16]	ZDT1, ZDT6, DTLZ2, KUR, COMET, and QV
Buche *et al* [32]	BSDK1
Bui *et al* [33]	ZDT1-ZDT6
Fieldsend and Everson [78]	DTLZ2
Goh and Tan [95]	ZDT1, ZDT4, ZDT6, FON and KUR
Hughes [127]	MOP3 [277]
Singh [241]	S1

a suite of multi-objective test problems with different features will allow us to examine the influence of noise on these features. Moreover, it should be noted that the different problem features will determine the extent and effect that noise has on the optimization process. For instance, we can expect problems with strong parameter dependencies and small isolated PF* to be more susceptible to noise as compared to those without these features. On the other hand, it has been reported that noise has a smoothing effect on the fitness landscape, which allows the EA to handle multi-modality with greater success [216]. The different test problems that have been extended for noise analysis are summarized in Table 2.1.

2.2 Performance Metrics for Noisy Multi-objective Optimization

Like deterministic multi-objective optimization, the optimization goal of noisy multi-objective optimization is to find a near-optimal, diverse, and uniformly distributed PF^A. To be precise, we are concerned about how good the PF^A truly is and not how it is perceived since it is the true objective values that matters during implementation. Therefore, performance metrics proposed for deterministic multi-objective optimization can be used directly to assess the performance of MOEAs in the presence of noise. The only difference between deterministic and noisy multi-objective optimization is that the objectives are perturbed by noise and the true values not be known. In this case, re-evaluation can be employed to compute the effective objective values before the results are assessed.

Visualization of the evolved PF^A is used in [33, 78, 127] to demonstrate the effectiveness of the proposed methods. Bui *et al* [33] employed GD and attainment surface to provide a quantitative measure of algorithmic performance. Teich [263] used coverage to compare the quality between different true PF^As, while Buche *et al* [32] measure that distance between PF^A and PF^* with respect to ten predefined points in the decision space.

Table 2.2 Parameter settings of the simulation study

Parameter	Settings
Chromosome	Binary coding; 15 bits per decision variable.
Population	Population size 100; Archive (or secondary population) size 100.
Selection	Binary tournament selection
Crossover operator	Uniform crossover
Crossover rate	0.8
Mutation operator	Bit-flip mutation
Mutation rate	$\frac{1}{chromosome\_length}$ for ZDT1, ZDT4 and ZDT6; $\frac{1}{bit\_number\_per\_variable}$ for FON and KUR;
Ranking scheme	Pareto ranking
Diversity operator	Niche count with radius 0.01 in the normalized objective space.
Evaluation number	50,000

However, noise-specific performance measures can also be found. Fieldsend and Everson [78] measured the Euclidean distance between true and noisy PF^A. Such a measurement provides an indication of how well the resampling technique performs. Basseur and Zitzler [16] proposed a probabilistic extension of the attainment function, which provides the visualization of the probablistic $k\%$ approximation set. This probablistic $k\%$ approximation set is defined as the set of evolved solutions which dominates objective vectors that have been attained with a probablity up to $k\%$.

In cases where the true PF^A can be determined, deterministic performance metrics should be used because it will provide a more accurate assessment of algorithmic performance. On a more practical side, it should be highlighted that the selection of final solution for implementation will be based on the corrupted PF^A. Therefore, apart from expending a certain amount of computational resource to reduce uncertainties, we can also expect noise-specific metrics, such as the probabilistic attainment function, to play a dual role in the evaluation of algorithmic performance as well as the solution selection.

2.3 Empirical Results of Noise Impact

Noise has a disruptive presence that is encountered in many industrial applications like control systems, medical instruments, communications, and multimedia networks. It is also present at all stages of implementation, such as from data acquisition to system modeling. This section examines the impact of noise on the dynamics of fitness and diversity in evolutionary multi-objective optimization.

The evolutionary model adopted in this section is based on the conceptual framework in Chapter 1. The algorithm employs a fixed-size population and an archive to store non-dominated solutions found during the evolution process. The archive is updated at each cycle, i.e. a candidate solution will be added to the archive if it is not dominated by any members in the archive. Likewise, any archive member dominated by this solution will be removed from the archive. When the predetermined archive size is reached, a recurrent truncation process based on niche count is used to eliminate the most crowded archive member. Although MOEAs have been implemented in different ways, most current state-of-the-art MOEAs include some form of elitism and diversity preservation mechanisms. In this paper, elitism is implemented by selecting individuals to a mating pool through a binary tournament selection of the combined archive and evolving population. Taking into account the study in [127], the selection criterion adopted in this chapter is based on the Pareto ranking scheme described in [84] and niche count [102] is used in the event of a tie. Note that the mechanism of niche sharing is used in the tournament selection and diversity maintenance in the archive.

Five benchmark problems, ZDT1, ZDT4, ZDT6, FON, and KUR, are selected to examine the effectiveness of MOEAs in converging and maintaining a diverse set of non-dominated solutions under the influence of noise. In this study, noise is implemented as an additive normal distributed perturbation with zero mean. It is assumed that noise has a disruptive influence on the value of each individual in the objective space [13, 27, 127, 128, 227], i.e.

$$\bar{f}(\mathbf{x}) = f(\mathbf{x}) + Normal(0, \sigma^2) \tag{2.2}$$

where σ^2 represents the level of noise present; $Normal$ denotes the normal distribution function; $\bar{f}$ and f denote the objective function with and without the additive noise, respectively. Investigations of other noise models are left for future work.

Experiments are conducted at noise levels of $\sigma^2 = \{0.2\%, 0.5\%, 1\%, 5\%, 10\%, 15\%, 20\%\}$ in order to study the impact of noise on evolutionary multi-objective optimization. Thirty independent simulation runs are performed for each of the test problems and the values of the various parameter settings in the algorithm are tabulated in Table 2.2. The mutation rates adopted in this chapter are based on experimental studies that have been successfully applied in [251]. A random initial population is created for each of the simulation runs in every test problem.

2.3.1 General MOEA Behavior under Different Noise Levels

The performance trace representing the mean of true values of GD and MS over 30 simulation runs for ZDT1, ZDT4, ZDT6, FON and KUR with

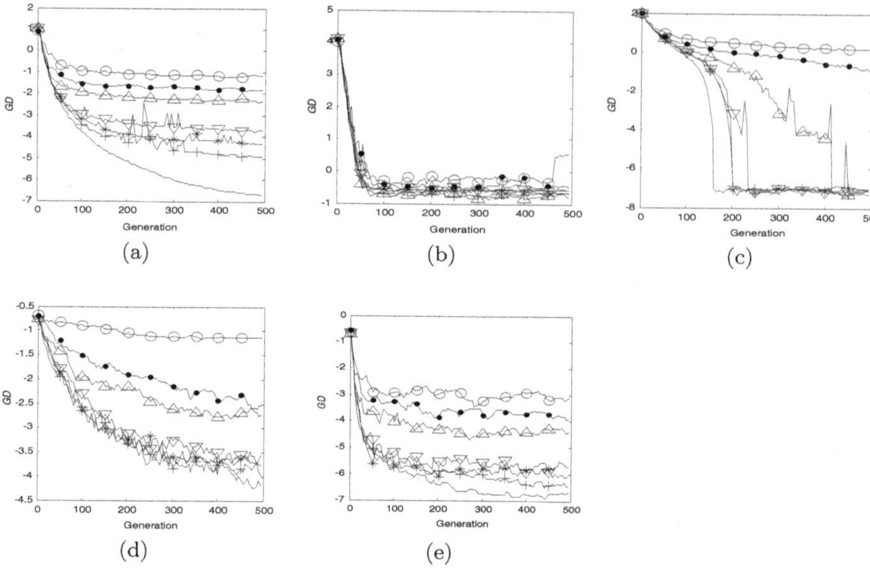

Fig. 2.2 Performance trace of GD for (a) ZDT1, (b) ZDT4, (c) ZDT6, (d) FON, and (e) KUR under the influence of noise level at 0.0%, 0.2%, 0.5%, 1.0%, 5.0%, 10% and 20%

different noise levels are shown in Fig. 2.2 and Fig. 2.3, respectively. The trace of GD and MS are sufficient to demonstrate the impact of noise on convergence and diversity.

According to Nissen and Propach [200], population-based EAs are inherently robust in SO optimization under low level of noise. It can be seen from Fig. 2.2 and Fig. 2.3 that MOEA is also capable of evolving satisfactory solutions in multi-objective optimization under the influence of low levels of noise.

In addition, the smoothing effect described by Rana *et al* [216] is also observable from the simulation results at low noise levels. An interesting finding is that the performance of MOEA actually improves with the introduction of low levels of noise. For instance, there is a high tendency to evolve better coverage for FON, which challenges the algorithm's ability to find and maintain the entire Pareto front uniformly. In the case of ZDT4, which challenges the algorithm's ability to deal with multi-modality, the presence of noise allows MOEA to reduce the gap between PF* and PF^A. As in the study conducted in [216], smoothing is achieved without resampling, probably an indication of implicit averaging.

In contrast, it can be observed that MOEA suffers from degenerate convergence properties and faces the problem of maintaining a diverse solution set under the influence of sufficiently high levels of noise. Fig. 2.4 shows that

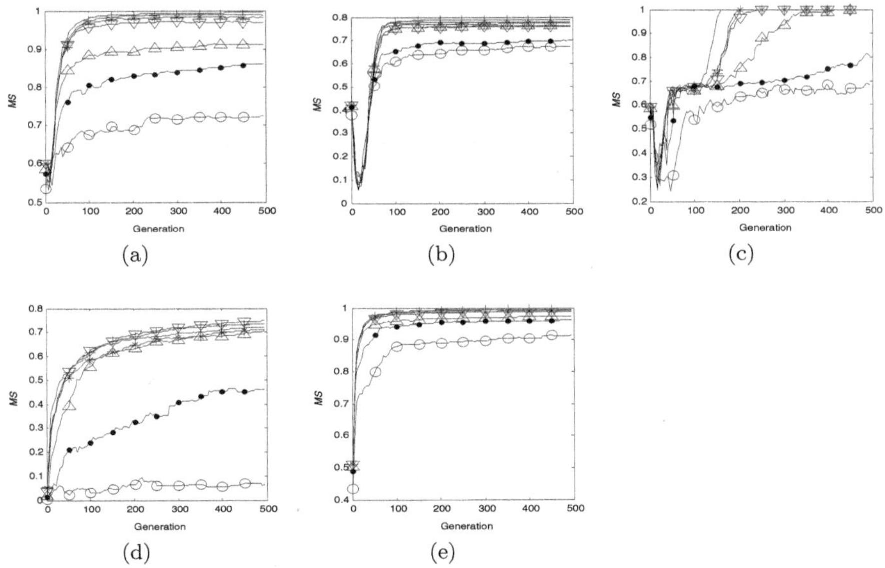

Fig. 2.3 Performance trace of MS for (a) ZDT1, (b) ZDT4, (c) ZDT6, (d) FON, and (e) KUR under the influence of noise level at 0.0%, 0.2%, 0.5%, 1.0%, 5.0%, 10% and 20%

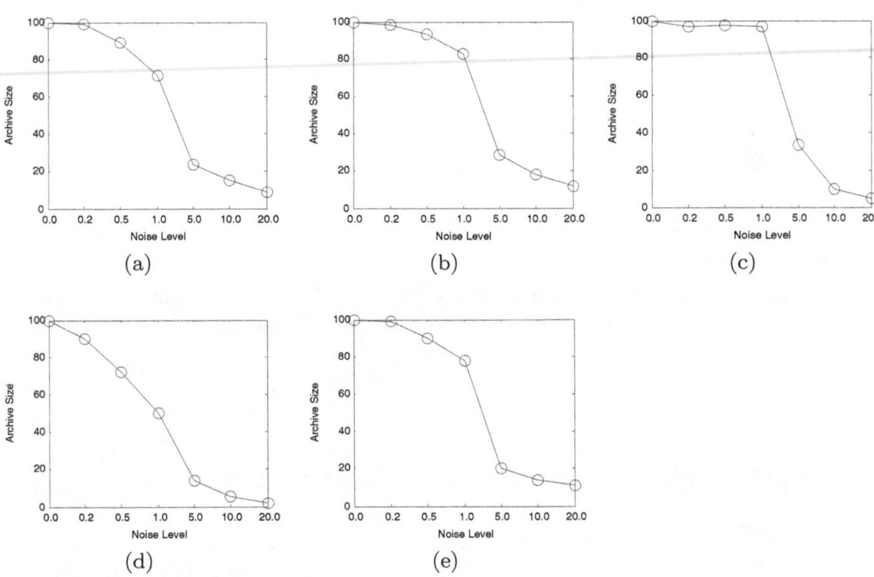

Fig. 2.4 Number of non-dominated solutions found for (a) ZDT1, (b) ZDT4, (c) ZDT6, (d) FON, and (e) KUR under the influence of different noise levels

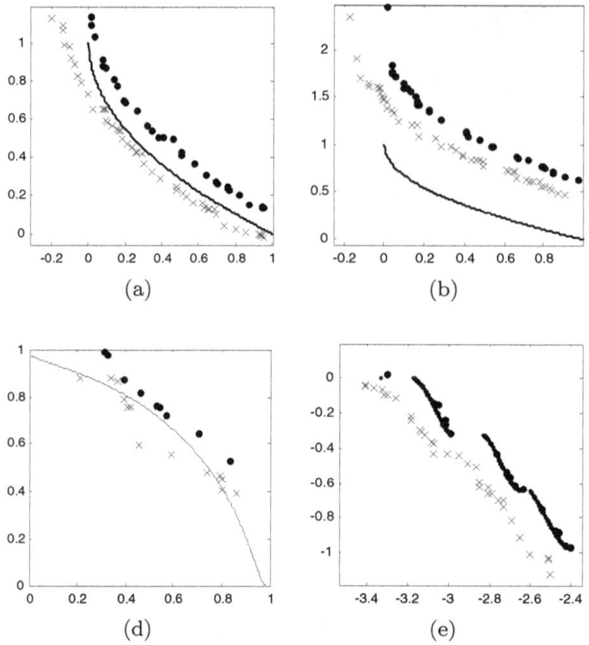

Fig. 2.5 The actual and corrupted location of the evolved tradeoff for (a) ZDT1, (b) ZDT4, (c) ZDT6, (d) FON, and (e) KUR under the influence of 5% noise. The solid line represents PF* while closed circles and crosses represent the actual and corrupted PF^A respectively.

the archiving process deteriorates with increasing noise levels and fails to maintain a stable archive of non-dominated solutions. Further investigations revealed that good solutions are actually kept out of the archive by the presence of noise-enhanced solutions. The impact of noise is also observed to be more severe on problems such as FON, ZDT1, and ZDT6. In particular, the MOEA is unable to improve in performance beyond the initial population for the problem of FON. Although the MOEA fails to escape the local optima of ZDT4, its performance for ZDT4 appears to be insignificantly affected by noise.

2.3.2 MOEA Behavior in the Objective Space

It is important to analyze the behavior of MOEA in the objective space since how it performs depend on the degree at which noise affects the fitness landscape. A straight-forward approach to examine algorithmic behavior in the objective space is, of course, to look at how the MOEA will perceive the perturbed solutions. On a more critical note, if the significance of the point on erroneous selection of solution for implementation made earlier had

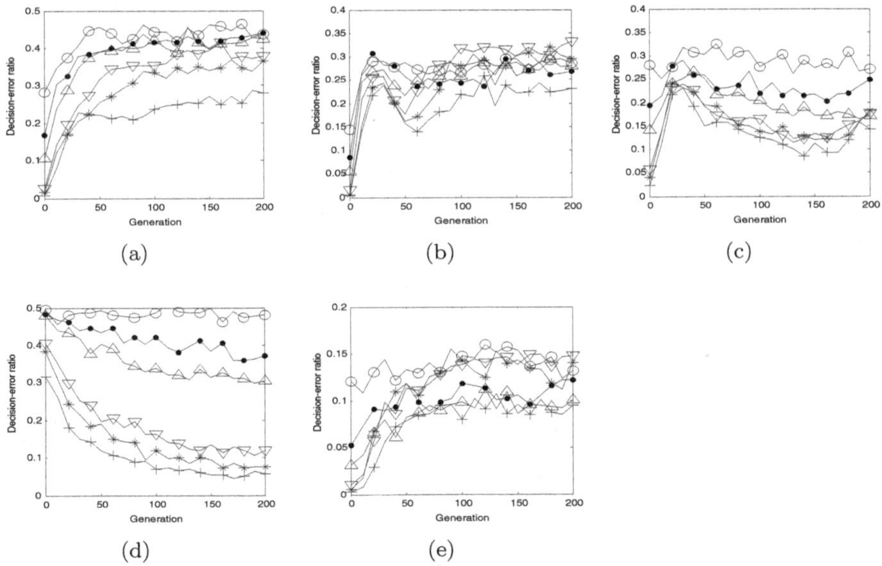

Fig. 2.6 Decision-error ratio for the various benchmark problems (a) ZDT1, (b) ZDT4, (c) ZDT6, (d) FON, and (e) KUR under the influence of different noise levels

not been fully appreciated, a quick inspection on the relationship between the actual and perceived locations of the final tradeoff illustrated in Fig. 2.5 should do the trick. Notice how the perceived PF^A of FON in Fig 2.5(d) seems to imply the presence of a knee solution. The situation will actually get worse with increasing noise levels. Therefore, it is definitely worth the extra computational effort required to perform re-evaluation to obtain the expected objective values for the final PF^* to get a better indication of solution quality.

For a more in-depth analysis of how this affects the MOEA, we will first consider how the MOEA makes decisions based on the perturbed fitness values. Fig. 2.6 shows the decision-error ratio against the number of generations for the five benchmark problems. The decision-error ratio is defined as the ratio between the number of incorrect decisions made for these operations and the total number of decisions made. From the figure, we can observe two trends, the first of which is the positive correlation between the decision-error ratio and noise. There is actually a special significance attached to the ratio at 0.5 because it provides an indication of the degree in which the evolutionary optimization process has degenerated into a random search process. Intuitively, the decision-error ratio should not exceed this 0.5 mark. In the event of such an interesting situation, then all we need to do is to "hard code" the MOEA to select the percieved inferior solution instead! True enough, with the exception of FON and ZDT1, the decision-error ratio did not exceed 0.4 even at $\sigma = 0.2$. This seems to imply that the algorithm should converge to

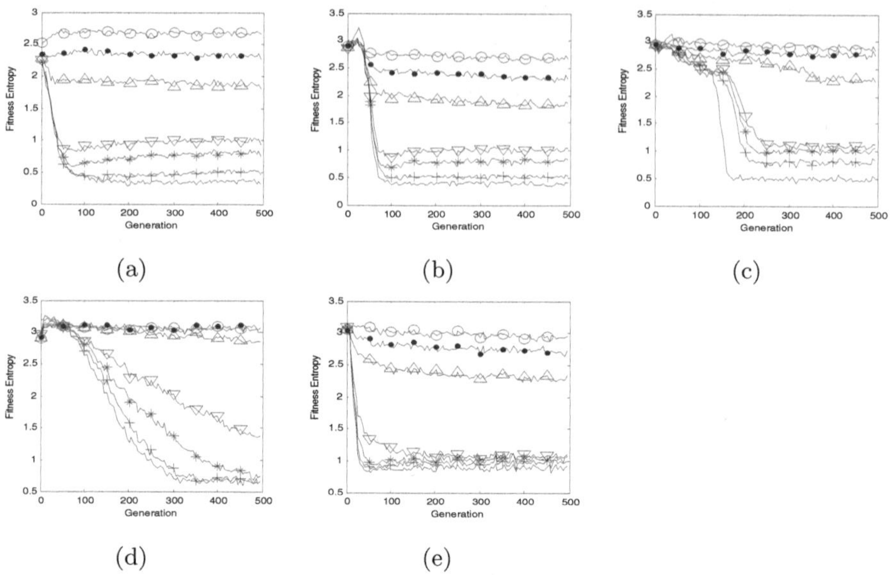

Fig. 2.7 The entropy value of individual fitness for (a) ZDT1, (b) ZDT4, (c) ZDT6, (d) FON, and (e) KUR under the influence of different noise levels

PF* eventually. Unfortunately, this may not happen due to a phenomenon which we term as the *curse of elitism*, because 1) the noise enhanced solutions in the archive are keeping out the good solutions (a point mentioned earlier) and 2) the optimization process is biased towards less promising areas due to the reinsertion of elitist solutions.

With the exception of FON, the second trend observed is that the decision-error ratio generally increases as PF^A approaches PF*, indicating a performance deterioration of optimization process along the evolution. For problems which demonstrate such characteristics, it is desirable to devise a mechanism that makes effective use of initial decisions to improve the convergence of evolutionary multi-objective optimization. Comparing Fig. 2.2, Fig. 2.3, and Fig. 2.6, it is apparent that the evolutionary optimization process stagnates as the error ratio saturates in the evolution. Such a degenerate convergence behavior of MOEA is due to the unreliability of selection, ranking, and archiving in the presence of noise. On the other hand, FON exhibits exactly the opposite behavior, with decision-error ratio improving with the number of generations. This is because the solutions of the initial population of MOEA are always located around a small region about $f_1 = f_2 = 1$ due to the high parameter interactions between the decision variables, which amplifies the effects of noise.

Another way of analyzing the impact of noise on the objective space is to examine the distribution of Pareto ranks assigned to the noisy solutions. Shannon's entropy [238] is applied to quantify the amount of uncertainty in

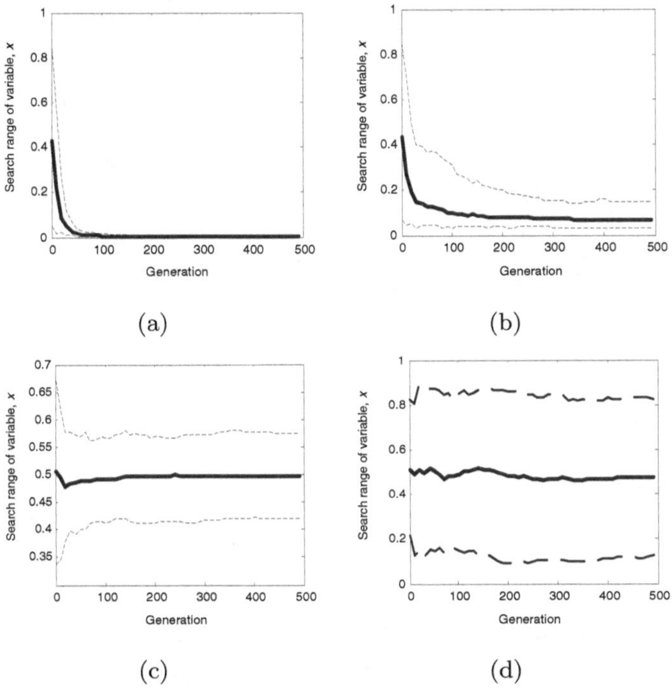

Fig. 2.8 Search range of an arbitrary decision variable for ZDT1 at (a) 0%, (b) 20% noise and FON at (c) 0% and (d) 20% noise. The thick line denotes the trace of the population mean along an arbitrary decision variable space, while the dashed line represents the bounds of the decision variable search range along the evolution.

the ranking process and the entropy of solution Pareto rank in the evolving population for all noise levels is shown in Fig. 2.7. It can be seen that only simulation runs with no or low noise levels exhibit behavior of a stable optimization process with a converged fitness distribution. This is because the ranks of these individuals should also converge to better rank values as the evolving population converges to a better set of individuals in a low-noise environment. In contrast, simulation runs at high noise levels demonstrated high levels of uncertainty in the evolutionary optimization process.

2.3.3 MOEA Behavior in Decision Space

We have observed that the optimization process tends to converge to sub-optimal PF even though the MOEA is making the correct decisions most of the time. So the natural question now is *what exactly are the effects of those right decisions?* Since the behavior of MOEA in the objective space seems to reveal little, we turn our attention towards how it behaves in the decision space. In order to examine algorithmic behavior in the decision space, Fig. 2.8(a)- (d) illustrates trace of the search range for ZDT1 and FON along the evolution for an arbitrary selected decision variable at 0% and 20% noise levels.

By comparing the search range depicted in Fig. 2.8(a)-(b) and Fig. 2.8(c)-(d), it is clear that a disciplined evolutionary search is lacking at high noise levels. Specifically, MOEA is capable of narrowing down the search range for better evolutionary search optimization in a noise-free environment. On the other hand, it can be observed that the mean location of individuals remains relatively the same despite a more diverse search space. This implies that the evolutionary process roughly knows where the promising regions are despite the presence of noise, most probably a consequence of the correct decision-making. More significantly, it also means that the impact of keeping out the true non-dominated solutions is greater than the reinsertion of inferior solutions. Therefore, apart from the need to improve the archive updating process of MOEAs, it is also desirable to devise a scheme that allows adaptation of search range based upon the mean location of individuals for better evolutionary search optimization in a noisy environment.

2.4 Conclusion

In this chapter, extensive studies have been performed to examine the impact of noisy environments in evolutionary multi-objective optimization, particularly for the population dynamics of fitness and diversity. It has been observed that the impact of noise on MOEA is different for the various benchmark problems, i.e. MOEA tends to evolve better solutions for some of the problems in the presence of low-level noise, while the evolutionary optimization process degenerates into a random search under increasing levels of noise. Furthermore, it seems that the selection process is more reliable in the early stages of evolution and the statistical analysis of online optimization behavior in the decision space has revealed that the evolution defines a population distribution with a mean value that remains relatively invariant in the decision space despite the different environmental conditions.

Chapter 3
Handling Noise in Evolutionary Multi-objective Optimization

In the previous chapter, we have shown empirically that the performance of MOEA deteriorates quickly with increasing noise intensities. As the results suggest, the canonical MOEA will face difficulties identifying non-dominated solutions, let alone maintaing a diverse set of near-optimal solutions.

In single-objective optimization, there are three basic approaches to suppress the detrimental effects of noise [144], 1) explicit averaging, 2) implicit averaging, and 3) selection modification. In explicit averaging, each solution is evaluated a number of times and averaged to compute the expected objective values. Increasing the number of samples (H) reduces the degree of uncertainty by a factor of $\sqrt{\mathrm{H}}$ at the expense of higher computational cost. Instead of re-evaluating and averaging the objective values over a number of samples, a large population is used in implicit averaging. When population size is large, there are many similar solutions and the influence of noise is compensated as the algorithm revisits the same region repeatedly. In selection modification, the ranking and selection procedures are modified such that a solution is judged better than another solution only if it satisfies certain conditions.

The ideas behind these three approaches can be easily applied to multi-objective optimization to suppress the effects of noise. But no matter which approach is adopted, it is necessary to consider how noise affects the selection, elitism, and diversity preservation processes in MOEA. There are two ways in which an inferior solution can be chosen over a better one in the selection process. Firstly, the selection mechanism can perceive an inferior solution to dominate a superior solution under the influence of noise. Secondly, as long as the inferior solution appears to be non-dominated, it can be selected by virtue of a better perceived degree of diversity measure. Similarly, the archive can be deceived into storing inferior solutions. Not only can these archived solutions drive out superior solutions, they can also prevent good solutions from entering the archive. Recall that most state-of-the-art MOEAs are elitist in nature. The worst problem is that these archived inferior solutions can mislead the entire optimization process, resulting in sub-optimal or unrealistic solution sets.

C.-K. Goh and K.C. Tan: Evolutionary Multi-objective Optimization, SCI 186, pp. 55–99.
springerlink.com

The straight-forward approach is to remove elitism completely. However, there is no indication that non-elitist MOEAs will perform significantly better for noisy problems and we can identify two difficulties in the absence of elitism. Firstly, good solutions will be lost along the evolution and we can expect slower convergence since elite solutions are not exploited. Furthermore, when noise intensity is sufficiently high, what will happen is that the MOEA will favor a particular set of solutions in one generation while favoring another set of solutions in another generation. This will result in the algorithm oscillating between different regions in the search space without ever converging.

Based on the above discussions, it is clear that much thought and effort must be devoted to the design of robust MOEAs for noisy multi-objective optimization. This chapter describes a number of existing MOEAs for noisy optimization, focusing on how archiving, selection, and diversity preservation are performed in the presense of noise.

3.1 Estimate Strength Pareto Evolutionary Algorithm

Teich [263] suggested the estimate strength Pareto evolutionary algorithm (ESPEA), which is based on the strength Pareto evolutionary algorithm (SPEA) developed by Zitzler and Thiele [300]. The main difference between the two algorithms is the use of probability theory to model the stochastic objective values. The major steps of SPEA and ESPEA within each iteration are shown below.

Strength Pareto Evolutionary Algorithm (SPEA)

Step 1: Evaluate all solutions in P_t.
Step 2: Update Archive:

- Add non-dominated solutions in P_t to A_t.
- Remove dominated solutions from updated A_t.
- Apply clustering if $|A_t|$ exceeds size limit.

Step 3: Fitness assignment:

- Assign fitness to archived solutions based on non-dominance.
- Assign fitness to solutions in P_t based on fitness assigned to archived solutions.

Step 4: Recombination process

Estimate Strength Pareto Evolutionary Algorithm (ESPEA)

Step 1: Evaluate all solutions in P_t.
Step 2: Update Archive:

- Calculate the probability of dominance for all solutions in P_t. Add the set of solutions with the best probability of dominance to A_t.
- Calculate the probability of dominance for all solutions in A_t. Remove the set of solutions with the worst probability of dominance from A_t.
- Apply clustering if $|A_t|$ exceeds size limit. Euclidean distance is calculated based on expected objective values.

Step 3: Fitness assignment:

- Assign fitness to archived solutions based on probabilistic non-dominance.
- Assign fitness to solutions in P_t based on fitness assigned to archived solutions.

Step 4: Recombination process

It can be easily observed that the general algorithmic stucture of SPEA is retained in ESPEA but the introduction of a probability of dominance brings about significant changes to how the actual archive updating and fitness assignment processes are being performed to improve algorithm robustness to noise.

In ESPEA, each noisy objective value is described by a property interval $[f_i^L, f_i^U]$, where f_i^L and f_i^U represent the lower and upper bounds of f_i. Noise is assumed to be uniformly distributed so f_i can take up any value within the interval $[f_i^L, f_i^U]$ with equal probability. Let us consider two solutions $\mathbf{F}_a$ and $\mathbf{F}_b$. The probability that solution $\mathbf{F}_a$ dominates solution $\mathbf{F}_b$ is given by

$$P(\mathbf{F}_a \preceq \mathbf{F}_b) = \prod_{i=1}^{M} P(f_{a,i} \preceq f_{b,i}) \tag{3.1}$$

where $P(f_{a,i} \preceq f_{b,i})$ is defined as

$$P(f_{a,i} \preceq f_{b,i}) = \begin{cases} 0, & \text{if } f_{b,i}^U < f_{a,i}^L \\ 1, & \text{if } f_{a,i}^U < f_{b,i}^L \\ \frac{1}{f_{a,i}^U - f_{a,i}^L} \cdot \left(\int_{f_{a,i}^L}^{f_{b,i}^L} dy + \int_{\max(f_{a,i}^S, f_{b,i}^S)}^{\min(f_{a,i}^U, f_{b,i}^U)} \frac{f_{b,i}^S - y}{f_{b,i}^U - f_{b,i}^S} \right), & \text{otherwise} \end{cases} \tag{3.2}$$

In this way, $\mathbf{F}_a \preceq \mathbf{F}_b$ only if the upper bound of the property interval $f_{a,i}^U$ is smaller than the lower bound $f_{b,i}^L$ for all objectives. In another words, we can say for certain that $\mathbf{F}_a$ dominates $\mathbf{F}_b$ when the worst case $\mathbf{F}_a$ is better than the best case $\mathbf{F}_b$. On the other hand, $\mathbf{F}_a$ does not dominate $\mathbf{F}_b$ as long as the upper bound $f_{b,i}^U$ is smaller than the lower bound $f_{a,i}^L$ for any one objective. When there is an overlap between the property intervals of both solutions, $\mathbf{F}_a \preceq \mathbf{F}_b$ with a certain degree of probability. If the noisy objective values of the solution $\mathbf{F}_a$ can be described by $[f_{a,i} - \delta_{a,i}, f_a + \delta_{a,i}]$, we can rewrite the third case of equation 3.2 as follows:

$$P(f_{a,i} \preceq f_{b,i}) = \frac{1}{2\delta_{a,i}} \cdot \Big[(f_{b,i} - \delta_{b,i} - \min(f_{b,i} - \delta_{b,i}, f_{a,i} - \delta_{a,i})) + \frac{1}{2\delta_{b,i}} \cdot (\max(f_{a,i} - \delta_{a,i}, f_{b,i} - \delta_{b,i}) - \min(f_{a,i} + \delta_{a,i}, f_{b,i} + \delta_{b,i})) \Big] \quad (3.3)$$

Intuitively, $P(f_{a,i} \preceq f_{b,i})$ is sensitive to the setting of $\delta_{a,i}$. Teich assumed a fixed $\delta_{a,i}$ which is set before each optimization run. However, *a priori* knowledge regarding $\delta_{a,i}$ is typically not available. One possible approach is to estimate $\delta_{a,i}$ during the clustering process when the expected cluster distances are calculated.

Archive Update: The algorithm maintains two populations, the population (P_t) of evolving individuals and the fixed-size archive (A_t) of non-dominated solutions. Non-dominated solutions found in P_t are copied and added to the archive. After which, dominated solutions found in the updated archive are removed. The challenge in this updating process is the identification of good solutions to be added and kept in the archive. This is where the probability of dominance comes into play in ESPEA, allowing solutions that would otherwise be rejected by the deterministic scheme a chance to enter the archive. The decision to add any solution to the archive is based on a mean probability that the solution is dominated by other solutions in P_t:

$$R(\mathbf{F}_a) = \frac{1}{(|\mathrm{P}_t|)} \sum_{\mathbf{F}_b \in \mathrm{P}_t} P(\mathbf{F}_b \preceq \mathbf{F}_a), \quad (3.4)$$

where a lower $R(\mathbf{F}_a)$ implies a higher probability that $\mathbf{F}_a$ is non-dominated. With each solution assigned a different probability of being dominated (implying varying degrees of non-dominance), it is not readily apparent which solutions belong to the first non-dominated front. In ESPEA, the criterion for solution entry to the archive can be governed by any of the three following rules:

- Add $\mathbf{F}_a$ to archive if the probability of being dominated $R(\mathbf{F}_a)$ is below a threshold α.
- Add β% of solutions with the lowest $R(\mathbf{F}_a)$ to the archive.
- Add all γ solutions with the lowest $R(\mathbf{F}_a)$ to the archive.

The idea is to ensure that all solutions with a decent probability of being non-dominated are copied into the archive. Thereafter, solutions with high probability of being dominated are removed from the archive and the probability of the stored solution being dominated by other solutions in A_t is calculated as follows:

$$R(\mathbf{F}_a) = \frac{1}{(|\mathrm{A}_t|)} \sum_{\mathbf{F}_b \in \mathrm{A}_t} P(\mathbf{F}_a \preceq \mathbf{F}_b), \quad (3.5)$$

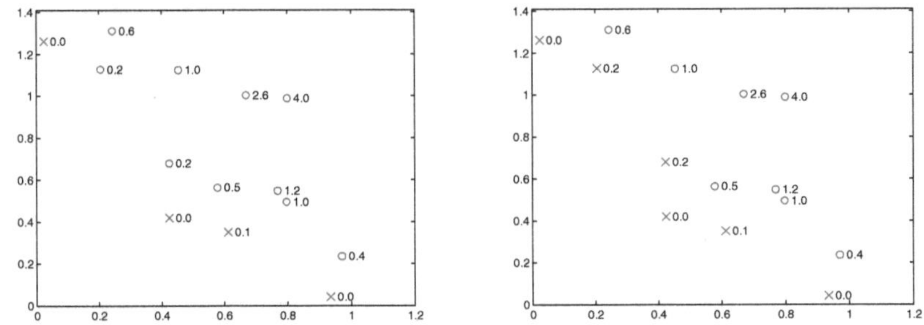

(a) (b)

Fig. 3.1 Archived solutions for (a) α=0.15 and (b) α=0.25 at δ=0.2. The probabilities of being dominated are appended next to the associated solutions. Archived and non-archived solutions are denoted by x and $\circ$, respectively.

As before, the decision to remove any solution from the archive is determined by any of the following heuristical rules:

- Remove $\mathbf{F}_a$ from A_t if the probability of being dominated $R(\mathbf{F}_a)$ is above a threshold α'.
- Remove β'% of solutions from A_t with the largest $R(\mathbf{F}_a)$).
- Remove all γ' solutions with the largest $R(\mathbf{F}_a)$) from A_t.

Therefore, what is left in the archive is a set of probably non-dominated solutions at the end of the updating process. Consider the set of solutions illustrated in Fig. 3.1. The first rule is applied in this particular instance to update appropriate solutions into the archive; archived solutions are denoted by x, while the remaining solutions are represented by $\circ$. The probabilities of being dominated $R(\mathbf{F}_a)$ are shown next to the associated solutions at δ=0.2. It is clear that the archiving process is sensitive to threshold settings. At $\alpha = 0.15$, some of the perceived non-dominated solutions are left out of the archive. This implies that a lower α may impose too strict a criterion and good solutions will still be denied the chance to survive. On the other hand, a higher α will result in more solutions being stored in the archive. While storing more solutions will increase the prospects of preserving good solutions, it will also increase the likelihood of updating more inferior solutions as well. Thus, the setting of α represents the tradeoff between preventing the potential loss of good solutions and exploiting solutions with high probabilities of non-dominance.

Fitness assignment: Fitness assignment is conducted after the archive update. Since the solutions from both P_t and A_t will participate in the recombination process, fitness is assigned to individuals from both populations. The fitness of the archived solutions, also called strength (S), are calculated first as follows:

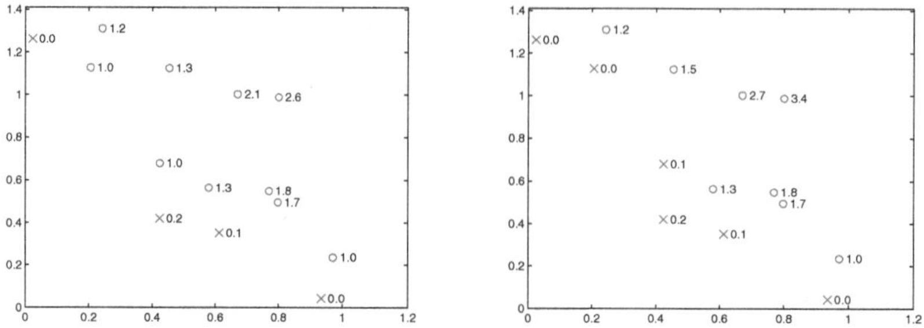

(a) (b)

Fig. 3.2 Fitness assignment for (a) α=0.15 and (b) α=0.25 at δ=0.2. Archived and non-archived solutions are denoted by x and ∘ , respectively.

$$S(\mathbf{F}_a) = \frac{1}{N+1} \sum_{\mathbf{F}_b \in \mathrm{P}_t} P(\mathbf{F}_a \preceq \mathbf{F}_b). \tag{3.6}$$

After which, the fitness of each of the individuals in P_t is equal to one plus the total strength of the archived solutions that dominate it with a probability greater than α.

$$F(\mathbf{F}_b) = 1 + \sum_{\mathbf{F}_b \in \mathrm{P}_t | P(\mathbf{F}_b \preceq \mathbf{F}_a) \geq \alpha} S(\mathbf{F}_a) \tag{3.7}$$

Fig. 3.2 shows the fitness of each solution at α=0.15 and α=0.25. All archived solutions have fitness values much smaller than 1.0, while solutions in the evolving population have fitness values greater than 1.0. The impact of the archiving process is readily apparent in these fitness values. Notice that some of the perceived non-dominated solutions have relatively poor fitness values in Fig. 3.2(a). If we are to increase α beyond 0.4, then the archive will also include some of the perceived dominated solutions. In this way, solutions perceived to be dominated will be assigned good fitness values and hence, increasing the chances of participating in the recombination process.

3.2 Multi-Objective Probabilistic Selection Evolutionary Algorithm

In an independent study, Hughes [127, 128] suggested a multi-objective probabilistic selection evolutionary algorithm (MOPSEA) which also employs probabilistic dominance to account for the effects of noise in the objective space. The probability of dominance introduced by Hughes is different from that described above in the sense that noise is assumed to be normally distributed. This assumption leads to a very different formulation and interpretation for the probabilistic dominance relationship between two solutions, $\mathbf{F}_a$ and $\mathbf{F}_b$. $P(\mathbf{F}_a \preceq \mathbf{F}_b)$ is actually the probability that the perception of $\mathbf{F}_a$ dominating $\mathbf{F}_b$ is wrong and it is calculated as follows:

$$P(f_{a,i} \preceq f_{b,i}) = 0.5 + 0.5\text{erf}(\frac{f_{a,i} - f_{b,i}}{2\sigma_n}) \quad (3.8)$$

$$P(\mathbf{F}_a \preceq \mathbf{F}_b) = \prod_{i=1}^{M} P(f_{a,i} \preceq f_{b,i}) \quad (3.9)$$

where σ_n is the standard deviation of the objective values and $P(f_{a,i} \preceq f_{b,i})$ is the probability that the judgement that the i-th objective of $\mathbf{F}_a$ is better than the i-th objective of $\mathbf{F}_b$ is wrong. One attractive feature of this formulation is that the probabilities of other dominance relationships between $\mathbf{F}_a$ and $\mathbf{F}_b$ can be calculated easily once $P(\mathbf{F}_a \preceq \mathbf{F}_b)$ is known.

$$P(\mathbf{F}_b \preceq \mathbf{F}_a) = 1 - P(\mathbf{F}_a \preceq \mathbf{F}_b) \quad (3.10)$$

$$P(\mathbf{F}_b \sim \mathbf{F}_a) = 1 - P(\mathbf{F}_a \preceq \mathbf{F}_b) - P(\mathbf{F}_b \preceq \mathbf{F}_a) \quad (3.11)$$

σ_n is typically not known *a priori* and it can only be estimated during the optimization process. If computational resources and time permits, σ_n can be estimated for each and every solution by evaluating the solution over a few samples. But this is not the case for many real-world applications. If variance is assumed to be constant throughout the search space, Hughes suggested estimating σ_n from a random solution at the start of the optimization process and using it for all subsequent comparisons.

Fieldsend and Everson [78] considered the efficient computation of probabilistic ranking and suggested an online Bayesian learning algorithm to estimate the noise variance σ_n. More significantly, it is demonstrated that the algorithm is capable of tracking noise variance under different conditions, such as unknown noise properties, independent noise for each objective, and etc.

Unlike most state-of-the-art MOEAs, MOPSEA maintains only one evolving population on which genetic operations are performed to generate new solutions. Elitism is implemented by retaining a significant portion of the best solutions. Nonetheless, Hughes suggested (but did not simulate) an unique approach of using a probabilistic tournament selection to update an archive in [129]. The basic algorithmic structure of MOPSEA is kept simple, with stochastic universal sampling, intermediate crossover, and uniformly distributed mutation operators, probably to highlight the contributions of probabilistic dominance in noisy multi-objective optimization. The main steps of the algorithm within an iteration are shown below:

Multi-objective Probabilistic Selection Evolutionary Algorithm (MOPSEA)

Step 1: Evaluate all solutions in P_t.
Step 2: Fitness assignment: Calculate the probabilistic Pareto ranking of all solutions.

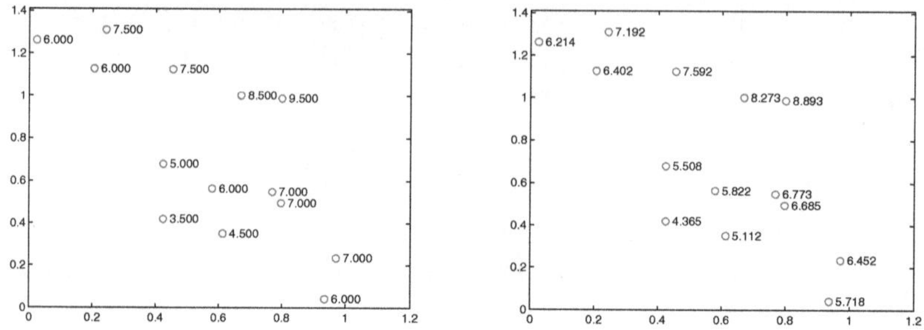

(a) (b)

Fig. 3.3 Probabilistic Pareto ranking for (a) $\sigma_n = 0$ and (b) $\sigma_n = 0.3$

Step 3: Elitism: Maintain the best 70% of the solutions in terms of fitness and discard the rest.

Step 4: Recombination process: Perform selection, crossover, and mutation on the population until it is filled to the original population size.

Fitness Assignment: In a certain sense, the probabilistic Pareto ranking scheme is rather similar to the scheme presented by Fonseca and Fleming [84] except that the rank of each solution is determined by the probabilities of other solutions dominating it. The rank of $\mathbf{F}_a$ is calculated as

$$R(\mathbf{F}_a) = \sum_{\mathbf{F}_b \in \mathrm{P}_t} P(\mathbf{F}_b \preceq \mathbf{F}_a) + 0.5 \sum_{\mathbf{F}_b \in \mathrm{P}_t} P(\mathbf{F}_a \sim \mathbf{F}_b) - 0.5 \tag{3.12}$$

The second term in equation 3.12 is introduced to account for the probability of non-dominance between the solutions. Notice that it is multiplied by a factor of 0.5 to maintain consistency in the sum of ranks. The third term allows a solution to compare with itself. Alternatively, we can rewrite equation 3.12 as

$$R(\mathbf{F}_a) = \sum_{\mathbf{F}_b \in \mathrm{P}_t} P(\mathbf{F}_b \preceq \mathbf{F}_a) + 0.5 \sum_{\mathbf{F}_b \in \mathrm{P}_t} P(\mathbf{F}_b \sim \mathbf{F}_a) \tag{3.13}$$

This ranking scheme is illustrated in Fig. 3.3 for $\sigma_n = 0$ and $\sigma_n = 0.3$. We can observe that although the perceived non-dominated solutions are still assigned better fitness, the probabilistic ranking scheme tends to favor well-spaced solutions. If you are to think about it a little deeper, you will realise that there is a higher degree of uncertainty regarding the dominance of a solution located in a denser region. Therefore, the probabilistic ranking scheme not only emphasizes on solutions with high probabilities of non-dominance but also on solutions with less uncertainties regarding their quality.

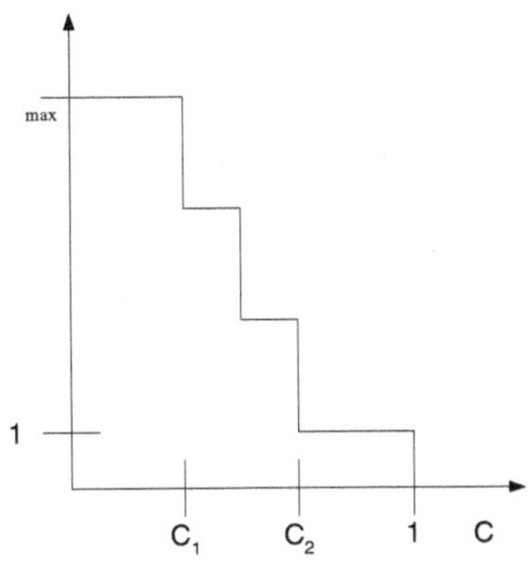

Fig. 3.4 Lifetime Dependence on ratio of fraction of archived solutions dominated

A probabilistic sharing mechanism is applied to encourage diversity and uniform distribution of solutions in the evolving population. The probabilistic niche count for individual $\mathbf{F}_a$ is defined as

$$nc(\mathbf{F}_a) = \sum_{\mathbf{F}_b \in \mathrm{P}_t} v(\mathbf{F}_a, \mathbf{F}_b) - \frac{h_d}{\sqrt{1+h_n^2}} + 1 \tag{3.14}$$

where $v(\mathbf{F}_a, \mathbf{F}_b)$ is the geometric mean between $\mathbf{F}_a$ and $\mathbf{F}_b$ in the objective space, $h_d = \sigma_s/\sqrt{2\sigma_n^2}$ and $h_d = \sigma_s/\sqrt{2\sigma_n^2}$. The shared fitness value of each individual is then calculated by dividing the fitness by its niche count.

3.3 Noise Tolerant Strength Pareto Evolutionary Algorithm

Buche *et al* [32] modified the archive updating mechanism of the SPEA [300] to reduce the detrimental effects of noise on the elitist process. The design of noise tolerant strength Pareto evolutionary algorithm (NTSPEA) is motivated by the authors' work on gas turbine combustion processes and they are particularly concerned about the influence of outliers on the optimization procedure. Outliers are the result of instrumentation failure, which can be of several magnitudes larger than measurement noise. Under the influence of an outlier, an inferior solution may be perceived to be unrealistically good, driving out the true non-dominated solutions and misleading the optimization process.

The NTSPEA introduces the concept of a lifetime to non-dominated solutions stored in the archive so the impact of inferior solutions on the optimization process will be limited. At every generation, non-dominated solutions with expired lifetime are removed from the archive, added to the evolving population, and re-evaluated. After the re-evaluation, it is unlikely that an outlier will remain non-dominated and hence, it will not be updated into

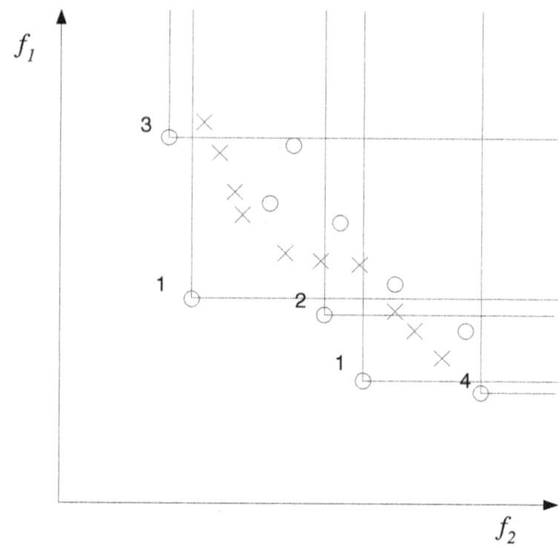

Fig. 3.5 Lifetime of solutions in population for $\kappa_{\max} = 4$, C1=0.1 and C2=0.3. Previously archived solutions and population individuals are denoted as x and ∘, respectively.

the archive again. On the other hand, a good solution is likely to remain non-dominated.

The lifetime of each non-dominated individual is dependent on the fraction of archived solutions it dominates and this relationship is shown in Fig. 3.4. Since the archive is empty initially, the first set of non-dominated solutions is assigned the maximum lifetime of $\kappa_{\max}$. In the subsequent generations, archived solutions with expired lifetimes are removed from the archive before the updating process. Each newly generated solution is assigned a lifetime which is inversely proportional to the fraction of previously archived solutions that it dominates. Fig. 3.5 shows the lifetime of solutions (denoted by ∘) found in the current generation with respect to the members of the previous archive (X) denoted by on a two-objective problem.

With the archive of non-dominated solutions in place, fitness are assigned to the solutions in archive and evolving population as in the case of SPEA. Binary tournament selection is then conducted to select solutions from the combined archive and evolving populations. Thereafter, crossover and mutation are performed on the selected solutions to create a new population. As mentioned before, archived solutions with expiring lifetimes are re-evaluated in NTSPEA. The archive is first checked for solutions with expiring lifetimes and these solutions are copied and added to the new population before the evaluation procedure. The main steps of NTSPEA are summarized below:

Noise Tolerant Strength Pareto Evolutionary Algorithm (NTSPEA)

Step 1: Evaluate all solutions in P_t.
Step 2: Update Archive:

- Delete expired solutions from A_t.
- Calculate lifetime for all solutions in P_t.

- Add non-dominated solutions in P_t to A_t.
- Remove dominated solutions from updated A_t.
- Apply clustering if $|A_t|$ exceeds size limit. Perform explicit averaging to calculate cluster distances.

Step 3: Fitness assignment:

- Assign fitness to archived solutions based on non-dominance.
- Assign fitness to solutions in P_t based on fitness assigned to archived solutions.

Step 4: Recombination process
Step 5: Add expiring solutions from A_t to P_t.

3.4 Modified Non-dominated Sorting Genetic Algorithm II

Babbar *et al* [12] introduced explicit averaging into NSGAII and modified the non-dominated sorting procedure to allow seemingly dominated solutions into the first non-dominated front. This modified NSGAII, which we shall call MNSGAII, also incorporates a procedure to remove unreliable solutions from the final set of non-dominated solutions. The main steps of MNSGAII are summarized below:

Modified non-dominated Sorting Genetic Algorithm II (MNSGAII)

Step 1: Perform explicit averaging to calculate expected objective values for all solutions in P_t. Update variance of each solution.
Step 2: Combine P_t and A_t.
Step 3: Perform modified non-dominated sorting on combined population.

- Perform non-dominated sorting on combined population as in NSGAII.
- Compute neighborhood distances between solutions from the first non-dominated front and the other fronts.
- Assign all solutions with neighborhood distance smaller than the neighborhood restriction factor to the first non-dominated front.

Step 4: Update neighborhood restriction factor.
Step 5: Create A_{t+1}: Perform crowding sort to truncate combined population size to $|A_t|$.
Step 6: Recombination process: Conduct tournament selection on A_{t+1} to form mating pool. Perform crossover and mutation to create P_{t+1}.

Interested readers are encouraged to read [61] on NSGAII. As in all dominance-based approaches, there is a bias towards the perceived non-dominated solutions. Since the solutions may seem better or worse than what they really are in noisy multi-objective optimization, it is highly probable that the selection pressure is directed towards inferior solutions, while the truly good solutions are lost quickly along the evolution. The first step taken in MNSGAII to reduce the effects of noise is to conduct explicit averaging for all solutions. Although explicit averaging can reduce the uncertainty in the selection process, it is not feasible to use large number of samples, particularly for problems with expensive objective functions.

Since the number of samples is limited, it is inevitable that the first non-dominated front will comprise of both dominated and non-dominated solutions. To prevent the loss of potentially useful solutions, Babbar *et al* suggested a clustering mechanism to induce solutions from the inferior non-dominated fronts into the first layer. This mechanism works by comparing the distances between solutions from the first non-dominated front and the other fronts. A higher ranked solution $\mathbf{F}_a$ is re-assigned to the first non-dominated front if the following criterion is satisfied for any arbitrary objective,

$$|f_{a,i} - f_{b,i}| < K\sqrt{\frac{v_{a,i} + v_{b,i}}{2}} \tag{3.15}$$

where $f_{b,i}$ is the i–th objective of solution $\mathbf{F}_b$ from the first non-dominated front, $v_{a,j}$ is the variance of $\mathbf{F}_a$ updated during the averaging process and K is the neighborhood restriction factor. The rationale is that it is very likely for a perceived inferior solution located in close proximity to a perceived non-dominated solution in the objective space to be a true non-dominated solution. By assigning the first rank to these otherwise higher ranked solutions, truely good solutions are given the reprieve necessary to survive the selection process.

Fig. 3.6 illustrates the modified non-dominated sorting procedure in MNSGAII. It is clear that the number of solutions present in the first non-dominated front is highly dependent on the setting of K; a large K setting will reduce selection pressure, while too small a K value will defeat the purpose of modifying the sorting procedure in the first place. MNSGAII applies a simulated annealing inspired adaptive scheme that reduces K over generations. The equation that governs the behavior of K is given by,

$$K = C(1 - \exp(\beta t)) \tag{3.16}$$

where C determines the largest setting of K and β controls the rate at which K reduces with t. Initially, K is large and more dominated solutions will be accepted into first non-dominated front. On the other hand, the number of solutions being accepted is reduced over the number of generations. The rationale is that the reliability of the solution will increase with time through explicit averaging, allowing the algorithm to sort the solutions into the different layers of non-dominated fronts with a greater degree of certainty.

At the end of the evolutionary process, the clustering mechanism is applied once again. However, in this instance, it is applied to remove solutions that are significantly different from the other archived solutions. Solutions that do not satisfy the following criterion for all M objectives are removed from the final archive.

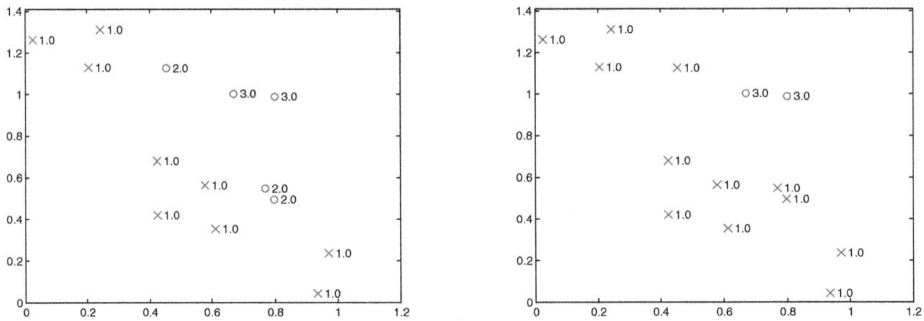

(a) (b)

Fig. 3.6 Solutions belonging to the first non-dominated front (x) for (a) K=0.2 and (b) K=0.3

$$|f_{a,i} - f_{b,i}| < K\sqrt{\frac{v_{a,i} + v_{b,i}}{2}} \; \forall i, ...M \tag{3.17}$$

3.5 Multi-objective Evolutionary Algorithm for Epistemic Uncertainty

Limbourg [183] extended a probabilistic theory known as the Dempster-Shafer framework of evidence to model epistemic uncertainties and suggested a set of guidelines for the design of selection and archive updating mechanisms for noisy multi-objective optimization. Associated with this probabilistic framework is the notion of belief and plausibility, which are used to characterize the lower and upper bounds of the k-nearest neighbour distances calculated during density estimation. We will call this algorithm MOEAEU and the major steps of the algorithm are shown below:

Multi-Objective Evolutionary Algorithm for Epistemic Uncertainty (MOEAEU)

Step 1: Evaluate all solutions in P_t.
Step 2: Update archive:

- Add solution to A_t as long as solution is not strong certain dominated.
- Remove solutions from A_t that are strong certain dominated.
- Apply truncation while $|A_t|$ exceeds size limit.
  - Calculate k-nearest neighbours belief and plausibility pairs for each solution.
  - Binary tournament selection:
    - Pick two solutions randomly from A_t.
    - Compare belief and plausibility pairs of selected solutions based on weak uncertain dominance.
    - Remove loser from archive.

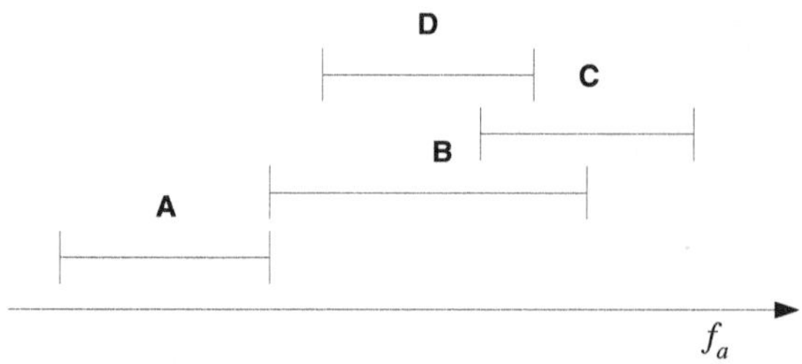

Fig. 3.7 Illustration of the Certain and Uncertain Dominance relation for 4 different solutions intervals

Step 3: Binary Tournament selection:

- Pick two solutions randomly from P_t.
- Compare selected solutions based on weak uncertain dominance.
- Add winner to mating pool.

Step 4: Recombination process

The Dempster-Shafer framework does not assume any probabilistic distribution and represents each noisy objective value as an interval, i.e. $f_a = [\underline{f}_a, \overline{f}_a]$. With such a representation, the manner in which comparisons are conducted between solutions is different. There are two degrees of dominance that an interval can be judged to be better than (or dominates) another:

Definition 3.1 Certain Dominance: The interval $f_a = [\underline{f}_a, \overline{f}_a]$ certain dominates $f_b = [\underline{f}_b, \overline{f}_b]$, denoted by $f_a \prec_c f_b$, $iff\ \overline{f}_a \leq \underline{f}_b$

Definition 3.2 Uncertain Dominance: The interval $f_a = [\underline{f}_a, \overline{f}_a]$ uncertain dominates $f_b = [\underline{f}_b, \overline{f}_b]$, denoted by $f_a \prec_{uc} f_b$, $iff\ \underline{f}_a \leq \underline{f}_b$ and $\overline{f}_a \leq \overline{f}_b$

f_a *certain* dominates f_b if the worst f_a is better than the best f_b. On the other hand, f_a *uncertain* dominates f_b if f_a is better than the best f_b and not worse than the worst f_b. The two relations can be illustrated using the four intervals (**A**, **B**, **C**, and **D**) shown in Fig. 3.7. **A** certain dominates **B**, **C**, and **D**. This follows that **A** uncertain dominates **B**, **C**, and **D** as well. **B** and **D** uncertain dominate **C** but are indifferent to **C** with respect to the certain dominance criterion. **B** and **D** are indifferent with respect to both dominance relations. It is clear that the certain dominance is a much stronger criterion as compared to uncertain dominance. We can make the decision that f_a is better than f_b with a much higher degree of confidence when certain dominance holds for f_a as compared to the case of uncertain dominance.

Extending these two interval relationships to multi-objective optimization, Limbourg suggested the following two dominance relations:

Definition 3.3 Weak Uncertain Dominance:$\mathbf{F}_a$ weakly and uncertain dominates $\mathbf{F}_b$, denoted by $\mathbf{F}_a \preceq_{uc} \mathbf{F}_b$, iff $f_{a,i} \prec_{uc} f_{b,i}$ $\forall i \in \{1, 2, ..., M\}$

Definition 3.4 Strong Certain Dominance: $\mathbf{F}_a$ strongly and certain dominates $\mathbf{F}_b$, denoted by $\mathbf{F}_a \prec_c \mathbf{F}_b$, iff $f_{a,i} \preceq_c f_{b,i}$ $\forall i \in \{1, 2, ..., M\}$ and $f_{a,j} \prec_c f_{b,j}$ $\exists j \in \{1, 2, ..., M\}$

Selection Process. The dominance relation described in Definition 3.3 is used as the decision-making criterion of the tournament selection. In another words, solutions are compared using the weak uncertain dominance relation. Note that this relation has a low degree of indifference and the probability of making a wrong decision based on this relation can be very high depending on the degree of overlap between the objective intervals. However, since decision errors are inevitable in the presence of noise, Limbourg argued that the tradeoff in accuracy must be made because a high degree of indifference in the comparison of solutions will result in a random search.

Archiving Process. The MOEAEU employs a fixed-size archive A_t to store all the non-dominated solutions found in P_t. In each generation, solutions that are not dominated in the strong certain sense, described by Definition 3.4, are added to the archive. Previously archived solutions that are strong certain dominated are removed from the archive. Notice that such an updating scheme allows all solutions that are probably non-dominated to be added to the archive and both perceived dominated and non-dominated solutions exist in the archive simultaneously. Since it is much harder to satisfy the strong certain dominance criterion, we can expect the archive to be filled up quickly with solutions, especially if the interval range is large. When the size of the archive reaches the limit, truncation is performed to keep the archive size under control. As in the case of MOEAs for deterministic optimization, only solutions located in the least crowded regions are kept. Density assessment is estimated using the *k-th* neighbour distance where the normalized distance between any two solutions $\mathbf{F}_a$ and $\mathbf{F}_b$ is calculated in the following manner:

$$d(\mathbf{F}_a, \mathbf{F}_b) = \sqrt{\sum_{i=1,\ldots,M} \hat{d}_i^2 \cdot (f_{a,i} - f_{b,i})} \tag{3.18}$$

$$\hat{d}_i = \frac{1}{\max_{\mathbf{F}_a \in \mathrm{A}_t}(\mathbf{F}_a) - \min_{\mathbf{F}_b \in \mathrm{A}_t}(\mathbf{F}_b)} \tag{3.19}$$

where $\hat{d}_i$ represents the maximum extend found in the archive for the i-th objective. Since the solutions $\mathbf{F}_a$ and $\mathbf{F}_b$ are represented by intervals, the normalized Euclidean distance between them is also an interval. In MOEAEU, $d(\mathbf{F}_a, \mathbf{F}_b)$ is represented by belief and plausibility pairs. The lower bound or belief is given by

$$bel\Big(d(\mathbf{F}_a,\mathbf{F}_b)\Big) = \min_{i=1,\ldots M}(f_{a,i}-f_{b,i})^2 \tag{3.20}$$

$$\min_{i=1,\ldots M}(f_{a,i}-f_{b,i})^2 = \begin{cases} 0, & \text{if } \underline{f}_{a,i}-\overline{f}_{b,i}\le 0 \\ & \text{or } \underline{f}_{b,i}-\overline{f}_{a,i}\le 0 \\ \min\Big((\underline{f}_{a,i}-\overline{f}_{b,i})^2,(\underline{f}_{b,i}-\overline{f}_{a,i})^2\Big), & \text{otherwise} \end{cases}$$

while the upper bound or plausibility is calculated as follows:

$$pl\Big(d(\mathbf{F}_a,\mathbf{F}_b)\Big) = \max_{i=1,\ldots M}(f_{a,i}-f_{b,i})^2 \tag{3.21}$$

$$\max_{i=1,\ldots M}(f_{a,i}-f_{b,i})^2 = \max\Big((\underline{f}_{a,i}-\overline{f}_{b,i})^2,(\underline{f}_{b,i}-\overline{f}_{a,i})^2\Big)$$

After calculating the (*bel*, *pl*) pairs for all the solutions in A_t, binary tournament selection is conducted to remove the more crowded solutions. During the tournament selection, two solutions are chosen at random and their *k-th* neighbour distance intervals are compared using the uncertain dominance criterion. The winner of the tournament remains in the archive, while the loser is removed. In the event that the (*bel*, *pl*) pair of the two solutions are indifferent, the survivor is selected randomly.

3.6 Indicator-Based Evolutionary Algorithm for Multi-objective Optimization

Basseur and Zitler [16] considered how noise can be handled in indicator-based evolutionary algorithms (IBEA). In this particular implementation of IBEA, the ϵ_+-indicator is used to guide the evolutionary process and the expected indicator values are used as fitness values to improve robustness. The main steps in each iteration of the algorithm is shown below:

Indicator Based Evolutionary Algorithm (IBEA)

Step 1: Evaluate all solutions in P_t.
Step 2: Fitness assignment: Assign expected $I_{\epsilon}+$ value as fitness to solutions in P_t.
Step 3: Remove individual with the smallest fitness value.
Step 4: Recombination process

- Perform binary tournament selection, crossover, and mutation.
- Add new individual to P_t.

Similar to MOPSEA, IBEA is a steady-state algorithm but only the worst solution is replaced by a new individual in each generation.

Fitness Assignment: Like Teich, the authors assumed noise to be uniformly distributed. But the way in which the solutions are modeled is significantly different from the other works discussed previously. Most researchers assume that a true objective vector exists for each solution. In IBEA, each solution is associated with a probability distribution over the objective space. In other words, each solution $\mathbf{x}_a$ is associated with a set of objective vectors $\{\mathbf{F}_a\}$ instead. As a consequence of such a model, the expected indicator value is not calculated based on the expected objective values but as the average of indicator values with respect to the set of objective vectors.

$$Fit(\mathbf{x}_a) = \frac{1}{|S(\mathbf{x}_a)|} \sum_{\mathbf{F}_a \in S(\mathbf{x}_a)} E\Big(I\big(F(\mathrm{P}_t), F_a\big)\Big) \tag{3.22}$$

$$E\Big(I\big(F(\mathrm{P}_t), F_a\big)\Big) = \sum_{\mathbf{F}_1 \in S(\mathbf{x}_1), \mathbf{x}_1 \in \mathrm{P}_t} \cdots \sum_{\mathbf{F}_j \in S(\mathbf{x}_j), \mathbf{x}_j \in \mathrm{P}_t} I_{\epsilon+}(\{\mathbf{F}_1, ..., \mathbf{F}_j\}, \{\mathbf{F}_a\}) \tag{3.23}$$

where $S(\mathbf{x}_a)$ is the set of samples drawn for the solution, F is the function that maps the solutions from the decision space to the objective space. Accordingly, the fitness value provides an indication of how much $\mathbf{x}_a$ contributes to the overall quality of the evolved solution set.

One drawback of this fitness scheme is the large computational cost involved in the calculation of expected $\mathrm{I}_\epsilon+$ for each solution. To this end, the authors exploited the fact that the minimum $\mathrm{I}_\epsilon+$ value determines the actual value and hence, not all combinations of objective vectors have to be considered. The pseudocode for this estimation is given as:

Expected $\mathrm{I}_\epsilon+$ Value Estimation

Step 1: Determine $\bar{\epsilon} = \min_{\mathbf{x}_j \in \mathrm{P}_t} \max_{\mathbf{F}_j \in S(\mathbf{x}_j)} I_{\epsilon+}(\{\mathbf{F}_j\}, \{\mathbf{F}_a\})$.
Step 2: Determine relevant $I_{\epsilon+}$ values:

- Create empty List, L.
- Compute $\epsilon = I_{\epsilon+}(\{\mathbf{F}_j\}, \{\mathbf{F}_a\})$ for all $\mathbf{x}_j \in \mathrm{P}_t$ and $\mathbf{F}_j \in S(\mathbf{x}_j)$.
- Add $(\epsilon, \mathbf{x}_j)$ to L if $\epsilon < \bar{\epsilon}$.

Step 3: Sort L in increasing order of ϵ.
Step 4: Set I=0 and initialize zero array N of size $|\ \mathrm{P}_t\ |$
Step 5: Compute expected $\mathrm{I}_\epsilon+$ value estimation: While L is not empty

- load first $(\epsilon', \mathbf{x}'_j)$ pair from L.
- $p = \frac{1}{|\mathrm{L}| - \mathrm{N}[\mathbf{x}'_j]} \cdot \prod_{\mathbf{x}_j \in \mathrm{P}_t} \frac{1 - \mathrm{N}[\mathbf{x}'_j]}{|\mathrm{P}_t|}$
- $\mathrm{I} = \mathrm{I} + p \cdot \epsilon'$
- $\mathrm{N}[\mathbf{x}'_j] = \mathrm{N}[\mathbf{x}'_j] + 1$. Remove first element from L.

Step 6: Return I as expected $\mathrm{I}_\epsilon+$ value estimation.

Basseur and Zitler suggested using a bucket sort to reduce computational complexity of the sorting performed in Step 4. Simulation results conducted showed that the adopted fitness assignment scheme is significantly more effective as compared to computing the expected indicator value based on the expected objective values.

3.7 Multi-Objective Evolutionary Algorithm with Robust Features

We suggested three noise-handling features [95] and incorporated them into a simple MOEA, which we call MOEA-RF. Unlike the other algorithms described so far, these noise-handling features are designed based on the observations made on population dynamics presented in Chapter 2. The experiential learning directed perturbation (ELDP) is designed upon the observation that MOEAs are better in decision-making at the early stages of evolution, while the gene adaptation selection strategy (GASS) exploits the observation that the mean location of archived individuals remains relatively constant under different noise conditions. Lastly, the possibilistic archiving methodology is developed using the concept of possibility and necessity measures, which helps to perform effective archiving of noisy objective vectors. The main steps of the algorithm within each iteration is summarized below:

Multi-Objective Evolutionary Algorithm with Robust Features (MOEA-RF)

Step 1: Evaluate and assign Pareto ranking to all solutions in P_t.
Step 2: Calculate niche count for all solutions in P_t with respect to archived solutions.
Step 3: Possibilistic archiving.
Step 4: Combine P_t and archive.
Step 5: Create mating pool:

- Binary tournament selection:
  - Pick two solutions randomly from combined population.
  - Compare selected solutions based on Pareto rank.
  - Add winner to mating pool.
- Perform GASS on all solutions in mating pool.

Step 6: Recombination process: Perform crossover and ELDP

Experiential Learning Directed Perturbation (ELDP): As mentioned earlier, ELDP makes use of the better decisions at early generations to improve performance. The ELDP is a deviation from conventional mutation paradigm in two aspects: 1) the change in chromosome is ordered instead of being by chance and 2) variation can be performed either in genotype or phenotype space. In particular, the actual adaptation in ELDP is based on posterior knowledge of favorable movements in the search space. For the ensuing discussion, an individual is represented as a vector $\mathbf{X} = (\mathbf{g}, \mathbf{p}, \mathbf{f})$, where the vectors $\mathbf{g}$ and $\mathbf{p}$ represent the decision vector in the genotype space $\mathbf{G} \in B^{CL}$ and the phenotype space $\mathbf{X} \in n_x$, respectively; $\mathbf{f}$ is the associated objective vector in the objective space, $\mathbf{F}^M$. The binary representation $\mathbf{g}$ of the decision variables is mapped by the function $f : \mathbf{G} \rightarrow \mathbf{X}$ from the genotype space to the phenotype space and there is a corresponding inverse function $f^{-1} : \mathbf{X} \rightarrow \mathbf{G}$.

The experiential learning strategy adopted by ELDP for directed perturbation in the phenotype space is inspired by the role of the momentum term in back-propagation for neural networks; accelerating movement in the direction of improvement, while restricting movement otherwise. The variation for each decision variable x_j can be described as follows,

$$\bar{\triangle} x_j(t) = \triangle x_j(t) + \alpha \cdot \bar{\triangle} x_j(t-1) \tag{3.24}$$

where α represents the learning rate; $\triangle$ refers to changes acquired through prior genetic operations such as crossover, while $\bar{\triangle}$ corresponds to changes including the effect of momentum. According to equation 3.24, the posterior knowledge comes in the form of past movements made by the individual in concern. The ELDP defines a two-mode operation to impose the necessary control for directed variation in the phenotype space and to perform bit-flip mutation in the genotype space for genetic diversity. The ELDP operation is given as follows,

$$x_j(t+1) = \begin{cases} p_j(t) + \bar{\triangle} x_j(t), & \text{if } \triangle_{min} < |\alpha \cdot \bar{\triangle} x_j(t-1)| < \triangle_{max} \\ f\Big(g_{BF}(\mathbf{g}(t) + \triangle \mathbf{g}(t))\Big), & \text{otherwise.} \end{cases} \tag{3.25}$$

where $\bar{\triangle} x_j(t)$ refers to the variation described in equation 3.24 and $g_{BF}()$ denotes bit-flip mutation for the j-th decision variable. Note that the corresponding changes will also be updated in the genotype for any variation in the phenotype space. From equation 3.25, the magnitude of directed perturbation is bounded by $\triangle_{min}$ and $\triangle_{max}$, which can be set by the user. The limiting bounds on directed perturbation for x_j ensure that a new search direction is initiated through bit-flip mutation to reduce the impact of outliers or whenever the evolutionary search process has stalled. For simplicity, $\triangle_{min}$ and $\triangle_{max}$ is set as 0.0 and 0.1, respectively, on the normalized decision space.

The operation of ELDP is illustrated in Fig. 3.8. By considering each and every decision variable, the ELDP provides a simple and efficient way for adaptation of the required variation associated with each parameter. From (3.24) and (3.25), the variation increases in magnitude in the direction of change and thus accelerates convergence when $\triangle x_j(t)$ and $\bar{\triangle} x_j(t-1)$ have the same sign. Likewise, the variation is small if $\triangle x_j(t)$ and $\bar{\triangle} x_j(t-1)$ are different in sign, implying that the ELDP performs local fine-tuning in the later stages of evolution where movements tend to fluctuate. Moreover, such properties are desirable in the context of noisy objective function optimization

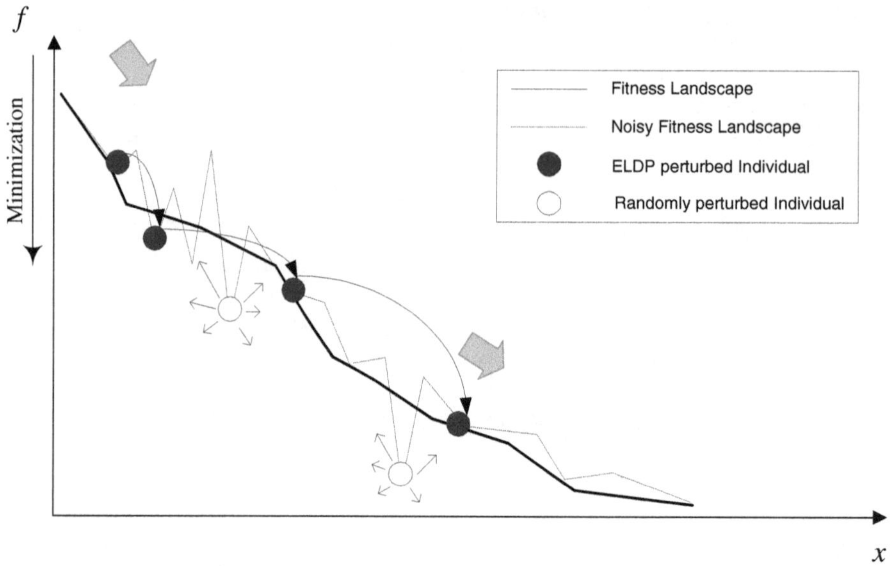

Fig. 3.8 Operation of ELDP

where inferior solutions are likely to participate in the recombination process. In such cases, the ELDP helps to reduce the stochastic influence of noise and prevents the individuals from changing haphazardly.

Gene Adaptation Selection Strategy (GASS): Recall that the mean of the population distribution remains relatively invariant in the decision space despite the different environmental conditions. In addition, the search range of the different variables tends to converge to a smaller region within the search space in a noise-free environment and remains relatively unchanged or even diverges in a noisy environment. It is thus useful to construct an approximate model of the ideal population behavior for guiding the evolutionary search process.

The proposed GASS attempts to manipulate population distribution so that the evolutionary algorithm exhibits certain desirable search characteristics. It first builds a posterior model of the desired population distribution and then adapts part of the selected individual's chromosome. Mathematically, the adaptation of gene structure is given as

$$x'_j(t) = \begin{cases} \mathrm{U}(a_j, b_j), & \mathrm{U}(0,1) < \frac{1}{n_x} \\ x_j(t), & \text{otherwise.} \end{cases} \tag{3.26}$$

Here $\frac{1}{n_x}$ is the probability of decision variable j being selected for adaptation. The GASS defines an operation in the phenotype space which is characterized by a uniformly distributed number U on the interval $[a_j, b_j]$ for each

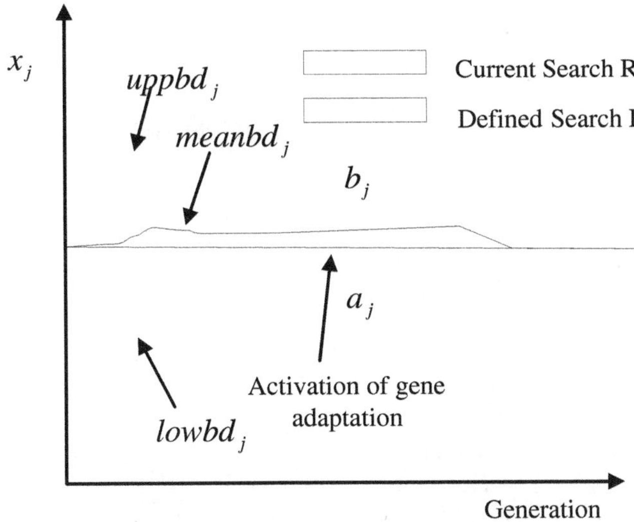

Fig. 3.9 Search range defined by convergence model

decision variable. After which, the corresponding genotypic adaptation will be updated. It adopts two different models to control the evolution for a better convergence, i.e. the interval $[a_j, b_j]$ is dependent on the state of evolution and the archival population distribution in the decision space.

Convergence model: The population distribution tends to converge as the evolving population approaches the final tradeoff. Since it is difficult to determine if PF^A corresponds to PF^*, the adopted model needs to define a space that is larger than the current search range along the j-th dimension to prevent a premature convergence. The corresponding interval is given as

$$\begin{aligned} a_j &= lowbd_j - w \cdot meanbd_j \\ b_j &= uppbd_j + w \cdot meanbd_j \end{aligned} \tag{3.27}$$

where w is a fixed parameter that controls the step change in the search range, $lowbd_j$, $uppbd_j$, and $meanbd_j$ correspond to the minimum, maximum and mean of x_j in the archive, respectively. The aim of the convergence model is to compel EA to look beyond the current search region as shown in Fig. 3.9. In the case where an individual corresponds to the global optimum, the overall quality of the evolving population is not adversely affected. This is because similar individuals have similar genetic information and the model creates individuals based on the converged search region.

Divergence model: A degenerate evolutionary search process is characterized by a non-convergent population distribution. In the situation where the evolutionary process degenerates into a random search, such as due to

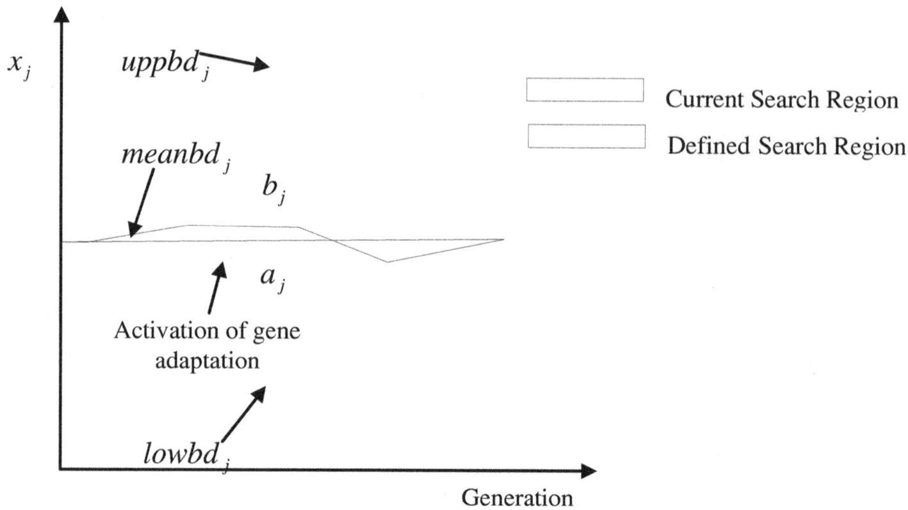

Fig. 3.10 Search range defined by divergence model

high level of uncertainty in the system, the interval for the j-th decision variable defines a small search region around its mean given as

$$\begin{aligned} a_j &= meanbd_j - w \cdot meanbd_j \\ b_j &= meanbd_j + w \cdot meanbd_j \end{aligned} \tag{3.28}$$

The aim of the model is to reduce stochastic change in gene structure due to random selection of individuals by providing a stable search range as shown in Fig. 3.10. Note that the interval specifying the location of new individuals is only a rough deduction of the search region based on the available information.

Intuitively, the utilization of statistical model can improve robustness of existing selection strategies, where individuals selected based on fitness are included directly in the evolving population. Note that such a procedure introduces another desirable effect. By focusing on a specific region of the search space, we are effectively performing implicit averaging. The selection of appropriate model is performed autonomously based on the condition of the evolutionary process. As shown by the experiments in Section 2.3, the search process degenerated by noise can hardly fill up more than 30% of the archive's capacity. Hence, the behaviors of convergence and random search can be determined based upon the growth rate of archive population and the gene adaptation strategy can be activated when there is sufficient indication of convergence or random search behavior in the evolution. We applied a simple scheme; divergence model is activated when 60% of the archive capacity is reached, while convergence model is activated when less than 60% of the archive is filled after 150 generations.

A Possibilistic Archiving Methodology: We also presented two archiving models, i.e. necessity-possible (NP-) archiving model and necessity (N-) archiving model, based on the concept of possibilistic Pareto dominance relation. Contrary to MOPSEA and ESPEA, where probabilities are employed to model uncertainty as part of the Pareto ranking procedure, fuzzy numbers are used here to represent the objective vectors. The proposed approach is based on the concept of possibility and necessity measures [67], which aims to rectify certain deficiencies present in the current Pareto-based updating strategy in handling noisy environments. Besides, a tagging system is proposed to allow both the models to co-exist in the situation where the uncertainty level is low.

The archive updating schemes adopted in existing MOEAs are largely based on the concept of Pareto optimality and some form of truncation process is usually applied to limit the number of good individuals stored in the archive due to the limitation of memory resource. Although such an updating scheme is simple and effective, it is not competent in dealing with individuals containing uncertainties in the objective functions since the dominance relationship for these individuals in the presence of noise is no longer deterministic. In the absence of a reliable decision maker, the standard archiving scheme can be easily deceived into removing non-dominated individuals from the archive or inserting dominated individuals to the archive, which could subsequently affect the performance of evolutionary multi-objective optimization in noisy environments.

The instance in Fig. 3.11 shows the distribution of archived individuals marked by closed circles and the newly evolved individuals marked by crosses in a two-dimensional objective space. From the definition of Pareto dominance, it is clear that **A**, **C**, and **D** will be selected to fill the archive in the evolution. However, **A** provides only marginal improvement for x_1 at a great expense of x_2, which gives little contribution to the overall quality of the solution set. In the face of limited archive storage, non-contributing individuals occupying valuable space that are usually located in isolated regions in the objective space are less unlikely to be removed during the truncation process. It is thus desirable if the updating function is capable of rejecting such non-dominated individuals according to some *a-priori* knowledge or user preference. In addition, it is also desirable if the updating mechanism can minimize removal of non-dominated individuals and provide a chance for individuals degraded by noise to survive in the evolution.

To understand the proposed archiving models, a number of definitions are given as follows:

Definition 3.5 *Necessity Condition*: Given that f_1 and f_2 are fuzzy numbers with membership functions μ_{f_1} and μ_{f_2}, respectively, the necessity that the largest possible value of f_1 is smaller than the smallest value of f_2 is given by

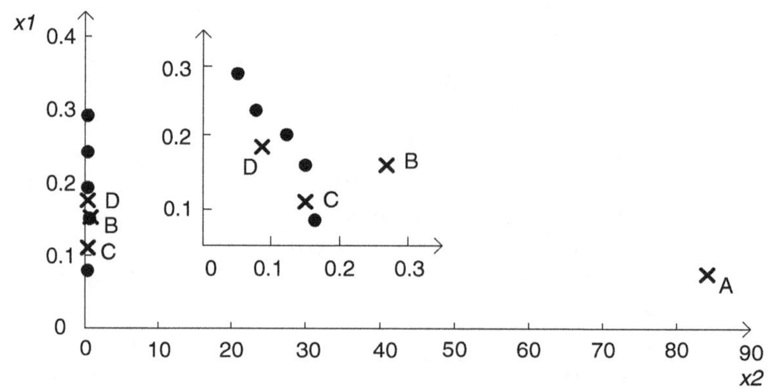

Fig. 3.11 Distribution of archived individuals marked by closed circles and the newly evolved individuals marked by crosses in a two-dimensional objective space

$$Nec(\overline{z}_1 < \underline{z}_2) = \inf_u \max\Big[1 - \mu_{f_1}(u), \inf_{v<u}\big(1 - \mu_{f_2}(v)\big)\Big] \tag{3.29}$$

Definition 3.6 *Possibility Condition*: Given that f_1 and f_2 are fuzzy numbers with membership functions μ_{f_1} and μ_{f_2}, respectively, the possibility that the smallest possible value of f_1 is smaller than the largest value of f_2 is given by

$$Pos(\underline{z}_1 < \overline{z}_2) = \sup_u \min\big[1 - \mu_{f_1}(u), \sup_{u<v}\mu_{f_2}(v)\big] \tag{3.30}$$

Definition 3.7 *NP-Dominance*: Given that $\mathbf{f}_1$ and $\mathbf{f}_2$ are M-dimensional objective vectors of fuzzy numbers with membership functions $\boldsymbol{\mu}_{f_1}$ and $\boldsymbol{\mu}_{f_2}$, respectively, $\mathbf{X}_1$ NP-dominates $\mathbf{X}_2$, denoted by $\mathbf{X}_1 \prec_{NP} \mathbf{X}_2$, *iff*

$$\begin{gathered} Pos(\underline{z}_{1,j} < \overline{z}_{2,j}) \geq Pos(\underline{z}_{1,i} < \overline{z}_{2,i}) \forall 1,2,...,M \\ \text{or} \\ Nec(\overline{z}_{1,i} < \underline{z}_{2,i}) = 1 \exists\ i \in 1,2,...,M \text{ and } Nec(\overline{z}_{1,j} < \underline{z}_{2,j}) < 1\ \forall j \in 1,2,...,M \end{gathered} \tag{3.31}$$

Definition 3.8 *N-Dominance*: Given that $\mathbf{f}_1$ and $\mathbf{f}_2$ are M-dimensional objective vectors of fuzzy numbers with membership functions $\boldsymbol{\mu}_{f_1}$ and $\boldsymbol{\mu}_{f_2}$, respectively, $\mathbf{f}_1$ N-dominates $\mathbf{f}_2$, denoted by $\mathbf{X}_1 \prec_N \mathbf{X}_2$, *iff*

$$Nec(\overline{z}_{1,i} < \underline{z}_{2,i}) = 1\ \forall i \in 1,2,...,M \tag{3.32}$$

Fig. 3.12 illustrates the different dominance relations for a minimization problem. The shaded region represents the area dominated by the individual marked by a circle. The NP-model behaves similarly to existing archiving models but allows decision-maker to reject certain non-dominated individuals in the evolution if necessary. This archiving model compares and updates

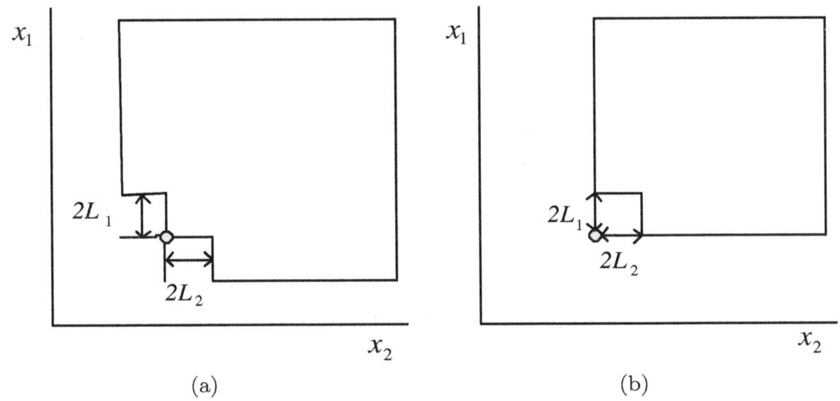

Fig. 3.12 Region of dominance based on (a) NP-dominance relation, and (b) N-dominance relation

individuals according to the NP-dominance relation. As shown in Fig. 3.12(b), the width of the fuzzy membership function associated with the i-th objective is denoted by L_i, which represents the tolerance level of inferiority for each objective. As L_i tends to zero, the behavior of NP-dominance approaches that of Pareto-dominance relation. The pruning criterion is based upon some degree of crowding or niche count, which helps to maintain population diversity in the archive.

The N-archiving model updates individuals according to the N-dominance relation, which stores a set of possibly non-dominated individuals. The membership function is a reflection of the uncertainty level present in the system, and the width L_i represents the possible values of the i-th objective. In order to minimize deletion of non-dominated individuals, the N-archiving model removes an archived individual only if it is N-dominated by an individual in the archive. In this model, an individual is selected if there is no archived individual that necessarily dominates it. Intuitively, the size of archive will grow exceedingly large with the increase of noise and any form of niche count or crowding comparison is of no practical meaning in the presence of noise. Therefore, the truncation criterion for the archive should be based upon the apparent ranking provided by the prior evaluation process.

It is clear that the proposed two archiving models operate at the two ends of the noise spectrum. A tagging system is thus proposed to provide a graceful integration of both models since it is often more desirable to incorporate both the model properties in the presence of low noise level. Each individual is assigned either a NP-tag or N-tag that defines the behavior it will experience during the archiving process, e.g. an individual assigned with the NP-tag is regarded as if only the NP-model is implemented. The assignment of tags is based on a probability distribution as shown in Fig. 3.13. If the noise level

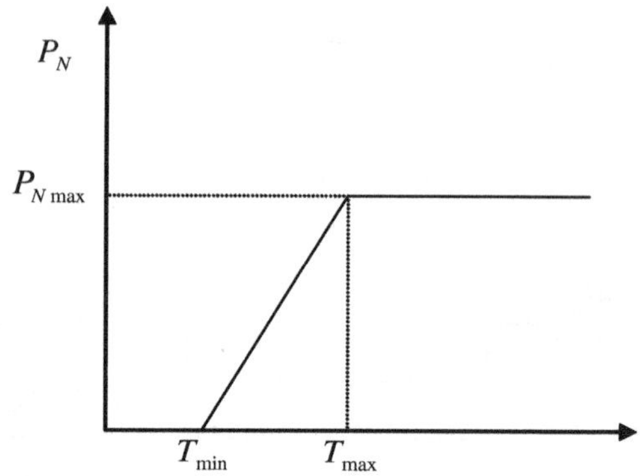

Fig. 3.13 Decision process for tag assignment based on the level of noise present

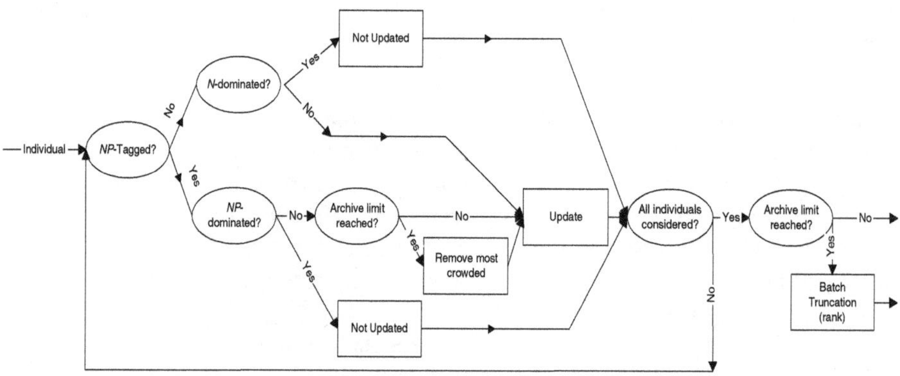

Fig. 3.14 Possibilistic archiving model

is below the minimum threshold of T_{min}, all individuals will be assigned the NP-tag. When the noise level is above the maximum threshold of T_{max}, all individuals will be assigned the N-tag with a probability of $P_N\text{max}$. If the noise level is between the two thresholds, the probability of P_N is a linear function of noise as depicted in Fig. 3.13. The *Possibilistic* archiving model is shown in Fig. 3.14.

3.8 Comparative Study

In this section, a comparative study is conducted between MOEA-RF, NTSPEA, MOPSEA, SPEA2, NSGAII, and PAES. Since re-sampling is probably the simplest and most common noise compensation technique, the baseline algorithm with a re-sampling rate of 10 (named as RMOEA) is also included in the study. The indices of the seven algorithms are listed in Table 3.1. In this study, different experimental setups with noise settings of

Table 3.1 Indices of the different algorithms

Index	1	2	3	4	5	6	7
Algorithm	MOEA-RF	RMOEA	NTSPEA	MOSPEA	SPEA2	NSGAII	PAES

Table 3.2 Parameter setting for different algorithms

Parameter	Settings
Populations	Population size 100 in NSGAII, SPEA2, NTSPEA, MOPSEA, RMOEA and MOEA-EF; Population size 1 in PAES; Archive (or secondary population) size 100.
Chromosome	Binary coding; 15 bits per decision variable.
Selection	Binary tournament selection
Crossover operator	Uniform crossover
Crossover rate	0.8
Mutation operator	Bit-flip mutation in NSGAII, SPEA2, NTSPEA, RMOEA and ELDP in MOEA-RF
Mutation rate	$\frac{1}{chromosome\_length}$ for ZDT1, ZDT4 and ZDT6; $\frac{1}{bit\_number\_per\_variable}$ for FON and KUR;
Hyper-grid size	2^5 per dimension
Niche Radius	Dynamic for MOEA-RF.
Evaluation number	50,000

$\sigma^2 = \{0\%, 5\%, 10\%, 20\%\}$ are applied to evaluate the performances of the algorithms.

Both the step size w for GASS and the learning rate α for ELDP are set as 0.3 in the algorithm. The *Possibilistic* archiving approach as shown in Fig. 3.13 is applied with triangular membership function for both the N- and NP-archiving models. Since the width of the membership function for the N-archiving model represents the noise level, it can be estimated by re-sampling one individual at the beginning of the evolution. The parameters for tag assignment, such as T_{min}, T_{max}, and $P_N\text{max}$, are set as 0.0, 0.1, and 1.0, respectively. In accordance to the original paper [32], k_{max} is set as 4, while c_1 and c_2 are set as 10% and 30%, respectively, for NTSPEA. The value of s is calculated by re-sampling ten individuals immediately after the first evaluation for MOPSEA.

ZDT1: Fig. 3.15 shows that the performances of the algorithms deteriorate with the increase in noise level; particularly there is a drastic performance change in PAES when the noise level is increased to 5%. Fig. 3.15(c) shows that the MOEA-RF, NTSPEA, and MOPSEA are capable of evolving better solutions in a noisy environment as compared to algorithms without any noise

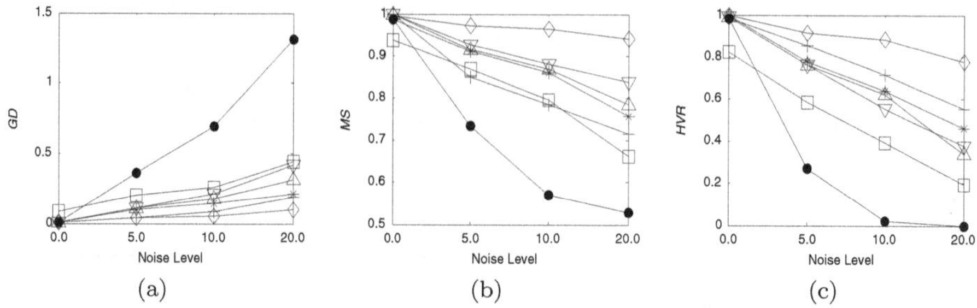

Fig. 3.15 Performance metric of (a) GD, (b) MS, and (c) HVR for ZDT1 attained by MOEA-RF (◇), RMOEA (□), NTSPEA(|), MOPSEA (∗), SPEA2 (▽), NSGAII (△) and PAES (•) under the influence of different noise levels

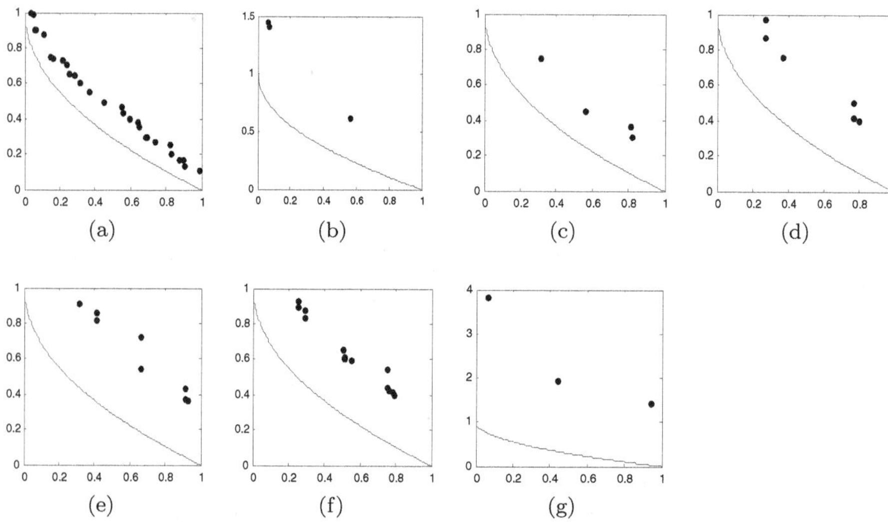

Fig. 3.16 The PF^A from (a) MOEA-RF, (b) RMOEA, (c) NTSPEA, (d) MOPSEA, (e) SPEA2, (f) NSGAII, and (g) PAES for ZDT1 with 20% noise

compensation techniques. With the exception of RMOEA, Fig. 3.17 shows that most algorithms encountered no problem in converging and maintaining a diverse set of solutions for ZDT1 under noiseless environment. As shown by the evolutionary trace of GD in Fig. 3.18, the poor performance of RMOEA can be attributed to the re-evaluation of candidate individuals. Although the performance of MOEA-RF for proximity is not the best, it has the fastest convergence for both GD and MS as can be seen from the evolutionary trace of GD and MS in Fig. 3.19. The metric of HVR in Fig. 3.17(d) indicates that the solutions evolved by MOEA-RF have the best overall quality. It

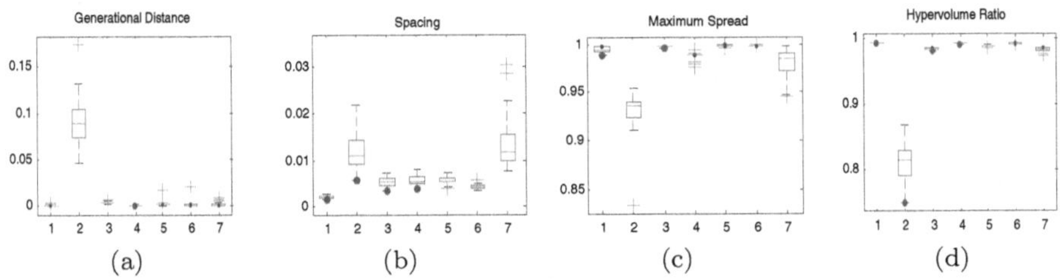

Fig. 3.17 Performance metric of (a) GD, (b) S, (c) MS, and (d) HVR for ZDT1 with 0% noise

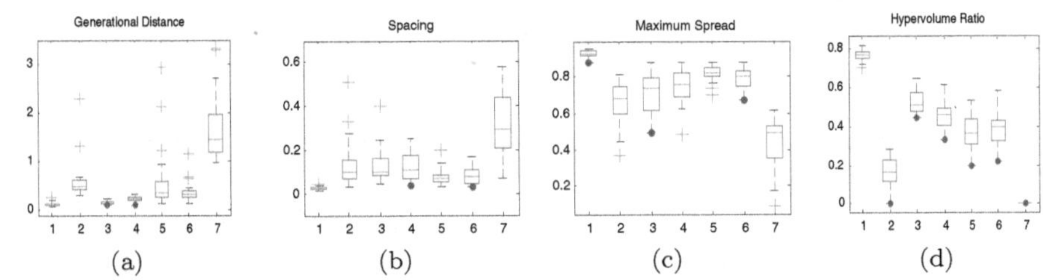

Fig. 3.18 Performance metric of (a) GD, (b) S, (c) MS, and (d) HVR for ZDT1 with 20% noise

also produces a more uniformly distributed Pareto front as shown by the low value of S. The KS test also revealed that MOEA-RF and other algorithms are statistically different in terms of S, MS, and HVR.

Among the conventional MOEAs, i.e., SPEA2, NSGAII, and PAES, it is apparent that the PAES is most affected by the noise. As can be seen from the distribution of GD in Fig. 3.18(a), MOEA-RF, NTSPEA, and MOPSEA produce competitive results since various features are included in these algorithms to deal with the noise. On the other hand, the performance of RMOEA is the worst among all algorithms except for PAES. As can be observed in Fig. 3.15 and Fig. 3.18(b)-(d), the MOEA-RF is capable of evolving a more diverse and uniformly distributed Pareto front for ZDT1 in the presence of noise as compared to other algorithms.

ZDT4: From the trend of GD over the various noise levels in Fig. 3.20(a), it is apparent that the smoothing effect of noise described in Section 2 is also present for the noise levels of 5% and 10%. In contrast to the algorithmic behaviors observed for ZDT1, this phenomenon enables some of the algorithms, such as SPEA2 and MOPSEA, to evolve better solutions as shown in Fig. 3.20(a) and Fig. 3.20(c). It can be observed from Fig. 3.21 and Fig. 3.23 that the local optima imposed by this benchmark appear to be a formidable barrier against the global convergence. At the end of 50,000 evaluations, RMOEA, MOPSEA, SPEA2, NSGAII and PAES only managed to discover one of the local Pareto fronts. On the other hand, Fig. 3.23(a) shows that

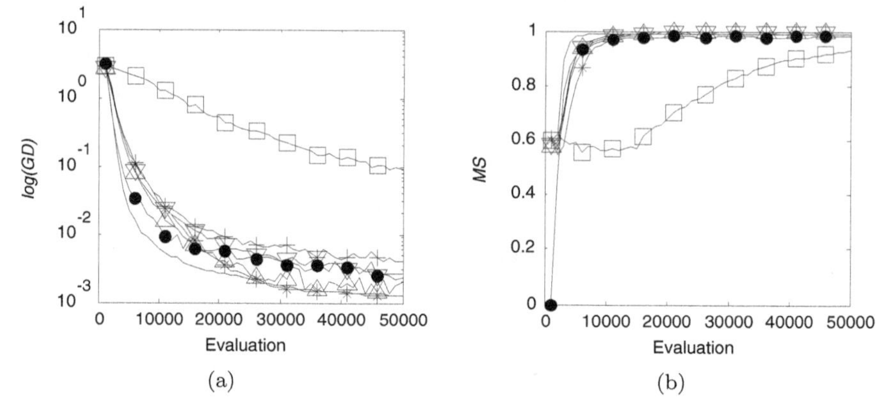

Fig. 3.19 Evolutionary trace of (a) GD and (b) MS for ZDT1 with 0% noise

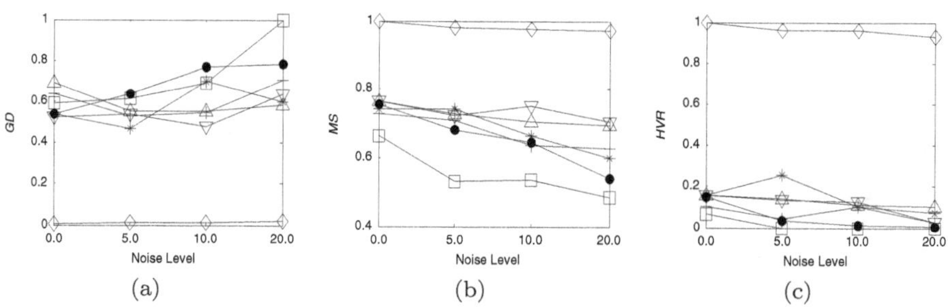

Fig. 3.20 Performance metric of (a) GD, (b) MS, and (c) HVR for ZDT4 attained by MOEA-RF ($\Diamond$), RMOEA ($\Box$), NTSPEA($|$), MOPSEA ($*$), SPEA2 (∇), NSGAII ($\triangle$) and PAES ($\bullet$) under the influence of different noise levels

MOEA-RF, incorporated with ELDP and GASS, is able to evolve individuals near to the global Pareto front consistently. From the convergence trace of GD in Fig. 3.25(a), it is clear that ELDP plays an important role in the algorithm to escape from the local optima. Moreover GASS is activated whenever the criterion of convergence is satisfied, which diverts the evolutionary search and avoids the local optima. The dips on the metric of MS, observed in Fig. 3.25(b), correspond to the effect of jumping from one local Pareto front to another during the evolutionary search. In this intermediate state of jumping, there is a transition from one relatively diverse set of individuals along a local Pareto front to another, which results in the effect of sudden dips. As can be observed in Fig. 3.23(b)-(d), MOEA-RF is capable of evolving a more diverse and uniformly distributed Pareto front under noiseless environment as compared to the other algorithms.

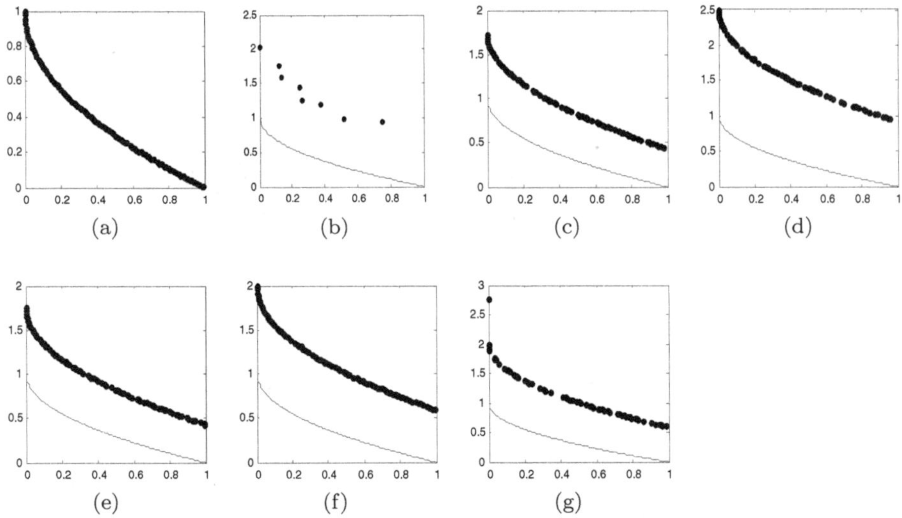

Fig. 3.21 The PF^A from (a) MOEA-RF, (b) RMOEA, (c) NTSPEA, (d) MOPSEA, (e) SPEA2, (f) NSGAII, and (g) PAES for ZDT4 with 0% noise

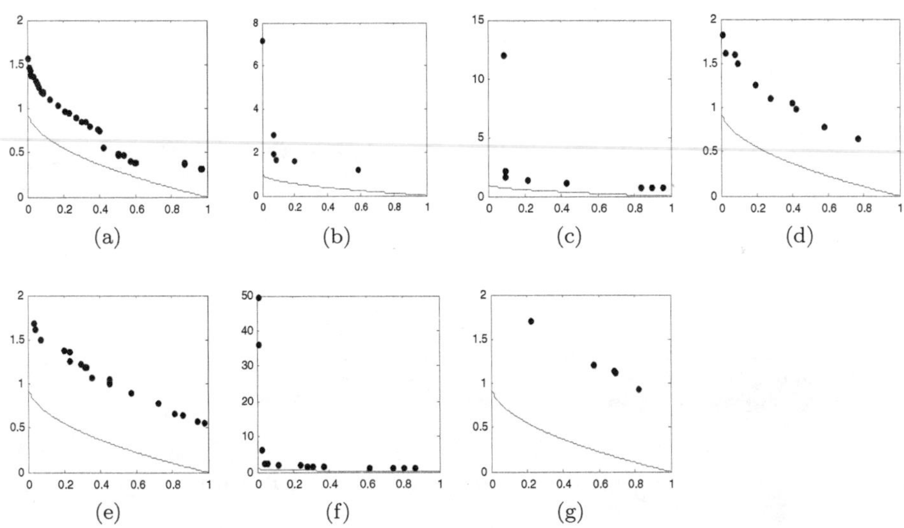

Fig. 3.22 The PF^A from (a) MOEA-RF, (b) RMOEA, (c) NTSPEA, (d) MOPSEA, (e) SPEA2, (f) NSGAII, and (g) PAES for ZDT4 with 20% noise

As can be seen in Fig. 3.22 and Fig. 3.24, the performances of NTSPEA, MOPSEA, SPEA2, NSGAII, and PAES are poor under the influence of noise. Furthermore, the number of non-dominated individuals discovered by these algorithms is also greatly reduced as shown in Table 3.3. By comparing Fig. 3.23(b)-(d) and Fig. 3.24(b)-(d), it is apparent that MOEA-RF is able

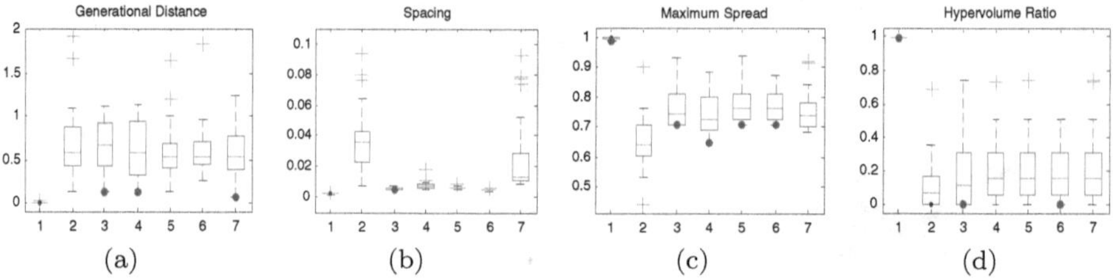

Fig. 3.23 Performance metric of (a) GD, (b) S, (c) MS, and (d) HVR for ZDT4 with 0% noise

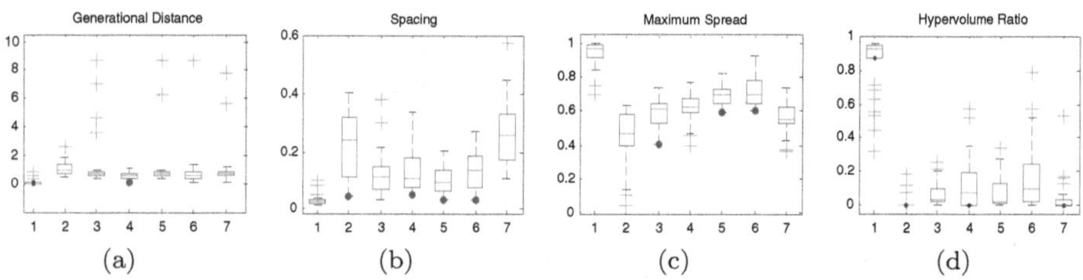

Fig. 3.24 Performance metric of (a) GD, (b) S, (c) MS, and (d) HVR for ZDT4 with 20% noise

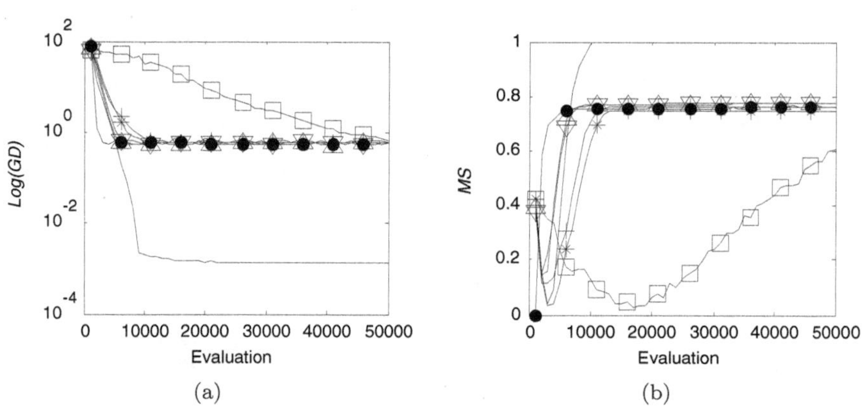

Fig. 3.25 Evolutionary trace of (a) GD and (b) MS for ZDT4 with 0% noise

to evolve individuals that are on or close to the global tradeoff for ZDT4, although its performance is generally affected by the presence of noise.

ZDT6: It can be observed from Fig. 3.26 that different algorithms behave differently, although their performances generally deteriorate with increasing noise levels. For instance, Fig. 3.26(c) shows that there are drastic drops in the performances of MOPSEA and PAES as reflected by the metric of HVR when the noise level is increased to 5%. On the other hand, the performances of NTSPEA, NSGAII, and SPEA2 seem unaffected for MS and GD over the

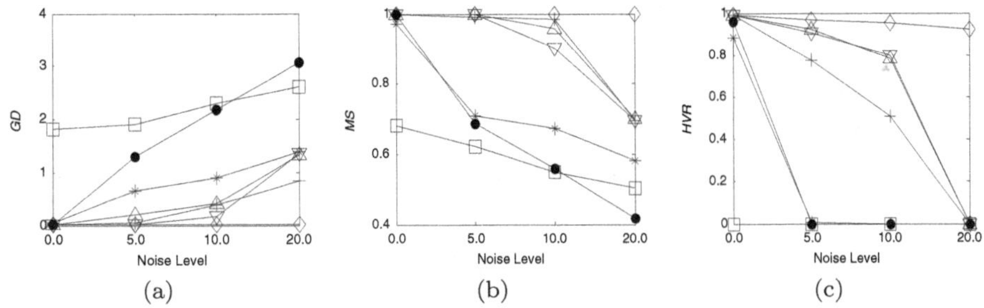

Fig. 3.26 Performance metric of (a) GD, (b) MS, and (c) HVR for ZDT6 attained by MOEA-RF (◊), RMOEA (□), NTSPEA(|), MOPSEA (∗), SPEA2 (∇), NSGAII (△) and PAES (•) under the influence of different noise levels

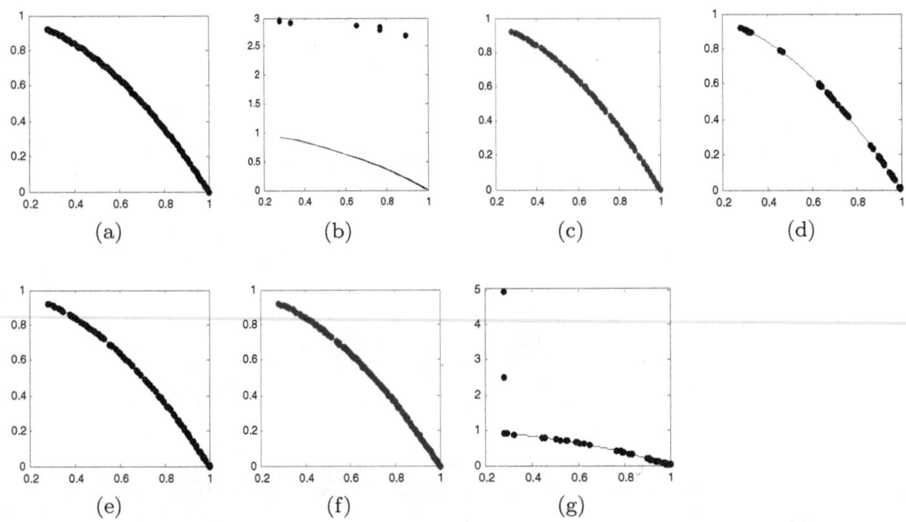

Fig. 3.27 The PF^A from (a) MOEA-RF, (b) RMOEA, (c) NTSPEA, (d) MOPSEA, (e) SPEA2, (f) NSGAII, and (g) PAES for ZDT6 with 0% noise

noise levels of 0%, 5%, and 10%, but deteriorate sharply when the noise level is increased to 20%. It can also be observed that the noise-handling algorithms of RMOEA, NTSPEA, MOPSEA, and MOEA-RF have different degrees of success in the presence of noise. For example, the re-sampling mechanism employed by RMOEA has a slight edge over only PAES at noise level of 20% for GD and MS, while MOEA-RF outperforms other algorithms on the various metrics of GD, MS, and HVR.

Although RMOEA, MOPSEA, and PAES can identify some parts of the tradeoff for ZDT6, Fig. 3.27(b), (d), and (g) show that these algorithms are unable to evolve a well-distributed Pareto front. Fig. 3.29(c) also shows that

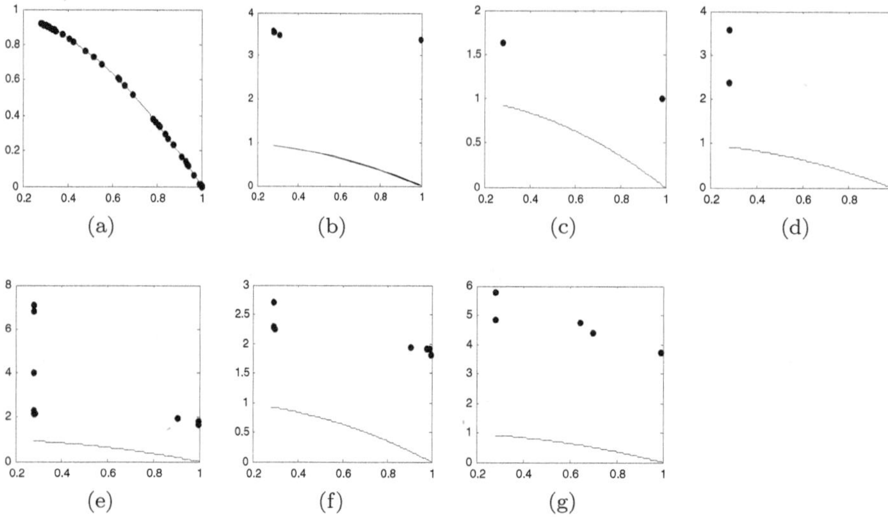

Fig. 3.28 The PF^A from (a) MOEA-RF, (b) RMOEA, (c) NTSPEA, (d) MOPSEA, (e) SPEA2, (f) NSGAII, and (g) PAES for ZDT6 with 20% noise

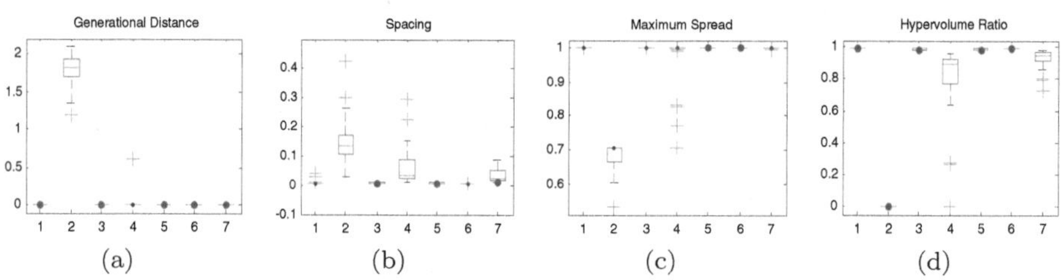

Fig. 3.29 Performance metric of (a) GD, (b) S, (c) MS, and (d) HVR for ZDT6 with 0% noise

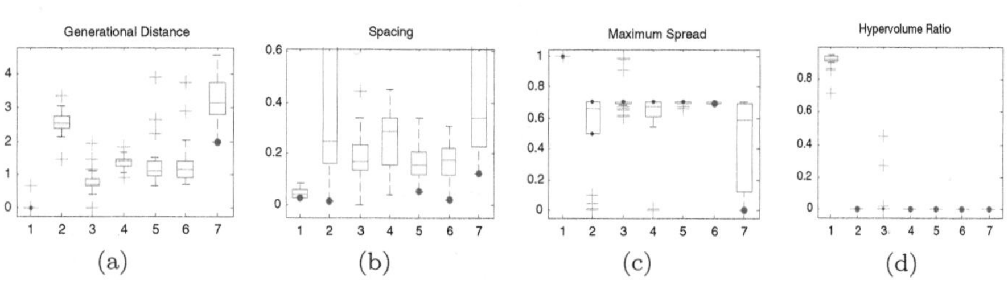

Fig. 3.30 Performance metric of (a) GD, (b) S, (c) MS, and (d) HVR for ZDT6 with 20% noise

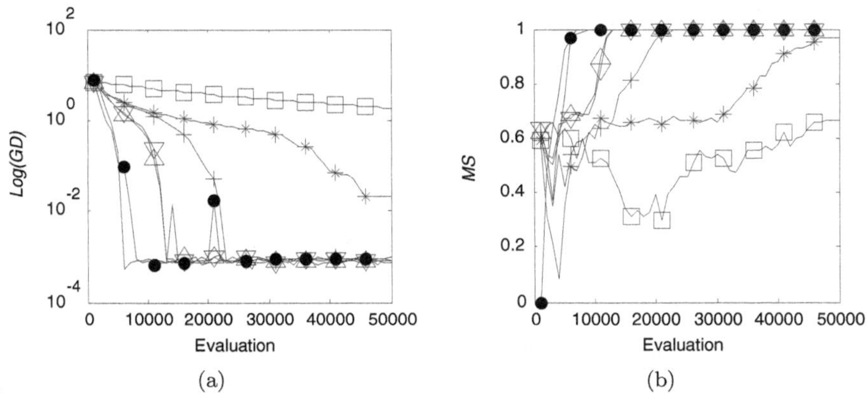

Fig. 3.31 Evolutionary trace of (a) GD and (b) MS for ZDT6 with 0% noise

RMOEA is unable to find a diverse solution set consistently. On the other hand, NSGAII, SPEA2, and MOEA-RF provide competitive results in all aspects, particularly for the metric of GD as shown in Fig. 3.29. In addition, the convergence traces of GD and MS in Fig. 3.31 show that MOEA-RF offers the fastest convergence among all algorithms due to the incorporation of ELDP.

Fig. 3.28 and Fig. 3.30 show that NTSPEA, MOPSEA, NSGAII, SPEA2, and PAES are unable to find any individuals along the global tradeoff under the influence of noise. The simple archiving technique employed in NTSPEA allows it to cope with noise better than MOPSEA, SPEA2, NSGAII and PAES. On the other hand, MOEA-RF is able to find a set of solutions near the global tradeoff consistently. With the exception of GD at 0% noise, the KS test indicates that the performance of MOEA-RF is statistically different from those of the other algorithms in all aspects of the MO optimization goals. The MOEA-RF also maintains a stable evolving environment through GASS that defines a concentrated search region. This results in a consistent algorithmic performance as reflected by the small variance of all metrics in Fig. 3.30. Conversely, RMOEA shows a large variance for the metric of S, MS, and HVR, despite the presence of re-sampling technique in the algorithm.
FON: It can be observed from Fig. 3.32 that none of the noise-handling MOEAs provides distinct advantage over NSGAII and SPEA2 for solving FON in noisy environments. In fact, only MOEA-RF is able to match the performances of NSGAII and SPEA2 in terms of convergence and diversity over the different noise levels. Conversely, the performances of other algorithms deteriorate drastically at noise levels of 10% and 20% as shown in Fig. 3.31. It can be observed from Fig. 3.33 and Fig. 3.34 that RMOEA are unable to find the final tradeoff, while other algorithms are capable of finding at least some parts of the optimal Pareto front. The results also show

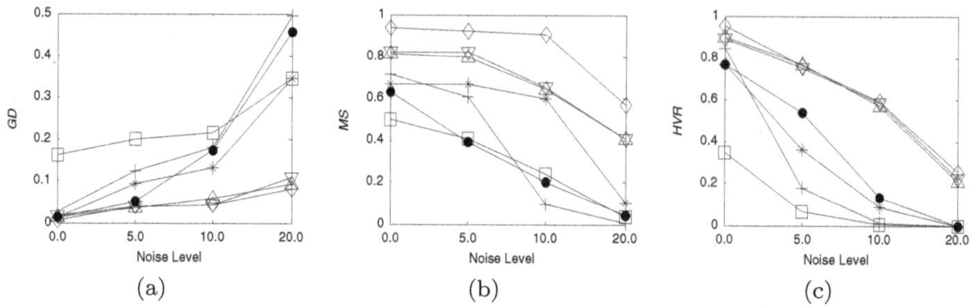

Fig. 3.32 Performance metric of (a) GD, (b) MS, and (c) HVR for FON attained by the algorithms under the influence of different noise levels

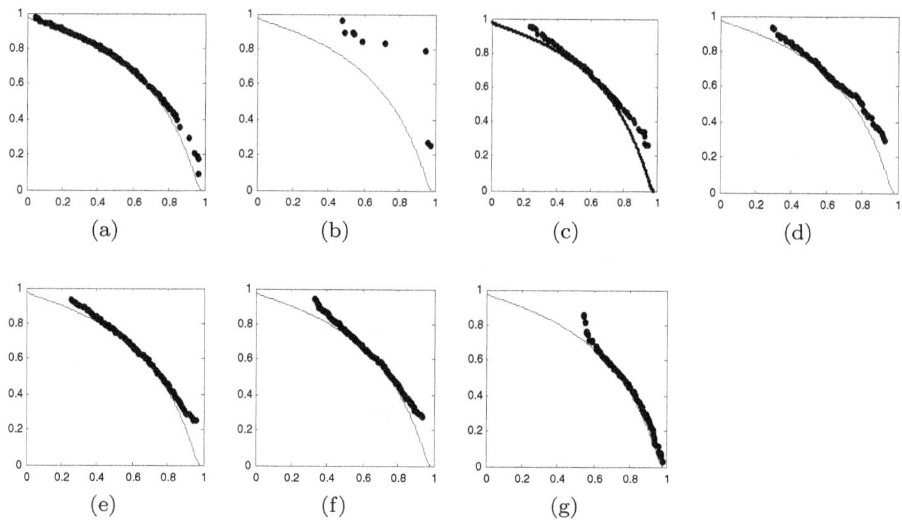

Fig. 3.33 The PF^A from (a) MOEA-RF, (b) RMOEA, (c) NTSPEA, (d) MOPSEA, (e) SPEA2, (f) NSGAII, and (g) PAES for FON with 20% noise

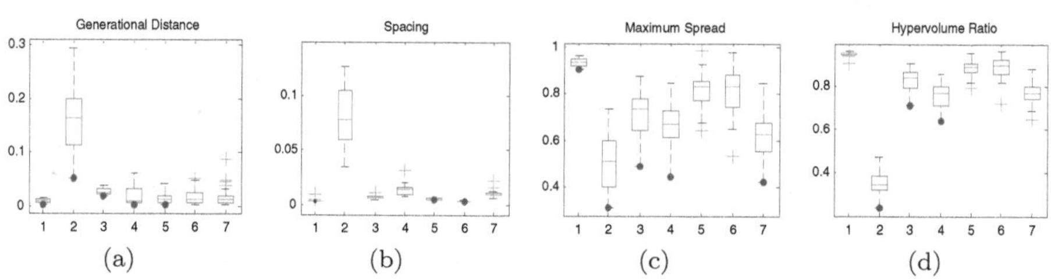

Fig. 3.34 Performance metric of (a) GD, (b) S, (c) MS, and (d) HVR for FON with 0% noise

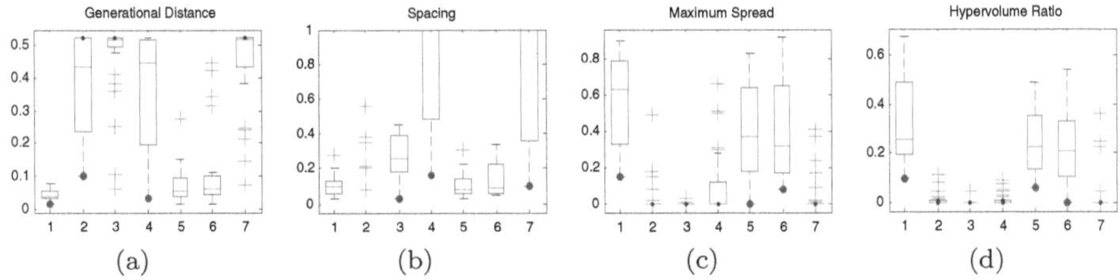

Fig. 3.35 Performance metric of (a) GD, (b) S, (c) MS, and (d) HVR for FON with 20% noise

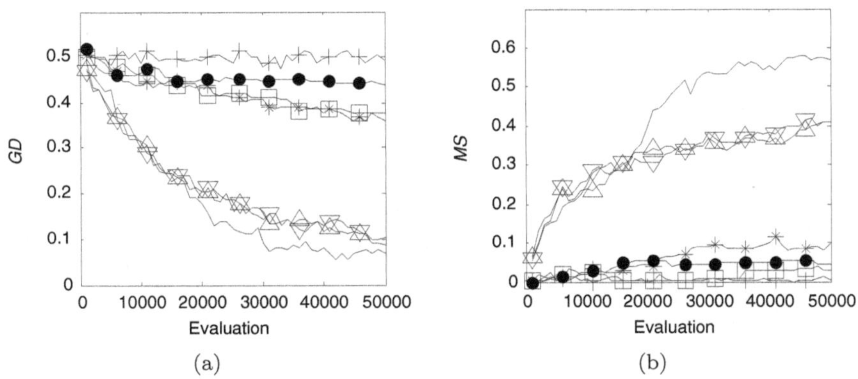

Fig. 3.36 Evolutionary trace of (a) GD and (b) MS for FON with 0% noise

that MOEA-RF offers the best performance in terms of spacing and spread and the KS test reveals that the performance of MOEA-RF is statistically different from those of the other algorithms in terms of MS, S, and HVR.

From the evolutionary trace in Fig. 3.36, it is obvious that RMOEA, NTSPEA, MOPSEA, and PAES are unable to improve beyond the initial candidate solutions at 20% noise. As a result, NTSPEA, MOPSEA, and PAES only manage to find one or two solutions that are far away from the final tradeoff for most of the 30 simulation runs, leading to the high values of GD and S as shown in Fig. 3.35(a)-(b). On the other hand, MOEA-RF, SPEA2, and NSGAII are able to discover some individuals that are near to the final tradeoff. The KS test indicates that the three algorithms are rather similar in performance for the various MO optimization metrics. Fig. 3.35(d) and Fig. 3.36 also show that MOEA-RF has a slight edge in producing better solutions as compared to the other algorithms, due to the proposed GASS that concentrates the evolutionary search to reduce the stochastic influence of noise as shown in Fig. 3.36(a) where the improvement of convergence for MOEA-RF coincides with the activation of GASS.

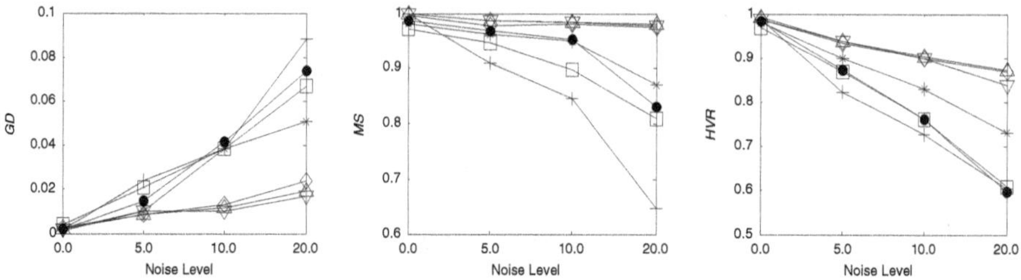

(a) (b) (c)

Fig. 3.37 Performance metric of (a) GD, (b) MS, and (c) HVR for KUR attained by the algorithms under the influence of different noise levels

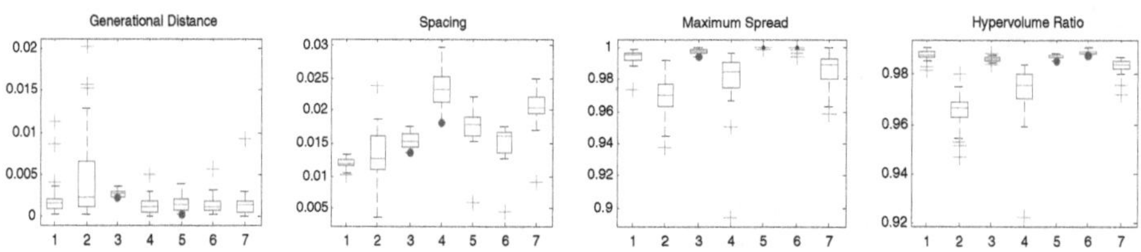

(a) (b) (c) (d)

Fig. 3.38 Performance metric of (a) GD, (b) S, (c) MS, and (d) HVR for KUR with 0% noise

KUR: Similar to FON, Fig. 3.37 shows that MOEA-RF, NSGAII, and SPEA2 are better for solving KUR in noisy environments as compared to the other algorithms. Fig. 3.38 shows that the global search mechanism of MOEAs generally responds well to the challenges of discontinuity and non-convexity posed by noiseless KUR. Among these algorithms, MOEA-RF, NTSPEA, SPEA2, and NSGAII are capable of finding a diverse and uniformly distributed Pareto front for most of the 30 simulation runs. It can be observed from Fig. 3.39(b)-(d) that RMOEA, NTSPEA, MOPSEA, and PAES have difficulty in distributing individuals uniformly along the discovered Pareto front in noisy environments. On the other hand, MOEA-RF, SPEA2, and NSGAII give good performance in terms of distribution and diversity under the influence of noise. Besides having similar results for GD and MS, Table 3.3 depicts that MOEA-RF, SPEA2, and NSGAII are capable of archiving more non-dominated individuals as compared to the other algorithms.

3.9 Effects of the Proposed Features

It can be observed from the comparative studies that MOEA-RF is capable of evolving near-optimal, diverse, and uniformly distributed Pareto fronts for

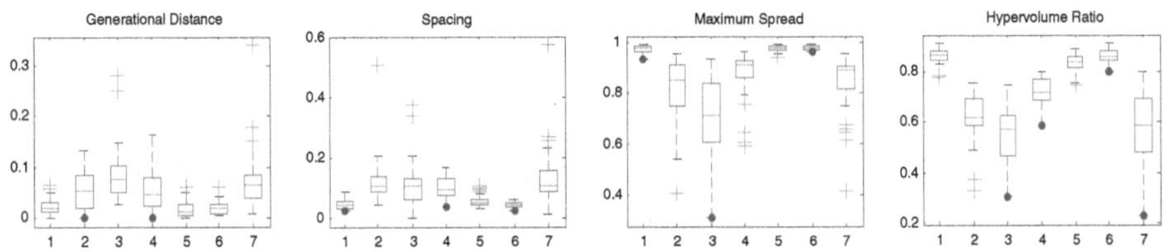

(a) (b) (c) (d)

Fig. 3.39 Performance metric of (a) GD, (b) S, (c) MS, and (d) HVR for KUR with 20% noise

Table 3.3 Number of non-dominated individuals found for the various benchmark problems at 20% noise level

		MOEA-RF	RMOEA	NTSPEA	MOSPEA	SPEA2	NSGAII	PAES
	1st quartile	28	6	5	7	18	17	4
ZDT1	Median	31	7.5	6	9	21.5	19.5	4.5
	3rd quartile	32	10	8	10	27	23	5
	1st quartile	29	5	6	9	12	20	5
ZDT4	Median	33	7	8	11	26.5	15	6
	3rd quartile	41	8	10	13	29	21	9
	1st quartile	82	2	4	3	8	8	2
ZDT6	Median	85	3	5	4	9	9	4
	3rd quartile	88	5	6	5	11	11	6
	1st quartile	9	1	1	1	6	6	1
FON	Median	11	2	1.5	2	8.5	8.5	2
	3rd quartile	17	2	2	3	12	12	3
	1st quartile	25	6	5	8	23	25	7
KUR	Median	27	8	5.5	9	25	27	9
	3rd quartile	30	9	7	11	28	30	10

the different benchmark problems. In this section, the dynamics and parameter settings of ELDP and GASS are examined in the presence of the possibilistic archiving model. Simulation results show that the proposed archiving model plays a complementary but crucial role in the preservation of good individuals discovered by ELDP and GASS, without which the potential of ELDP and GASS may not be easily exploited. Note that ZDT4 and FON are used in the study here since it has been observed in the previous section that most algorithms are unable to deal with these two benchmark problems effectively across the different noise conditions.

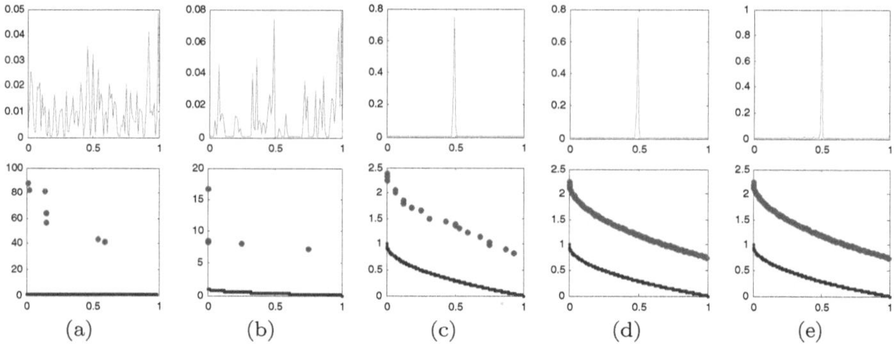

Fig. 3.40 The first row represents the distribution of one decision variable and the second row shows the associated non-dominated individuals of baseline MOEA at generation (a) 0, (b) 10, (c) 60, (d) 200, and (e) 350 for ZDT4

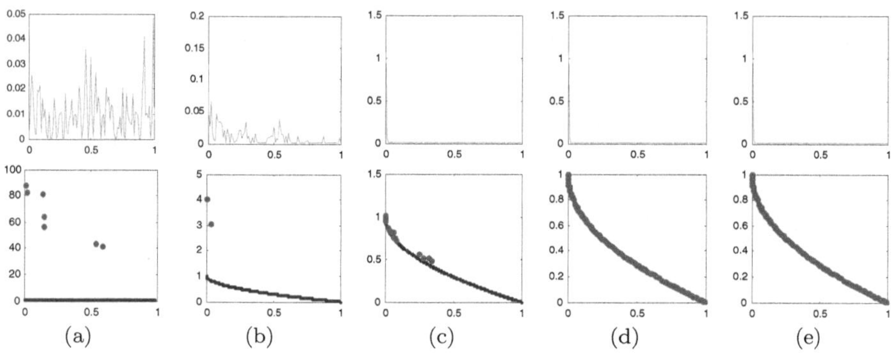

Fig. 3.41 The first row represents distribution of one decision variable and the second row shows the associated non-dominated individuals of baseline MOEA with ELDP at generation (a) 0, (b) 10, (c) 60, (d) 200, and (e) 350 for ZDT4

The Parzen window density estimation [210] is used to estimate the distribution of individuals in the decision space. Fig. 3.40(a)-(e) shows the distribution of one decision variable and the associated non-dominated individuals of baseline MOEA at the generation of 0, 10, 60, 200, and 350 for ZDT4. Similarly, the effects of ELDP and GASS are shown in Fig. 3.41 and Fig. 3.42, respectively. To illustrate the working dynamic for the problem of FON, the distribution of one decision variable and the associated non-dominated individuals for baseline MOEA without and with the proposed features at the generation of 0, 50, 150, 350, and 500 are shown in Fig. 3.43 to Fig. 3.45. Note that the possibilistic archiving model behaves like the standard archive in the absence of any preference or noise. The distribution and the associated non-dominated individuals demonstrate how the different features influence and

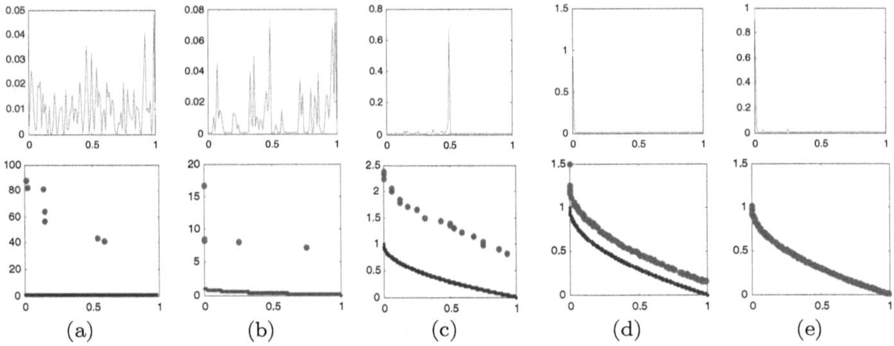

Fig. 3.42 The first row represents distribution of one decision variable and the second row shows the associated non-dominated individuals of baseline MOEA with GASS at generation (a) 0, (b) 10, (c) 60, (d) 200, and (e) 350 for ZDT4

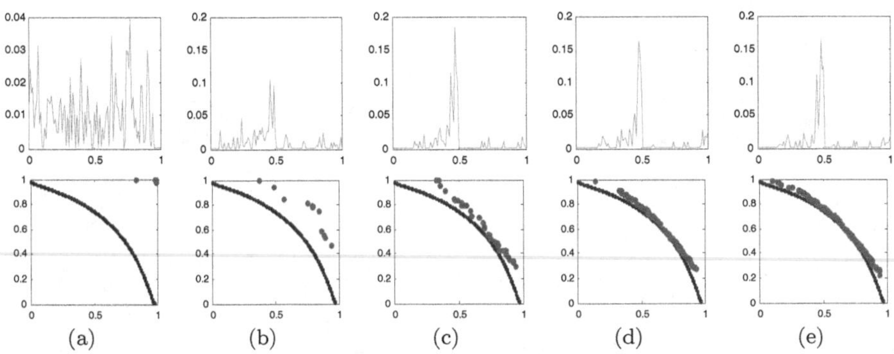

Fig. 3.43 The first row represents the distribution of one decision variable and the second row shows the associated non-dominated individuals of baseline MOEA at generation (a) 0, (b) 50, (c) 150, (d) 350, and (e) 500 for FON

improve the optimization process. In addition, it shows whether the proposed features are behaving in accordance to the design specifications.

It can be seen from the figures that ELDP and GASS have a distinct advantage in overcoming local optimality for ZDT4 as well as in finding a diverse tradeoff for FON. By comparing the decision variable distribution and the evolved non-dominated solutions across the different generations, it is evident from Fig. 3.41(a)-(c) and Fig. 3.44(a)-(c) that the population distribution converges faster when ELDP is incorporated. The slight divergence of the decision variable distribution about the main peak in Fig. 3.44(d)-(e) illustrates the local fine-tuning capability of ELDP, which is important in leading the evolution towards the global tradeoff. By comparing the decision variable distribution between Fig. 3.40(c) and Fig. 3.42(c) as well as between

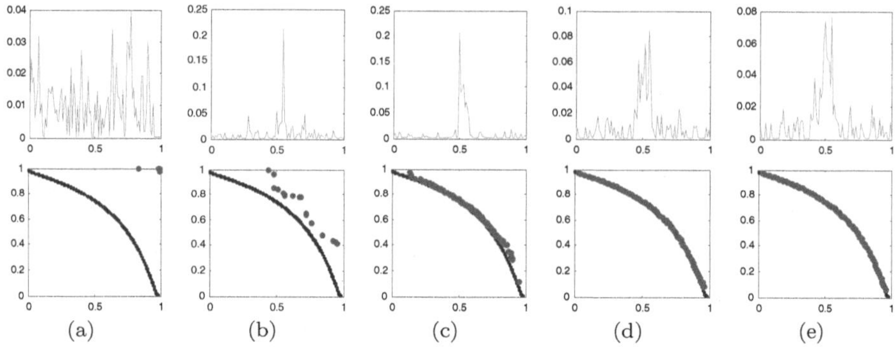

Fig. 3.44 The first row represents the distribution of one decision variable and the second row shows the associated non-dominated individuals of baseline MOEA with ELDP at generation (a) 0, (b) 50, (c) 150, (d) 350, and (e) 500 for FON

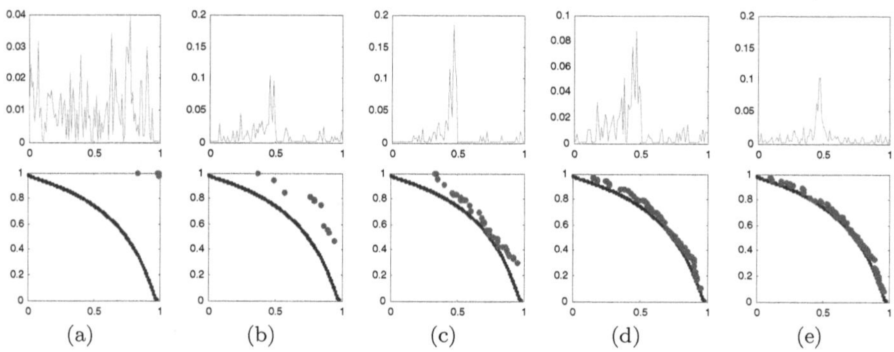

Fig. 3.45 The first row represents the distribution of one decision variable and the second row shows the associated non-dominated individuals of baseline MOEA with GASS at generation (a) 0, (b) 50, (c) 150, (d) 350, and (e) 500 for FON

Fig. 3.43(c)-(e) and Fig. 3.45(c)-(e), it can be seen that the incorporation of GASS results in a diverse distribution of individuals in the decision space. This shows that GASS is capable of diverting the evolution to other search regions upon the detection of a convergence, thus allowing the algorithm to discover the global tradeoff for ZDT4 as well as to achieve a good spread of non-dominated individuals for FON.

To examine the effect of parameter sensitivity for ELDP and GASS, a number of simulations are performed with different settings of α=\{0.0,0.05,0.1,0.3,0.5\} for ELDP and w=\{0.05,0.1,0.3,0.5\} for GASS at noise levels of 0% and 20%. The setting of α=0 for ELDP is equivalent to the operation of bit-flip mutation. Apart from demonstrating that ELDP provides better performances over the bit-flip mutation, it is observed that ELDP and GASS are capable of performing consistently and effectively within a large range of α and w settings for ZDT4 and FON at different noise levels.

3.10 Further Examination

The results in Sections 3.8 and 3.9 reveal that the proposed features can improve the performance of multi-objective optimization in terms of proximity, diversity, and distribution under the influence of noise. In this section, the features of ELDP and GASS are applied to SPEA2 and NSGAII to examine if their effects can be reproduced in conventional MOEAs. The ELDP is used in place of the bit-flip mutation operator in SPEA2 and NSGAII, while the GASS is implemented in conjunction with existing selection schemes. The possibilistic archiving model is not implemented here since the archiving strategies of SPEA2 and NSGAII play an important role in defining the behaviors of the algorithms.

It has been observed in the previous section that SPEA2 and NSGAII can neither discover the global tradeoff for ZDT4 nor maintain a well-distributed set of individuals for FON. The performances of these two algorithms are also largely affected by noise in ZDT4 and FON. Hence, these two benchmark problems are used in the study here. NSGAII-RF and SPEA2-RF denote the respective algorithms incorporated with the proposed features. The metric distributions of the simulation results for noiseless and noisy ZDT4 are shown in Fig. 3.46(a)-(d) and Fig. 3.47(a)-(d), respectively. Similarly, the performance of the algorithms for noiseless and noisy FON is shown in Fig. 3.48(a)-(d) and Fig. 3.49(a)-(d), respectively.

It can be observed from Fig. 3.46 - Fig. 3.49 that ELDP and GASS are capable of improving the performances of SPEA2 and NSGAII in terms of convergence and diversity of individuals along the tradeoff for ZDT4 and FON. In the case of ZDT4, the incorporation of the proposed features allows NSGAII-RF and SPEA2-RF to escape the local optima in ZDT4. In the case of FON, Fig. 3.48 shows that the incorporation of ELDP and GASS improves the performance in terms of GD, MS, and HVR. It can also be observed from

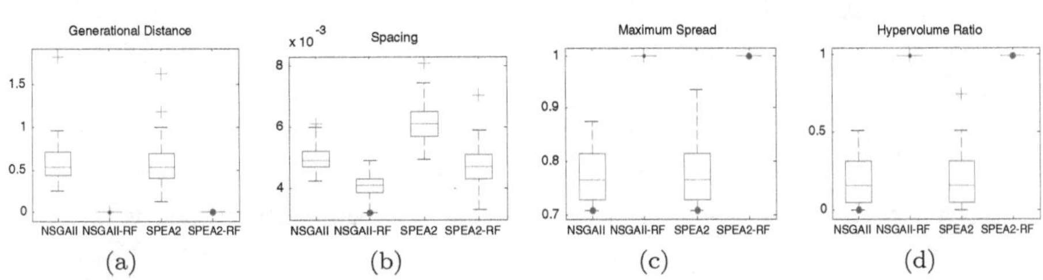

Fig. 3.46 Performance metric of (a) GD, (b) S, (c) MS, and (d) HVR for ZDT4 with 0% noise

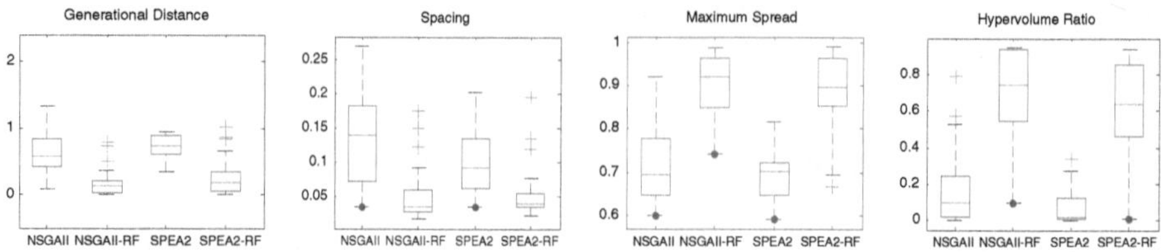

(a) (b) (c) (d)

Fig. 3.47 Performance metric of (a) GD, (b) S, (c) MS, and (d) HVR for ZDT4 with 20% noise

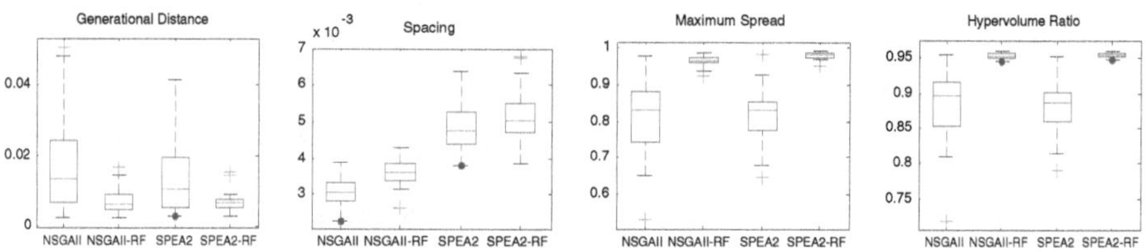

(a) (b) (c) (d)

Fig. 3.48 Performance metric of (a) GD, (b) S, (c) MS, and (d) HVR for FON with 0% noise

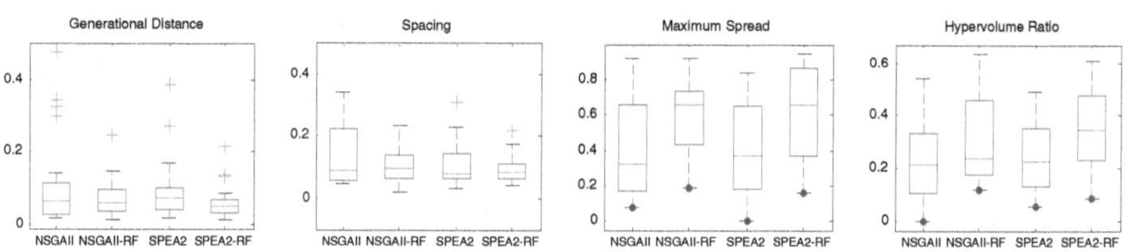

(a) (b) (c) (d)

Fig. 3.49 Performance metric of (a) GD, (b) S, (c) MS, and (d) HVR for FON with 20% noise

Fig. 3.49 that NSGAII-RF and SPEA2-RF have a slight edge over NSGAII and SPEA2 in almost all aspects of the multi-objective optimization goals.

3.11 Conclusion

Noise has a detrimental effect on the selection, elitism, and diversity presevation processes in MOEAs. Hence, it is not surprising that most of the algorithms presented in this chapter involve different schemes to improve the performance of these mechanisms in the presence of noise.

The simplest approach to reduce the impact of noise is to conduct explicit averaging. In this regard, the modified non-dominated sorting genetic algorithm II (MNSGAII) and the indicator-based evolutionary algorithm (IBEA) worked on the expected fitness values. Furthermore, in MNSGAII, dominated solutions are inducted into the first non-dominated front using a clustering algorithm to prevent the loss of good solutions due to errors made during the non-dominated sorting procedure. To improve the reliability of the selection process, algorithms, such as estimate strength Pareto evolutionary algorithm (ESPEA), multi-objective probabilistic selection evolutionary algorithm (MOPSEA), and multi-objective evolutionary algorithm for epistemic uncertainty (MOEAEU), introduce the concept of probability to model the noisy objective vectors. To minimize the impact of noise on the optimization process, the noise tolerant strength Pareto evolutionary algorithm (NTSPEA) assigns a finite lifetime to all archived solutions. The MOEA-RF incorporates two heuristics to guide the optimization process in the presence of noise and applies a possibilistic archiving to minimize the removal of true non-dominated individuals and provide a chance for individuals degraded by noise to survive in the evolution.

Chapter 4
Handling Noise in Evolutionary Neural Network Design*

In this chapter, we consider the design of artificial neural networks (ANNs) as an instance of noisy design problem. In the context of ANN design, evolutionary optimization has led to the development of evolutionary artificial neural networks (EANN) in which adaptation is performed primarily by means of evolution. Given that the intrinsic relationship between the architecture and the associated synaptic weights can be quite complex, the design methodology would be flawed if we were to decouple these two properties during the training phase of the network. The design of ANN has two intrinsical noise sources:

- The same network structure can give rise to different fitness values due to different weight instantiations.
- Excess hidden layer neurons tend to fit the observed features of the training samples which are not representative of the intrinsic underlying distribution of observations, obstructing the characterization of the true properties of the system or problem.

To minimize the effects of these two forms of noise, we developed a hybrid multi-objective evolutionary neural network (HMOEN) in [98]. A micro-hybrid genetic algorithm (μHGA) is applied to optimize the synaptic weights with respect to any new ANN structure introduced to reduce the former effects of noise, while singular vector decomposition (SVD) is incorporated in an architectural recombination operator to handle the latter. The role of the SVD operator is important, particularly in obtaining the geometrical relevancy of neurons in hidden layer space when attempting to prune a neural network to finally arrive at a parsimonious network architecture.

C.-K. Goh and K.C. Tan: Evolutionary Multi-objective Optimization, SCI 186, pp. 101–121.
springerlink.com

4.1 Singular Value Decomposition for ANN Design

As mentioned above, excess hidden neurons will lead to the phenomenon of over-fitting of data that are not characteristic of the underlying problem. However, it is difficult to quantify the contribution of additional neurons in the hidden layer without the use of an independent validation set of data. In this chapter, we suggest the use of a simple, yet, robust information measure based on the SVD operator in the framework of EANNs to achieve this purpose, in removing neurons in the hidden layer of the evolved single hidden layer feedforward neural network.

Computationally, the SVD is very robust and allows the discrimination against noise contamination. Typically, the SVD is utilized in computing the pseudoinverse (Moore-Penrose generalized inverse) of a rectangular, possibly singular, matrix. The SVD has also been extensively applied in problems of least squares, spectral estimation, and system identification. In signal processing, the SVD plays a central role in subspace modeling or low-rank approximation (similar to our problem of estimating the number of hidden layer neurons) of signals.

4.1.1 Rank-Revealing Decomposition

Consider the output matrix H of the hidden layer corresponding to the N training samples and n hidden neurons. The actual rank (say k) of H may be different from its numerical rank (say n), where $k \leq n$. Such a situation usually arises when the original matrix H is contaminated by E resulting in a matrix, $\tilde{H}$

$$\tilde{H} = H + E \tag{4.1}$$

with rank(H)=k and rank($\tilde{H}$)=n. This contamination, commonly referred to as noise, obstructs the characterization of the true properties of the system or problem given the observed training samples. This phenomenon actually corresponds to the marginal role played by additional hidden layer neurons that tend to fit the features of the training samples which are not representative of the intrinsic underlying distribution of observations.

Given a real matrix $H \in R^{Nxn}$, applying the SVD results in the orthogonal transformation,

$$\begin{aligned} U^T HV &= [\sum; 0] \\ \sum = \mathrm{diag}(\sigma_1, ...\sigma_n) \end{aligned} \tag{4.2}$$

where $U \in R^{NxN}$ and $V \in R^{nxn}$ are known as the left and right singular vectors of H. $\sum \in R^{Nxn}$ is a diagonal matrix with unique, non-negative entries ordered in decreasing magnitude. This decoupling technique of the SVD allows the expression of the original matrix as a sum of the first n columns of u and u^T, weighted by the singular values. The rank of H is determined by observing the n largest singular values that are non-zero.

While SVD does not reveal the actual rank of full-ranked $\tilde{H}$ [159, 245], through the structure of its zero elements, it provides information of the actual rank through the structure of small elements. Let the singular values of H and $\tilde{H}$ be $(\sigma_1, ...\sigma_n)$ and $(\tilde{\sigma_1}, ...\tilde{\sigma_n})$, respectively. From Schmidts Sub-Space Theorem, we have

$$\tilde{\sigma}_{k+1}^2 + ... + \tilde{\sigma}_n^2 \leq ||E||_F^2 \tag{4.3}$$

where $|| \ ||_F$ denotes the Frobenius norm, revealing the rank of $\tilde{H}$ such that its $n - k$ smallest singular values are bounded by the Frobenius norm of E.

4.1.2 Actual Rank of Hidden Neuron Matrix

Every neuron in the hidden layer constructs a hyperplane in the input feature space [124] and the contribution of each hidden neuron to the separating capability of the ANN depends on its uniqueness. In a geometrical sense, the rank of H denotes the space in which the columns of H occupy, representing the number of separating hyperplanes in the system. In the case where $n = H - 1$ hidden neurons are used with a suitable non-linear activation function such that H is full-ranked, the rank requirement [124, 229] is satisfied, giving rise to $N - 1$ separating hyperplanes for N training samples. This follows that, using simple matrix inverse, we are guaranteed perfect reconstruction for the training set. Intuitively, if H is of higher rank, H better fits the training data, at the expense of generalization when $\lim_{n \to N} \text{rank}(H)$. This full-rank condition also ensures that the patterns projected onto the hidden layer space are linearly independent; accordingly this is also known as ϕ-general position [46].

However, it should be noted that even if the transformed patterns produced in the hidden layer space are in general position, some of these patterns may be degenerate in the sense that certain patterns can be represented as a linear combination of other patterns. This leads to the issue of whether this additional hidden neuron is contributing to the actual separability of the samples or merely compensating for the presence of noise in the observations. Clearly, there is a limit, for which introducing additional hidden neurons will tend to over-fit the training data. Therefore, the actual rank of the matrix H is more useful in estimating the appropriate number of hidden neurons in the Single-Hidden Layer Feedforward Network (SLFN) for a given problem. In the context of representing the input patterns in hidden layer space, we can think of additional hidden layer neurons as causing degeneracy in this hidden layer space, for increasing the number of hidden layer neurons is akin to introducing noise into the system thus perturbing the hidden layer space such that the hidden layer space is now being represented by n hidden neurons which are of marginal benefit.

In Fig. 4.1, we illustrate the problem of hyperplane construction in hidden layer space and its corresponding relationship with the singular values of

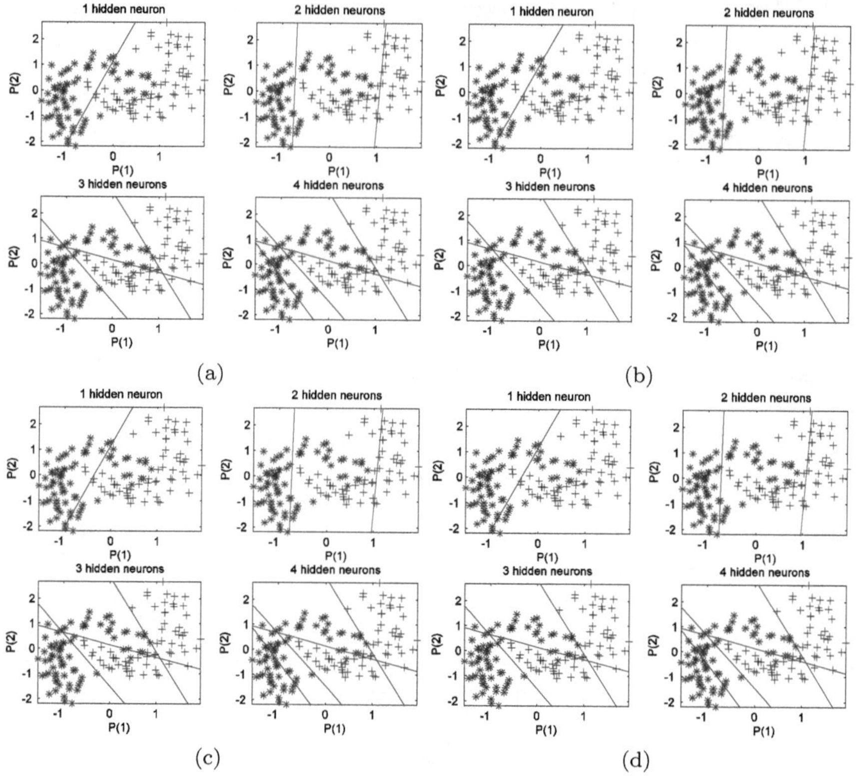

Fig. 4.1 Illustration of constructed hyperplanes in hidden layer space with (a) 1-4, (b) 5-8, (c) 9-12 hidden neurons and (d) corresponding decay of singular values as number of hidden layer neurons is increased

the hidden layer output matrix H using a rwo-dimensional toy problem that is easily visualizable. From observation, we know that three appropriately placed hyperplanes should provide us with a good balance of network capacity and complexity without sacrificing the generalization capability of the resulting classifier. The placement of these hyperplanes is achieved through the use of learning algorithms (e.g. EA or backpropagation using gradient descent) on the set of training data. These learning algorithms will usually attempt, to the best of their abilities, to position these hyperplanes such that their construction is as linearly independent as possible for the given set of training data. This usually requires that the learning algorithm has converged prior to using the SVD to decompose the matrix H and obtain the set of singular values. The singular values confirm that the use of three hidden layer neurons should be sufficient for the network capacity given the complexity of the problem. We arrive at this conclusion from the presence of a noticeable gap in the decay of singular values, indicating that most of the spectral energy of H can be attributed to the first three hidden layer neurons. Note that while we may not know the identities of these three most linearly independent hyperplanes (unless we carry out a recalculation of the SVD using a permutation of all possible combinations of hidden neurons), we find that it is more efficient to retrain the whole network using the found number of hidden neurons, which in this case is three.

Theorem 4.1: Define the numerical ε-rank k_ε of the matrix H with respect to some tolerance ε [115] by

$$k_\varepsilon = k_\varepsilon(H, \varepsilon) \equiv \min_{||E||_2 \leq \varepsilon} \{\text{rank(H+E)}\} \tag{4.4}$$

which states that if there is a gap between the k_ε-th and the $k_{\varepsilon+1}$-th singular values of size ε, then H has actual rank (ε-rank) k_ε. The larger this gap ε is, the more robust the matrix is to perturbation. To avoid possible problems when is itself perturbed, the definition of actual rank is refined by introducing δ as an upper bound for ε for which the numerical rank remains at least equal to k.

Theorem 4.2: The matrix H has a numerical rank of (δ, ε, r) with respect to the norm $|| \cdot ||$ if δ, ε, and r satisfy the following:

$$\begin{aligned} k &= \inf\{\text{rank}(B) : ||A - B|| \leq \varepsilon\} \\ \varepsilon &< \delta \leq \sup\{\eta : ||A - B|| \leq \eta \Rightarrow \text{rank}(B) \geq k\} \end{aligned} \tag{4.5}$$

σ_k provides an upper bound for δ, while σ_{k+1} provides a lower bound for ε.

The above definitions are equivalent to saying that the matrix H is linearly-independent when perturbed by E up to a threshold determined by $||E||^2 \leq \varepsilon$. This result also means that the singular values of H satisfy $\sigma_{k_\varepsilon} > \varepsilon > \sigma_{k\varepsilon+1}$. As described in [115], a well-determined gap between the singular values of σ_{k_ε} and $\sigma_{k_{\varepsilon+1}}$, represented by ε should exist in order for the above definition to make much sense; k_ε should be, in other words, well-defined for small perturbations of the threshold ε and the singular values. Alternatively, the numerical ε-rank is the smallest integer k for which (Schmidts Sub-Space Theorem),

$$\sum_{j=k+1}^{n} \sigma_j^2 \leq \varepsilon^2 \tag{4.6}$$

This result suggests that as more neurons are added to the hidden layer, the contribution of each additional hidden neuron decreases after a certain threshold. From a geometric point of view, additional hyperplanes constructed by these newly introduced hidden neurons are not unique, or different as compared to existing hyperplanes (these new hyperplanes may be almost parallel to existing ones). The significance of these new hyperplanes can be quantified by examining the singular values of the matrix H as more hidden layer neurons are added. For detailed proofs of these theorems, the interested reader is directed to [159, 245].

In our context, this suggests that we should use a SLFN with a fewer number of hidden neurons since a higher number of neurons in the hidden layer may unnecessarily fit the noise that is inherently present in the samples. As noted in [115], E is typically unknown but what we do have knowledge of is the source of E, which in turn can be used to estimate the norm of E.

These singular values indicate the degree of mutual correlation between features in the hidden layer space with column degeneracy resulting when these hidden space features are highly correlated, which in turn leads to the conclusion that these additional neurons are redundant. While the singular values do not provide information on which of these features are correlated (the identities of the neurons are not explicitly known), the presence of small singular values would indicate that these additional hidden neurons can be removed without affecting the performance of the SLFN significantly.

4.1.3 Estimating the Threshold

A long-standing problem in the use of the SVD as a tool in determining the actual or effective rank of a perturbed matrix is in the distinguishing of significantly small and insignificantly large singular values [164].

Suppose $H_n \in R^{Nxn}$ represents the hidden layer output activation matrix of a SLFN with n neurons in the hidden layer. As n increases, the input-output space mapping that is discovered by the MLP better approximates the training data. Increasing n can be seen as increasing the complexity (and hence capacity) of the network. Since all problems would have an inherent degree of complexity, which is essentially unknown, estimating this complexity characterized by k that is in some sense close to the inherent complexity of the problem is our objective. However, as n is increased, the better fitting of the training data by the MLP gives rise to a lesser ability to generalize on unseen examples (i.e. the training set). Let $\sigma_i(H_n)$ denote the i-th singular value of the SLFN with n hidden neurons. A way to measure the contribution of the i-th singular value to say, the separability of classes, is to relate its value to the other singular values.

If ε is large, we can assume that the matrix H is relatively robust to perturbations; conversely, if ε is selected to be small (but not too small such that the numerical ε-rank does not make sense), external noise that is introduced to the system may cause the matrix H to be rank-degenerate. Often, there is no clear value for k where $\sigma_k - \sigma_{k+1}$ is obvious. If the SLFN has been well-trained and has converged (there is little change in its weight values), the decay of the singular values is gradual and not very distinct and hence, we cannot conclude confidently that the numerical rank of matrix H is less than its actual rank. This has been explored in further detail in [262].

4.1.4 Moore-Penrose Generalized Pseudoinverse

To resolve a linear system of the general form $H\beta = T$ is straightforward if the matrix H is square and non-singular. However, under many practical circumstances, the matrix H is usually singular and likely to be rectangular. The Moore-Penrose generalized pseudoinverse simplifies the treatment by providing the solution to the linear system in a least-squares sense. The pseudoinverse of H is defined differently depending on the rank and dimensionality of H.

In most practical problems, the system is over-determined and hence, would want to find the least-square error of $||H\beta - T||_2$ in the presence of the inconsistencies introduced by the additional equations. Thus, β is obtained from $||H\beta - T||_2$. The pseudoinverse can be shown to be the minimum norm least-squares solution of the system, i.e. the pseudoinverse of β, which is $\beta^{\dagger}$, minimizes $||H\beta - T||_2$. For further details on the pseudoinverse, readers are directed to [237].

4.2 Hybrid Multi-Objective Evolutionary Neural Networks

4.2.1 Algorithmic Flow of HMOEN

To design an evolutionary ANN that is capable of evolving the architecture and weights of the ANN simultaneously, a few features, such as variable-length chromosome representation, specialized genetic operator in the form of the SVD-based Architectural Recombination (SVAR), and micro-hybrid genetic algorithm (μHGA) for effective local search, are incorporated in HMOEN. The pseudocode for HMOEN is given below:

Hybrid Multi-Objective Evolutionary Neural Networks (HMOEN)

Step 1: Initialize population of ANNs, P_t.
Step 2: Evaluation of ANNs on training set and Pareto ranking.
Step 3: Update archive:

- Add non-dominated solutions in P_t to A_t.
- Remove non-dominated solutions from updated A_t.
- Remove most crowded solutions based on niche count if $|A_t|$ exceeds size limit.

Step 4: Combine updated A_t and P_t. Perform binary tournament selection to form mating pool.
Step 5: Calculate *epr* and *gen_req* for all selected solutions.
Step 6: Perform SVAR and mutation.
Step 7: Implement μHGA.
Step 8: If stopping criterion is not satisifed, repeat process from Step 2. Otherwise output A_t.

The design process begins with the initialization of population. All individuals will be evaluated according to the objective functions and ranked according to their dominance relationship in the population. The objective functions and ranking scheme will be described in Section 4.2.2. After the ranking process, non-dominated solutions will be updated into the archive. HMOEN applies the fixed-sized archive applied in Chapter 2. Elitism is implemented by a two-stage process; 1) the archive and evolving population are combined and 2) binary tournament selection of this combined population is conducted to fill up the mating pool. In binary tournament selection, two individuals are selected randomly from the combined population and compared based on the Pareto rank. The individual with a lower rank, i.e. better, is selected for genetic variation. In the event of a tie in Pareto ranks, niche count is used as the tie-breaker. The selected ANNs will then undergo the process of SVAR, which adapts the network architecture, and the mutation process. In order to reduce the noise presented by the change in network architecture as well as to improve convergence, the offspring are allocated to the μHGA for local exploitation. The evolution process repeats until the stopping criterion is satisfied. The mechanisms of SVAR and μHGA are described in Sections 4.2.4 and 4.2.5, respectively.

4.2.2 Multi-objective Fitness Evaluation

ANN design is cast as a multi-objective optimization problem where a number of objectives such as training accuracy and degree of complexity can be specified. The conflicting objectives of maximizing network capacity and minimizing network complexity is manifested in the tradeoffs between training and test accuracies. As before, the Pareto ranking scheme [84] assigns the same smallest cost for all non-dominated individuals.

One of the primary reasons why a weighted objective is not favored is due to the fact that it is difficult to properly apportion the weights that should be associated with each objective in converting a multi-objective problem into a SO problem. Most objectives that are considered, such as training accuracy and size of neural network weights, are not commensurable (not of the same dimensional quantity), which makes it rather difficult to place these two objectives on a similar platform for comparison.

In this chapter, we consider the simultaneous evolution of both the neural architecture as well as the synaptic weights. Further, this problem is distinguished from previous works by formulating the problem as a multi-objective problem where the twin objectives of classification accuracy and network complexity are conflicting in nature. Therefore, the optimization problem for the ANNs generalization on unseen data can be written as

$$\begin{aligned} f_1 &= \min\left\{\sum_{\theta=1}^{N}\sum_{k=1}^{C}\left(y_k(\theta) - d_k(\theta)\right)^2\right\} \\ f_2 &= \min\{NH_k\} \\ f_3 &= \min\{||W_k||_2\} \end{aligned} \tag{4.7}$$

where f_1 refers to the sum-of-squared (SSE) errors of the classification errors. In our problem formulation, we use only one hidden layer of neurons (k=1), as dictated by the Universal Approximation Capability (UAC) theorem for neural networks. N is the number of samples, C is the number of classes, and d_k is the desired output.

The two other objectives that we consider in our multi-objective approach are firstly, the minimization of the number of neurons in the hidden layer (f_2), and secondly, the minimization of the L_2-norm of the hidden layer weights (f_3). We consider each in turn, as will be demonstrated later in our experimental results. The use of f_2 and f_3 in addition to f_1 typically leads to improved generalization performance (as compared to the use of f_1 as the sole objective to be maximized) as the size, or complexity of the network is now controlled. There is little distinction, empirically, of which of the two additional objectives to be minimized (f_2 or f_3) leads to better testing accuracies. We will address this issue in a later section of this chapter.

4.2.3 Variable-Length Representation for ANN Structure

EAs process a set of encoded parameters, providing EA designers with the flexibility to design an appropriate representation of the potential solutions. Appropriate representation implies that it fulfils some criteria, such as ease of implementation or exploitation of the problem structure. For simplicity, the chromosome is often represented as a fixed-structure and the embedded variables are usually assumed to be independent and context insensitive. In EANNs, a hybrid structure between binary and real-number representation is commonly used for the simultaneous optimization of weights and architecture. However, such a representation is not suitable for ANN design problems where flexibility is essential for the simultaneous evolution of architecture and connection weights.

In this chapter, a variable-length chromosome representation is adopted to represent the ANN topology, including the number of neurons in the hidden layers and the connection weights linking the input, hidden, and output layers as illustrated in Fig. 4.2(a). Fig. 4.2(b) is the instantiation of the representation in Fig. 4.2(a). Each neuron is coupled with its associated weights, thus allowing easy manipulation by search operators for the addition or deletion of neurons. The chromosome may consist of different number of neurons, which reflects the complexity of the ANN, but the number of connections is fixed by the number of input attributes. Such a representation is efficient and facilitates the design of problem-specific genetic operators.

4.2.4 SVD-Based Architectural Recombination

In EANNs, the recombination process between two ANNs is unlikely to produce a good offspring due to the lack of a clear definition of a building block

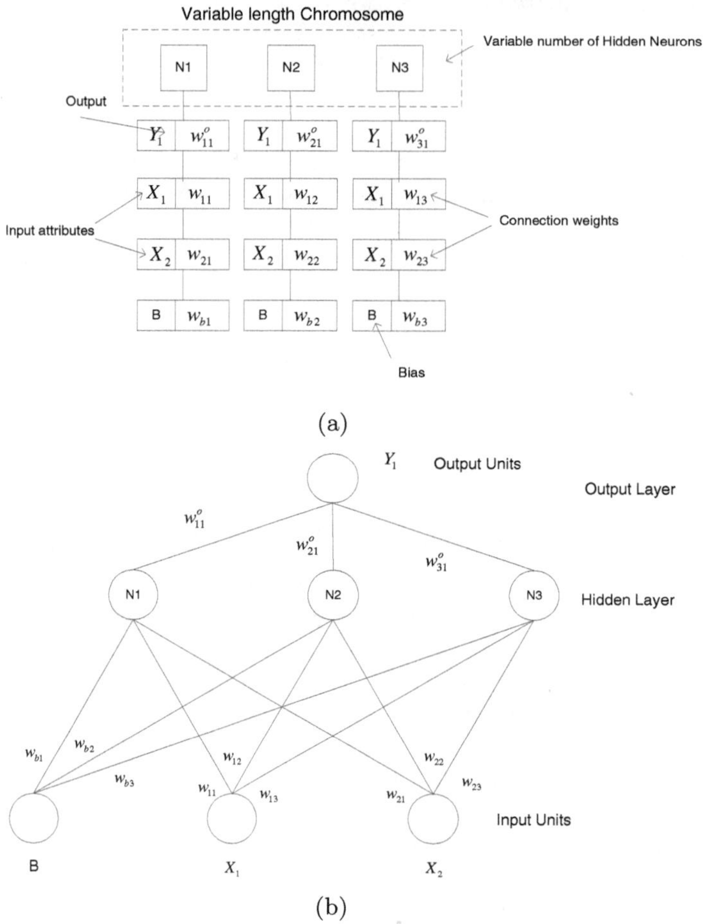

Fig. 4.2 An instance of the variable chromosome representation of ANN and (b) the associated ANN

in the framework of ANN [284]. However, the lack of recombination to facilitate the exchange of information between candidate solutions implies that each individual is expected to adapt independently by making best use of all available local information. This motivates the development of the SVAR approach which is based on the fact that each neuron constructs a hyperplane in the input feature space and hence, contributes to the resulting separating capability of the ANN. It follows that each neuron and its associated connections define a building block which contributes to the capacity of the ANN.

The issues considered in the design of the SVAR operator include the selection of the appropriate neuron and its associated weights for recombination as well as the decision to remove or add an appropriate neuron to the candidate ANN design. SVAR is performed between two parent ANNs and the procedure is outlined in the pseudocode below:

SVD-Based Architectural Recombination (SVAR)

Step 1: Apply SVD operator to determine presence of redundant neurons for both parents with a corresponding threshold ε and to determine the number of (redundant) hidden layer neurons to remove.
Step 2: Selection of neurons for removal or exchange.
Step 3: For both parents, perform either removal, exchange, or addition of neurons:

- If U(0,1) < 1/3,
    Perform exchange of marked neurons.
  Else if there are redundant neurons as indicated from Step 1,
    Perform removal of marked neurons.
  Else add neurons marked for exchange to both parents

For our proposed approach, the building blocks of each network are the neurons (together with their incoming weights from the previous layer). We call these neurons building blocks as they are the smallest units for which we operate on (such as performing crossover). The SVD is used as the tool to determine the presence of redundant neurons, while the calculation of inter- and intra- subspace angles is used for the selection of neurons to be removed or exchanged. The SVD operator will decide the number of hidden layer neurons to be removed.

The number of neurons in the hidden layers that are deemed redundant by the SVD operator is in effect a function of the threshold that is used, with ε assuming the role of the SVD threshold. In deciding *which* hidden layer neuron to remove or prune, we use a geometrical approach, where the algorithm examines the subspace spanned between the hidden layer neurons such that the neuron(s) which is (are) most linearly correlated with the other hidden layer neurons is (are) consequently removed. This is to prevent unnecessary removal of a neuron at the initial stages when the weights are not yet adapted to the problem.

The rationale for utilizing subspace angles as the selection criterion for pruning and exchange is to encourage the linear independency between the neurons. In [262], the authors used the SVD operator to first determine the appropriate, or necessary, number of hidden layer neurons on an initially large structured feedforward neural network. After which, the network is retrained using the same learning algorithm but using the reduced set of hidden layer neurons. In our approach, however, we adopt an online method such that during the evolutionary process, the identity of the hidden layer neuron(s) to be removed are determined geometrically by the subspace approach as described above. From the flowchart, it is observed that it is possible for a candidate ANN with redundant neurons to retain the same structure via the exchange of neurons from the other parent.

4.2.5 Micro-Hybrid Genetic Algorithm

As mentioned in the introduction, the optimization of neural network structures is an inherently noisy problem, i.e. the immediate neural network fitness after the recombination process may not be a good indicator of the new network structure due to an inappropriate set of weights. Thus, it is necessary to optimize the synaptic weights with respect to the new ANN structure after any structural changes. The mutation operator offers a simple option for local fine-tuning [139]. However, domain information cannot be easily incorporated and its stochastic nature tends to render the search operation inefficient. Intuitively, the change in genetic structure should be ordered instead of being left to chance in order for the local search to be robust. While the well-known back-propagation (BP) algorithm is a directed search, by means of gradient descent, it is prone to being trapped in local optima. In view of these concerns, the EANN is hybridized with the μHGA .

μHGA

THe μHGA exploits the synergy between a μGA [150] and the pseudoinverse operator to decompose the large and complex search space. Specifically, the μGA performs local fine-tuning of the hidden layer weights, while the pseudoinverse operator optimizes output weights in the least squares sense based on the weights found by the μGA. The pseudocode of the μHGA is shown below:

Micro-Hybrid Genetic Algorithm (μHGA)

Step 1: Initialize POP by creating $|POP| - 1$ variants of selected ANN.
Step 2: Evaluate POP and store best ANN.
Step 3: Perform binary tournment to select $|POP| - 1$ ANNs from POP. Insert best ANN into mating pool.
Step 4: Perform crossover on selected neurons.
Step 5: Perform mutation on selected neurons.
Step 6: Apply pseudoinverse operator.
Step 7: If stopping criterion is not satisifed, repeat process from Step 2. Otherwise, output best ANN.

In μHGA, simulated binary crossover (SBX) and uniform mutation (UM) is applied to evolve the desired set of connection weights. The mutation strength of UM is adapted as,

$$s = 0.1 \cdot (uppbd_w - lowbd_w) \tag{4.8}$$

where $uppbd_w$ and $lowbd_w$ correspond to the minimum and maximum of the associated weights in the population, respectively.

Balance between Evolution and Learning

While the incorporation of local search can accelerate the convergence of the evolutionary optimization process, hybrid EAs also give rise to issues pertinent to the tradeoffs between evolution and learning. Apart from the obvious challenge posed by limited computational resources, balance between exploration and exploitation is necessary to maintain diversity in the evolving population for the approximation of the Pareto optimal front. Consequently, these concerns have led to the recent development of resource utilization schemes, such as local search probability [139] and simulated heating [14].

While local search probability can reduce the computational time utilized for local fine-tuning, the exploration-exploitation dilemma is not explicitly considered. The fundamental idea behind simulated heating is based on simulated annealing, where the intensity of local search increases with time. Although it is intuitive that more computational time for local search should be allocated in the later stages, online search requirements are not considered in simulated heating. In contrast to existing methods which allocate resources based on a predetermined schedule, the allocation of resources here is based on the feedback of an online performance measure, the evolutionary progress rate [252]. The evolutionary progress rate ($\mathrm{epr}(t)$) can be defined as the ratio of the number of new non-dominated solutions discovered in generation t, $new\_nondomsol(t)$, to the total number of non-dominated solutions, $total\_nondomsol(t)$,

$$\mathrm{epr}(t) = \frac{new\_nondomsol(t)}{total\_nondomsol(t)} \tag{4.9}$$

The set of new non-dominated individuals discovered at each generation is basically composed of individuals that dominate the non-dominated individuals of the previous generation and individuals that contribute to the diversity of the solution set.

In this adaptive scheme, the number of individuals allocated for local search is adapted based on the $\mathrm{epr}(t)$ in every generation. Mathematically, the adaptation of computational resource allocation can be written as,

$$gen\_req(t) = \big(1 - \mathrm{epr}(t)\big) \cdot \big(upp\_bd_{com} - low\_bd_{com}\big) + low\_bd_{com} \tag{4.10}$$

where gen_req is the number of generations performed by μHGA, while upp_bd_{com} and low_bd_{com} denote the upper and lower limits of available computational resource, respectively. The rationale is that a high value of $\mathrm{epr}(t)$ means that the algorithm is still in the exploratory stage and the need for local fine-tuning is low. Likewise, a low value of $\mathrm{epr}(t)$ is an indication of convergence and more resources are required to meet the requirements of local fine-tuning. In this chapter, upp_bd_{com} and low_bd_{com} are set as 20 and 10 of the total population size, respectively.

4.3 Experimental Study

4.3.1 Experimental Setup

In order to evaluate the effectiveness of the proposed methods, a detailed empirical study is carried out on seven different datasets. HMOEN is implemented using the MATLAB technical computing platform and corresponding simulations are performed on an Intel Pentium 4 2.8 GHz computer. Thirty independent runs are performed for each of the datasets to obtain statistical information such as consistency and robustness of the algorithms. The various parameter settings of HMOEN are tabulated in Table 4.1.

In the training phase for the classifiers, we use 30-fold cross-validation, partitioning the data into two independent training and testing sets. 60% of the available samples are randomly selected as training data, with the remaining 40% as testing data. Prior to training, pre-processing is carried on the samples of each dataset. All input features are scaled and transformed such that the resulting input features have a mean of 0 and a variance of 1, as it has been shown that convergence is usually faster if the average of each input variable over the training set is close to zero [178]. For the outputs, since we consider classification problems, we use binary target values with a 1-out-of-C encoding where for a C-class problem, the largest output i is assigned to class i, with i=$\{1,2,..,C\}$. For the k-th training sample, the desired class output c where $d_k(\theta) = \{0, 1\}$ is 1 for $k = c$ and 0 otherwise. This is essentially a winner-take-all approach for the output layer neurons, and is a common approach used for classification purposes. Hidden layer neurons

Table 4.1 Parameter settings of HMOEN for the simulation study

Parameter	Settings
Population	Main population size: 20; Archive size: 20 μHGA: 4.
Chromosome	HMOEN: Variable-length real number representation; μHGA: Real number representation;
Selection	Binary tournament selection
Crossover operator	HMOEN: SVAR μHGA: SBX
Crossover rate	0.9
Distribution index (SBX)	10
Threshold, ε (SVAR)	0.995
Mutation operator	Normally distributed mutation
Mutation rate	0.1
Mutation strength	Adaptive
Nice radius	Dynamic.

Table 4.2 Characteristics of Data Set

Dataset	Samples	Attributes	Classes	Remarks
Cancer	699	9	2	Determine the patients for whom the tumour is benign or malignant
Pima	768	8	2	Determine whether a patient shows sign of diabetes according to World Health Organization Criteria
Heart	297	13	2	The learning task is to predict the presence or absence of heart disease given the results of various medical tests carried out on a patient.
Hepatitis	155	19	2	The hepatitis problem is a complex and noisy dataset as it contains a large number of missing data (there are 167 missing values in total in this dataset). The learning task is to predict whether a patient with hepatitis will live or die.
Horse	368	27	2	The objective here is to determine, based on the physical ailments and attributes of a particular horse, if it should have surgery performed on it.
Iris	150	4	3	This dataset is perhaps the best-known database to be found in pattern recognition literature. One class is linearly separable from the other two; the latter are NOT linearly separable from each other.
Liver	345	7	2	The learning task for this dataset is to determine, if the adult male that is tested using blood tests suffer from liver disorders that might arise from excessive alcohol consumption.

use a hyperbolic tangent non-linearity, while the output nodes use a linear output activation function.

The real-world datasets used in this simulation study, represent some of the most challenging problems in machine learning, were obtained from the UCI machine learning database (http://www.ics.uci.edu/~mlearn/MLRepository.html). Many researchers have used these datasets in validating the performances of their algorithms, and thus these datasets provide a good test suite of problem for evaluation of the proposed approach. The key characteristics of these problems and their associated learning tasks are summarized in Table 4.2.

4.3.2 Analysis of HMOEN Performance

Experimental results: Table 4.3 and Table 4.4 show the results of HMOEN over 60 independent runs on the seven problems. HMOEN_HN denotes that ANNs are evolved based on the criteria of f_1 and f_2, while HMOEN_L2 denotes that ANNs are evolved based on the criteria of f_1 and f_3. The test and training results are based on the ANN with the best training accuracy on the dataset from each run. Network size refers to the mean of the number of hidden layer neurons of the associated ANNs.

With the exception of Hepatitis, the networks evolved by HMOEN_HN and HMOEN_L2 have comparable sizes. For instance, the mean network size evolved by HMOEN_HN and HMOEN_L2 are 9.8667 and 9.6833, respectively. This is probably because architectural adaptation is governed by the same mechanism of SVAR. On the other hand, apart from Liver, the paired-T test reveals that the different optimization criteria have a significant impact on test accuracies. In particular, HMOEN_L2 performs significantly better for the problems of Pima, Hepatitis, Horse, and Iris, while HMOEN_HN only fares slightly better for Cancer and Heart.

Effects of proposed features: In this section, the effects and contributions of multi-objectivity, and the proposed features of SVAR and μHGA are examined for the different datasets. Note that only HMOEN_L2 is used in the study here since it has been observed in the previous section that f_1 and f_3 are generally the better optimization criteria. Here, different case setups are used to represent different combinations of features in HMOEN_L2. The different cases are summarized in Table 4.5. In Case 1, f_1 and f_3 are aggregated with the weight vector [0.8 0.2], i.e. $F = 0.8 \cdot f_1 + 0.2 \cdot f_2$. HMOEN_L2 is

Table 4.3 HMOEN_HN Performance on the Seven Different Datasets. The Table Shows the Mean Classification Accuracy and Mean Number of Hidden Neurons for all Datasets.

Method		Data Set						
		Cancer	Pima	Heart	Hepatitis	Horse	Iris	Liver
HMOEN_HN								
Training	Mean	0.9816	0.8075	0.9021	0.9668	0.9991	1	0.797
	Std	0.0015	0.0053	0.0098	0.0113	0.002	0	0.0098
Testing	Mean	0.9682	0.7536	0.8106	0.7551	0.977	0.9103	0.6894
	std	0.0058	0.0182	0.0261	0.0441	0.0171	0.0313	0.0271
Network size	Mean	4.8	8.0833	9.6833	10.7	9.15	2.6833	6.7667
	Std	1.8485	1.8712	3.1218	3.5668	5.3673	1.0813	1.1552

Table 4.4 HMOEN_L2 Performance on the Seven Different Datasets. The Table Shows the Mean Classification Accuracy and Mean Number of Hidden Neurons for all Datasets.

Method		Data Set Cancer	Pima	Heart	Hepatitis	Horse	Iris	Liver
HMOEN_L2								
Training	Mean	0.9825	0.7954	0.9035	0.9498	0.9991	0.9839	0.8059
	Std	0.0017	0.0052	0.0088	0.0164	0.0022	0.0056	0.0119
Testing	Mean	0.9626	0.7845	0.7969	0.803	0.9838	0.98	0.6800
	std	0.011	0.0122	0.0294	0.048	0.0131	0.0184	0.0294
Network size	Mean	4.6667	7.5167	9.8667	11.3833	7.2	3.0833	6.8333
	Std	1.5367	2.446	3.5865	3.2682	5.0583	0.8693	1.2374

Table 4.5 Different Case Setups to Examine Contribution of the Various Features

Case	Optimization Goal	Settings SVAR	μHGA
1	Weighted	Yes	Yes
2	Multi-objective	No	No
3	Multi-objective	Yes	No
4	Multi-objective	No	Yes
5	Multi-objective	Yes	Yes

represented as Case 5. The variable-length chromosome, Gaussian mutation operator, and archive mechanism remain fixed for the five cases.

The distributions of the classification accuracy and network size are represented by boxplots in Fig. 4.3 and Fig. 4.4, respectively. By comparing Case 1 (which is the SO version of HMOEN) and Case 5 in Fig. 4.3, we can note that the multi-objective approach is generally better with similar structural complexities. The paired-T test conducted also indicates that the performance between Case 1 and Case 5 is statistically different in all problems except Cancer. The purpose of conducting variants of HMOEN_L2 with and without SVAR is to ascertain the contribution of the proposed operator. The effects of SVAR can be observed by comparing the performances between Cases 2 and 3 and Cases 4 and 5 in Fig. 4.3. Clearly, without the use of the SVD as a form of capacity control in SVAR, the performances demonstrated in Case 2 and Case 4 are inferior to those in Case 3 and Case 5, respectively, for most of the problems. In addition, comparable, if not better, classification accuracies are achieved with smaller networks as evident in Fig. 4.4. These results substantiate our earlier hypothesis that each (hidden) neuron, together with

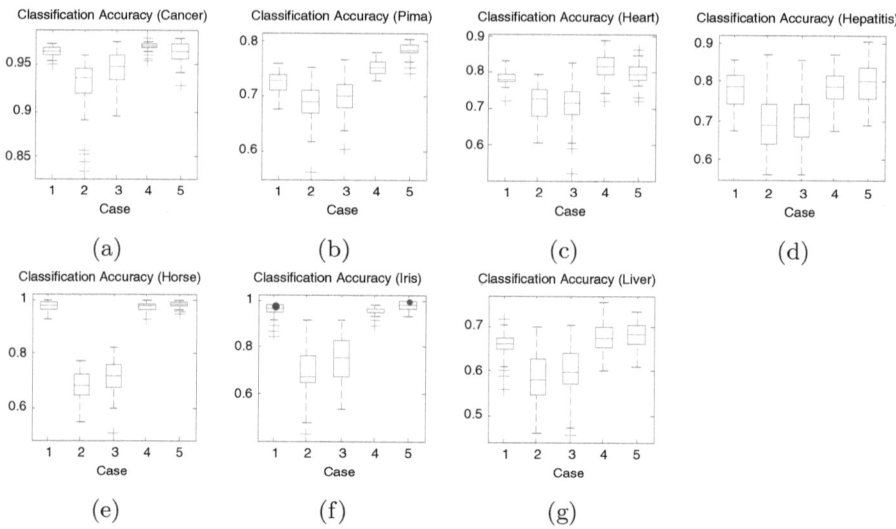

Fig. 4.3 Test Accuracy of the Different Cases for (a) Cancer, (b) Pima, (c) Heart, (d) Hepatitis, (e) Horse, (f) Iris, and (g) Liver

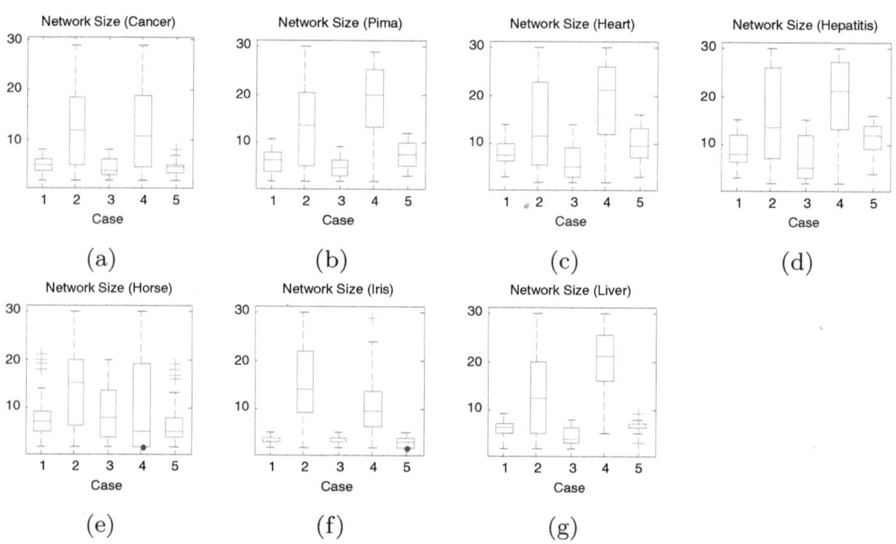

Fig. 4.4 Network Sizes of the Different Cases for (a) Cancer, (b) Pima, (c) Heart, (d) Hepatitis, (e) Horse, (f) Iris, and (g) Liver

its corresponding hidden layer weights (leading from the input layer to the hidden layer), functions as a building block for an EANN. The specialized recombination operator acts specifically on these neuronal building blocks. This notion is intuitively appealing because when viewed from the perspective of hidden layer space, each hidden neuron and its input set of weights (for the single hidden layer) construct a separating hyperplane in hidden layer space; thus, each hidden neuron together with its corresponding set of input weights, which are treated as a set of building blocks, are accountable for determining the separation of the training samples in hidden layer space. By comparing the performances between Cases 2 and 4 and Cases 3 and 5, it is clear that the introduction of μHGA provides substantial improvements to the classification performance on the testing sets of all the datasets. It can be noted that the local search ensures that the final network is sufficiently well-trained such that the SVD is able to operate on the hidden layer activation matrix effectively. Recall that the use of the SVD, as described earlier, requires the network to be well-trained. In other words, without μHGA, the SVAR tends to remove neurons excessively as reflected in the generally lower number of hidden neurons (large number of neurons are pruned). Therefore, it is also evident that SVAR and μHGA are complementary mechanisms in HMOEN.

Effects of SVD threshold settings: It can be observed in the previous section that the SVAR allows HMOEN to evolve smaller networks with comparable, if not better, classification accuracies. In this section, experiments are conducted for the various datasets over SVD threshold settings of $\{0.9, 0.95, 0.98, 0.99\}$ to investigate its effects on network structure and classification performance. Trends of testing accuracies and network sizes over the four threshold settings for the different datasets are plotted in Fig. 4.5 and Fig. 4.6, respectively.

Theoretically, the size of the network is expected to increase with the SVD threshold. This can be explained from the fact that the pruning mechanism, implemented through the proposed SVD-based crossover, becomes stricter so as to maintain more of the spectral energy of the singular values. This requires that more hidden layer neurons need to be kept. This conjecture is reinforced from Fig. 4.6 where a monotonically increasing trend for the architectural or structural complexity is observed as the SVD threshold is progressively increased from 0.95 to 0.99.

From a conventional and theoretical perspective, the trends for the training and testing accuracies are usually strongly positively correlated up to a certain point, beyond which the performance of the classifier on the test set degrades. The correlation between these two sets becomes negative when the training accuracy steadily increases while the testing accuracy drops. Continuing the training process beyond this point would result in the classifier being overtrained and any subsequent training, while increasing the training accuracy, will always degrade the performance of the classifier on the testing set.

Overtraining occurs in two ways, either with excessively many training epochs or when the architecture of the network is overly complex. The former applies to the traditional learning route for neural networks, while the latter

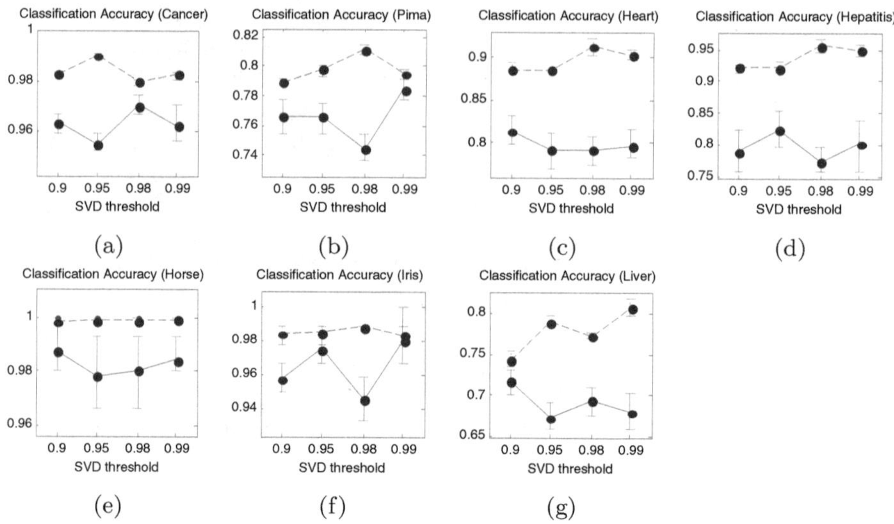

Fig. 4.5 Trend of Training Accuracy (- -) and Testing Accuracy (-) over different SVD threshold settings for (a) Cancer, (b) Pima, (c) Heart, (d) Hepatitis, (e) Horse, (f) Iris, and (g) Liver. The trend is connected through the mean while the upper and lower edges represent the upper and lower quartiles respectively.

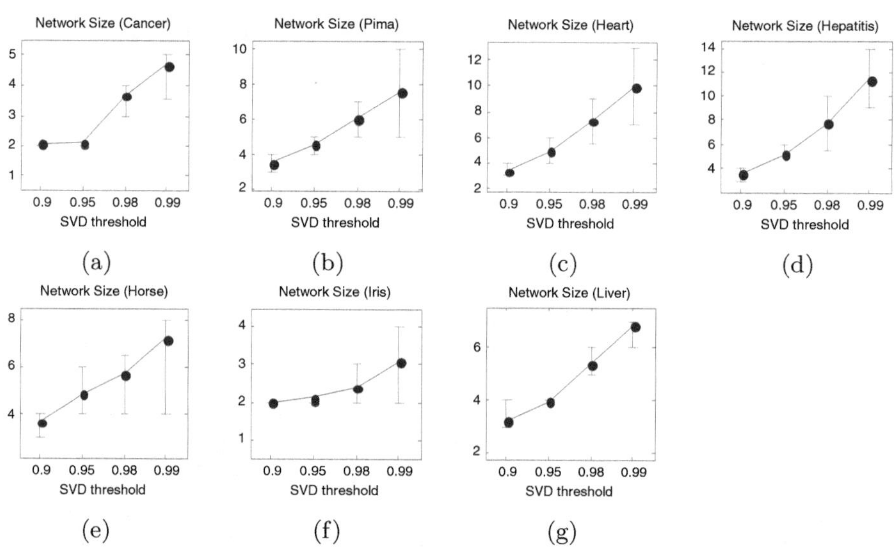

Fig. 4.6 Trend of Network Size over different SVD threshold settings for (a) Cancer, (b) Pima, (c) Heart, (d) Hepatitis, (e) Horse, (f) Iris, and (g) Liver. The trend is connected through the mean while the upper and lower edges represent the upper and lower quartiles respectively.

applies to our approach, where the SVD threshold is increased during the pruning mechanism which in turn is implicitly implemented via the SVD-based crossover. We can understand the SVD threshold from the perspective of architectural complexity. A higher SVD threshold means larger networks with more hidden layer neurons being evolved. However, using the proposed approach, we clearly see from Fig. 4.5 that using a larger SVD threshold does not necessarily lead to a lower performance in generalization as measured by the accuracy on the testing set. Alternatively, this can be understood from the perspective of the proposed method being able to prevent over-training from occurring.

4.4 Conclusion

In this chapter, we highlighted the two sources of noise in the design of neural networks, 1) the same network structure giving rise to different fitness values due to different weight instantiations and 2) excess hidden layer neurons over-fitting the observed features of the training samples which are not representative of the intrinsic underlying distribution of observations. We described a hybrid multi-objective evolutionary approach which applies a robust information measure based on the singular value decomposition to estimate the necessary number of neurons to reduce the second form of noise. Subsequently, the SVD-based architectural recombination is presented for the purpose of facilitating the exchange of neuronal information between candidate neural network designs and adaptation of the number of neurons for each individual based on a geometrical approach in identifying hidden layer neurons to prune. In addition, two other problem-specific operators comprising a variable-length representation and a micro-hybrid genetic algorithm with adaptive local search intensity are also proposed to handle the fundamental issues of structural adaptation and local fine-tuning. The use of μHGA effectively removes the first source of noise. Experimental studies are also performed to examine the effectiveness of the proposed methods with respect to real-life datasets to illustrate that both the SVAR and μHGA models assume different, but nonetheless significant, roles in the evolution of effective ANN designs. While we have demonstrated the effectiveness of our proposed approach for classification problems, we believe that the methods that we have employed in this article are sufficiently flexible and robust to be extended to handle a variety of problem domains, such as regression, prediction, as well as system identification problems, all of which we hope to investigate in future works.

Part II
Tracking Dynamic Multi-objective Landscapes

Chapter 5
Dynamic Evolutionary Multi-objective Optimization

Many real-world systems include time-varying components and, very often, the environment in which they operate is in a constant state of flux. For problems involving such dynamic systems, the fitness landscape changes to reflect the time-varying requirements of the systems. Examples of such problems can be found in the areas of control, scheduling, vehicle routing, and autonomous path planning.

In dynamic single-objective problems, it is unlikely that the optimal solution will remain invariant in face of the changing fitness landscape. Previously found solution becomes obsolete and it must be updated to meet the new requirements. Dynamic multi-objective problems are certainly much more complex than their single-objective counterparts because we are dealing with a set of solutions instead of just one single solution. Not only are we required to adapt the obsolete set of Pareto solutions when the problem changes, we are expected to find a new set of solutions that satisfies the three multi-objective optimization goals quickly.

In a certain sense, the dynamic multi-objective problem can be considered as the consecutive optimization of different time-constrained multi-objective problems with varying complexities. In this case, the simplest approach to handle dynamic problems is to restart the optimization process everytime a landscape change is detected. However, one of the challenges of evolutionary dynamic optimization is to exploit past information to improve tracking performance; it is simply too inefficient to restart the optimization process everytime a change in landscape is detected, especially when the new optimal solution set is somewhat similar to the previous solutions. It is also imperative that the MOEA must be capable of high speed convergence to find the optimal solution set before it changes. On the other hand, when MOEA converges to the new Pareto optimal set, the difficulty is that there will be a lack of diversity necessary to explore the search space for the new Pareto optimal front and Pareto optimal set when landscape changes.

In this chapter, we will present a taxonomy of dynamic problems and describe a number of existing test functions for dynamic multi-objective

C.-K. Goh and K.C. Tan: Evolutionary Multi-objective Optimization, SCI 186, pp. 125–152.
springerlink.com

optimization that can be found in the literature. We will discuss the issues of performance assessment and present a number of performance metrics in dynamic multi-objective optimization. We will also present a number of MOEAs that are designed to handle the difficulties posed by the dynamic landscapes.

5.1 Dynamic Multi-objective Optimization Problems

Dynamism in real-world problems may arise from a variety of factors, some of which are due to human intervention, while the rest are inherent to the problem; a change in preference by the decision-maker, a new job in the production line, an obstacle in the path of a robot, and breakdown of vehicle in vehicle routing, etc. In certain cases, the number of objectives or decision variables to be optimized may change requiring a drastic change in the ranking or representation on the part of the MOEA. In this work, we will focus on dynamic multi-objective problems with fixed objective and design space dimensionality and which require the MOEA to track the changing fitness landscape.

In contrast to noisy fitness functions, the fitness topology of dynamic multi-objective problems may change but the objective values are deterministic at any one time. In this context, the term *static* is more appropriate than deterministic for denoting multi-objective problems without explicit consideration of its dynamism. For dynamic multi-objective problems, the PF^* and the PS^* are unlikely to remain invariant and the optimization algorithm must be capable of tracking the PS^* over time. The dynamic multi-objective problem can be described as

$$\min_{\mathbf{x}\in\mathbf{X}^{n_x}} \mathbf{F}(\mathbf{x},t) = \{f_1(\mathbf{x},t), f_2(\mathbf{x},t), ..., f_M(\mathbf{x},t)\} \tag{5.1}$$

where t is typically measured in terms of solution evaluations. For subsequent discussions, we will affix the time variable to the multi-objective optimization notations described in Chapter 1 to distinguish dynamic multi-objective optimization from static multi-objective optimization. The terms PF_t^* and PS_t^* refer to the desired Pareto front and solution set at time t, while the set of tradeoffs and non-dominated solutions evolved by the dynamic MOEA at time t will be termed as PF_t^A and PS_t^A, respectively.

5.2 Dynamic Multi-objective Problem Categorization

In dynamic single-objective problems, a solution can either deteriorate due to a shift in the fitness landscape or become obsolete due to the emergence of a new optimum. Likewise, such traits can be found in dynamic multi-objective problems, except that we are now dealing with a set of solutions, which makes the tracking process a lot trickier. Another distinct characteristic of dynamic multi-objective problems is that the shape and distribution of PF^*

are susceptible to change as well. This makes it necessary to consider the dynamics of both feature spaces in the investigation of dynamic MOEAs.

Accordingly, Farina *et al* [73] identified four different types of dynamic multi-objective problems according to the changes affecting the optimal Pareto front and the optimal Pareto set,

- Type I where PS_t^* changes, while PF_t^* remains invariant,
- Type II where both PS_t^* and PF_t^* changes,
- Type III where PF_t^* changes, while PS_t^* remains invariant and
- Type IV where both PS_t^* and PF_t^* remain invariant.

Farina *et al* further noted that even though both PS_t^* and PF_t^* are time-invariant in Type IV problems, it is possible that the fitness topology is changing with time. Continuous changes in the fitness landscape can easily mislead the optimization process, posing a challenge to the MOEA finding the desired solutions.

The above classification scheme is only applicable to dynamic multi-objective optimization problems. There also exist other appropriate but more general categorizations of dynamic problems. Deb *et al* [58] pointed out that the changes in a dynamic multi-objective problem can take place in the objective functions, the constraint functions, and the variable boundaries. In [145], dynamic problems are classified according to how the optimal solutions move after a landscape change. Jin and Sendoff stated that the location of the optimum can 1) move linearly, 2) move non-linearly, 3) oscillate among a few points, or 4) move randomly in the decision space. Another different but important perspective of dynamic problems can be found in the single-objective domain. Branke [26] proposed the categorization of dynamic problems based on 1) frequency of change, 2) severity of change, 3) predictability of change, and 4) periodicity of change.

These classifications are complementary schemes. The first and second schemes categorize the dynamic multi-objective problem based on the physical aspects of change, the third considers how the optimum behaves with time, while the fourth considers how the changes are effected. A more general approach would be to decompose the dynamic problem into its spatial and temporal components. Table 5.1 shows the list of spatial features and their attributes, while Table 5.2 summarizes the different temporal features. Note that the spatial component is further decomposed into physical and non-physical attributes of change. Physical attributes refer to *physical aspects* of problem change such as PS_t^* and PF_t^*. Non-physical attributes refer to the *manner* in which these spatial physical attributes are changed. Further, note that these physical spatial attributes are unique to dynamic multi-objective problems since we are dealing with a set of solutions in contrast to a single solution in single-objective optimization.

A dynamic multi-objective problem may be characterized by more than one specific instance of spatial and temporal attributes. For example, a dynamic

Table 5.1 Spatial Features of Dynamic Multi-Objective problem

	Type	Attributes
Physical	PS_t^*	The whole or part of PS_t^* moves to a new location
	PF_t^*	The shape of PF_t^* changes or a part of PF_t^* disappears
	Fitness landscape	Fitness landscape changes without affecting PS^* and PF^*
	Random	Random PS_t^*, PF_t^* and fitness landscape changes. Aspect of change may occur at the same time or may not happen at all.
Non-Physical	Random	Changes to physical attributes are random
	Trend	Changes to physical attributes follow a fixed pattern. Past physical topology may or may not be revisited again
	Periodic	Changes to physical attributes are periodic. Changes within each period may or may not follow a fixed pattern

Table 5.2 Temporal Features of Dynamic Multi-Objective problem

Type	Attributes
None	No change is triggered at all
Random	Change is triggered randomly
Fixed	Change is triggered at a fixed interval
Scheduled	Change is triggered based on a predetermined schedule such that it may follow a trend or even appear random.
Conditional	Change is triggered after some predefined condition is satisfied

multi-objective test problem can exhibit PF_t^* and PS_t^* changes simultaneously as in the case of Type II problem described earlier. At the same time, PF^* changes may follow a fixed trend that is triggered randomly, while PS^* changes may be periodic that is triggered based on a fixed schedule. In the event of random physical changes, it is still possible that whenever a particular aspect, such as PF_t^*, changes, the change may be following a certain trend.

5.3 Dynamic Multi-objective Test Problems

Dynamic multi-objective problems are essentially multi-objective problems. Therefore, fitness landscape characteristics such as multi-modality, high-dimensionality, and test suites suggested in the evolutionary multi-objective optimization literature are applicable and should be taken into account when constructing any test function or test suite. Some features specific to the dynamic domain that should be considered in a dynamic multi-objective test suite include

- Cyclic spatial changes;
- Predictable spatial changes;
- Landscape changes, such as emergent landscapes, that are difficult to detect;
- Landscape properties that allow very fast convergence or no exploitable spatial changes at all, i.e. memory has no significant advantage at all;

In general, any dynamic test suite should include features that challenge the dynamic MOEA capability to 1) detect a change in the environment, 2) maintain or generate the necessary diversity to explore the search space upon any changes, 3) exploit past information, and 4) converge to the new PS_t^* quickly.

The spatial and temporal features described in Table 5.1 and Table 5.2 provide different challenges in the design of dynamic MOEAs. For example, the storage of past PS_t^* in the form of memory will improve algorithmic performance for problems exhibiting periodic non-physical attributes. In cases where spatial and temporal features follow some trend, the presence of predictive elements can prepare the evolutionary process in anticipation of the problem's future behavior. On the other hand, we can expect that these mechanisms will have little or no significant advantage for problems which do not revisit previous PS_t^* or exhibit any trend. Furthermore, it is always possible that the reintroduction of previous solutions or prediction strategy may mislead the optimization process instead.

5.3.1 TLK2 Dynamic Test Function

TLK2 [257] is one of the earliest dynamic multi-objective test problems to be suggested in the literature. It is based on the moving peaks function [26], which provides an artificial multi-dimensional landscape consisting of several peaks, where the height, width, and position of each peak are varied as the environment changes. The problem of TLK2 is given as

$$\min f_1(\mathbf{x}) = x_1 \tag{5.2}$$

$$\min f_2(\mathbf{x}) = \frac{1}{x_1 \cdot g(t)} \tag{5.3}$$

$$g(t) = \max_{i=1,\ldots,5}\left[\frac{H_i(t)}{1 + W_i(t)\sum_{j=1}^{5}\left(x_{j+1} - X_{ij}(t)\right)^2}\right] \tag{5.4}$$

At a predefined frequency τ_T, the height and width of each peak are changed by adding a random Gaussian variable. The location of each peak is moved by a vector v of fixed length s in a random direction, i.e. the parameter s controls the severity of change. A change in the peak is governed by the following equations

$$\sigma \in N(0,1) \tag{5.5}$$
$$H_i(t) = H_i(t-1) + 7 \cdot \sigma \tag{5.6}$$
$$W_i(t) = W_i(t-1) + 0.01 \cdot \sigma \tag{5.7}$$
$$X_i(t) = X_i(t-1) + v \cdot \sigma \tag{5.8}$$

The change in multi-modal function g may result in a shift of the optimum to a different location. In this case, the evolutionary search needs to jump or cross a valley in order to find the new optimum.

5.3.2 FDA Dynamic Test Functions

The FDA test suite proposed by Farina *et al* [73] is built upon the ZDT and DTLZ frameworks described in Chapter 1. This test suite has been used in [58, 117, 192, 289]. Formally, the two-objective FDA test problems have the following functional structure.

$$\min f_1(\mathbf{x_I}, t) = \sum_{x_i \in \mathbf{x_I}} x_i^{F(t)}, F(t) > 0 \tag{5.9}$$
$$\min f_2(\mathbf{x_{II}}, \mathbf{x_{III}}, t) = g(\mathbf{x_{II}}, t) \cdot h(\mathbf{x_{III}}, f_1, g) \tag{5.10}$$
$$g(\mathbf{x_{II}}, t) = 1 + G(t) + \sum_{x_i \in \mathbf{x_{II}}} \left(x_i - G(t)\right)^2 \tag{5.11}$$
$$h(\mathbf{x_{III}}, t) = 1 - (\frac{f_1}{g})^{H(t)} \tag{5.12}$$

where F, H, and G are time-dependent functions which control how the density of Pareto solutions, shape of the PF_t^* and PS_t^* change with time.

The M-objective FDA test problems have the following functional structure.

$$\min f_1(\mathbf{x}, t) = \left(1 + g(\mathbf{x}_{II})\right) \cdot \cos(0.5\pi y_1) \cdots \cos(0.5\pi y_{M-1}) \tag{5.13}$$
$$\min f_2(\mathbf{x}, t) = \left(1 + g(\mathbf{x}_{II})\right) \cdot \cos(0.5\pi y_1) \cdots \sin(0.5\pi y_{M-1}) \tag{5.14}$$
$$\vdots$$
$$\min f_M(\mathbf{x}, t) = \left(1 + g(\mathbf{x_{II}})\right) \cdot \sin(0.5\pi y_1) \tag{5.15}$$
$$g(\mathbf{x}_{II}, t) = G(t) + \sum_{x_i \in \mathbf{x}_{II}} (x_i - G(t))^2 \tag{5.16}$$
$$y_i = x_i^{F(t)} \tag{5.17}$$

where F controls how the density of Pareto solutions changes over time and G controls the changes in the PF_t^* and PS_t^* over time.

The dynamics of the FDA test functions are governed by the equation,

$$t = \frac{1}{n_T}\lfloor\frac{\tau}{\tau_T}\rfloor \tag{5.18}$$

where n_T and τ_T specify the severity and frequency of landscape change, respectively. Interestingly, due to the sinusoidal behavior of $G(t)$ and $H(t)$, n_T also determines the periodicity of similar solution sets emerging. In particular, a smaller n_T implies that the number of different PS_t^* is small. Both n_T and τ_T has a lower bound of 1.0. Setting a value of $n_t < 1.0$ will result in a magnitude change that is out of range, while increasing values of n_T produce decreasing magnitudes of change. Likewise, decreasing values of τ_T will result in increasingly static environments.

Mehnen *et al* [192] also suggested a generalization of the FDA framework, which allows for the number of Pareto front segments and the number of local optimal fronts to be dynamic as well. This extended framework is defined as follows,

$$\min f_1(\mathbf{x_I}, t) = x_1^{F(t)} \tag{5.19}$$

$$\min f_2(\mathbf{x_{II}}, t) = g(\mathbf{x_{II}}, t) \cdot h(f_1, g) \tag{5.20}$$

$$g(\mathbf{x_{II}}, t) = 1 + \sum_{x_i \in \mathbf{x_{II}}} \left(x_i - G(t)\right)^2 - \cos\left(\omega(t)\pi(x_i - G(t)\right) + 1 \tag{5.21}$$

$$h(f_1, t) = 2 - (\frac{f_1}{g})^{H(t)} - (\frac{f_1}{g}) \cdot |\sin(t \cdot \pi f_1)|^{H(t)} \tag{5.22}$$

Based on the FDA framework, Farina *et al* suggested five different problems and the definitions of these dynamic multi-objective test functions are summarized in Table 5.3. The first three problems are bi-objective problems. FDA1 is a Type I problem where only the PS_t^* is dynamic. FDA2 is a Type III problem where only the PF_t^* changes, from a convex to non-convex front. FDA3 is a type II problem where both the PF_t^* and PS_t^* changes. FDA4 and FDA5 have scalable number of objectives. These two test functions are Type I and Type II problems, respectively.

5.3.3 dMOP Test Functions

In [94], we developed three test functions based on the construction guidelines provided by Farina *et al* described above. The definitions of these dynamic multi-objective test functions are summarized in Table 5.4. The first test function, dMOP1, is a Type III problem where only the PF_t^* is dynamic, while dMOP2 is a Type II problem where both PS_t^* and the PF_t^* change with time. Like FDA1, dMOP3 is a Type I problem. However, the variable that controls the spread of the PF_t^* changes as well.

Table 5.3 Definition of FDA Dynamic Test Functions

	Test	Definition
1	FDA1	$f_1(\mathbf{x}_I) = x_1,$ $f_2(\mathbf{x}_{II} m) = g \cdot h,$ $g(\mathbf{x}_{II}) = 1 + \sum_{i=2}^{m} (x_i - G(t))^2,$ $h(f_1, g) = 1 - \sqrt{\frac{f_1}{g}}$ $G(t) = \sin(0.5\pi \cdot t)$ where $\lvert\mathbf{x}_{II}\rvert = 9,\ \mathbf{x}_I \in [0,1],\ \mathbf{x}_{II} \in [-1,1]$
2	FDA2	$f_1(\mathbf{x}_I) = x_1,$ $f_2(\mathbf{x}_{II}) = g \cdot h,$ $g(\mathbf{x}_{II}) = 1 + \sum_{i \in \mathbf{x}_{II}}^{m} {x_i}^2,$ $h(\mathbf{x}_{III}, f_1, g) = 1 - (\frac{f_1}{g})^{H(t) + \sum_{x_i \in \mathbf{x}_{III}} (x_i - H(t))^2}$ $H(t) = 0.75 + 0.7 \cdot \sin(0.5\pi \cdot t)$ where $\lvert\mathbf{x}_{II}\rvert = \lvert\mathbf{x}_{III}\rvert = 15,\ \mathbf{x}_I \in [0,1],\ \mathbf{x}_{II}, \mathbf{x}_{III} \in [0,1]$
3	FDA3	$f_1(\mathbf{x}_I) = \sum_{x_i \in \mathbf{x}_I} x_i^F(t),$ $f_2(\mathbf{x}_{II}) = g \cdot h,$ $g(\mathbf{x}_{II}) = 1 + G(t) + \sum_{x_i \in \mathbf{x}_{II}} (x_i - G(t))^2,$ $h(f_1, g) = 1 - \sqrt{\frac{f_1}{g}}$ $G(t) = \lvert\sin(0.5\pi \cdot t)\rvert$ $F(t) = 10^{2\sin(0.5\pi t)}$ where $\lvert\mathbf{x}_I\rvert = 5, \lvert\mathbf{x}_{II}\rvert = 25,\ \mathbf{x}_I \in [0,1],\ \mathbf{x}_{II} \in [-1,1]$
4	FDA4	$f_1(\mathbf{x}) = (1 + g(\mathbf{x}_{II})) \prod_{i=1}^{M-1} \cos(\frac{x_i \pi}{2}),$ $f_k(\mathbf{x}) = (1 + g(\mathbf{x}_{II})) \prod_{i=1}^{M-1} (\cos(\frac{x_i \pi}{2})) \sin(\frac{x_{M-k+1}\pi}{2}),$ $f_M(\mathbf{x}) = (1 + g(\mathbf{x}_{II})) \prod_{i=1}^{M-1} \sin(\frac{x_1 \pi}{2}),$ $g(\mathbf{x}_{II}) = \sum_{x_i \in \mathbf{x}_{II}} (x_i - G(t))^2,,$ $G(t) = \lvert\sin(0.5\pi \cdot t)\rvert$ where $\mathbf{x} \in [0,1]$
5	FDA5	$f_1(\mathbf{x}) = (1 + g(\mathbf{x}_{II})) \prod_{i=1}^{M-1} \cos(\frac{y_i \pi}{2}),$ $f_k(\mathbf{x}) = (1 + g(\mathbf{x}_{II})) \prod_{i=1}^{M-1} (\cos(\frac{y_i \pi}{2})) \sin(\frac{y_{M-k+1}\pi}{2}),$ $f_M(\mathbf{x}) = (1 + g(\mathbf{x}_{II})) \prod_{i=1}^{M-1} \sin(\frac{y_1 \pi}{2}),$ $g(\mathbf{x}_{II}) = G(t) + \sum_{x_i \in \mathbf{x}_{II}} (x_i - G(t))^2,$ $y_i = x_i^{F(t)},$ $G(t) = \lvert\sin(0.5\pi \cdot t)\rvert$ $F(t) = 1 + 100\sin^4(0.5\pi \cdot t)$ where $\mathbf{x} \in [0,1]$

Table 5.4 Definition of dMOP Dynamic Test Functions

	Test	Definition
1	dMOP1	$f_1(x_1) = x_1,$ $f_2(x_2, ...x_m) = g \cdot h,$ $g(x_2, ...x_m) = 1 + 9 \cdot \sum_{i=2}^{m} {x_i}^2,$ $h(f_1, g) = 1 - (\frac{f_1}{g})^{H(t)}$ $H(t) = 0.75 \cdot \sin(0.5\pi \cdot t) + 1.25$ where $m = 10, \quad x_i \in [0, 1]$
2	dMOP2	$f_1(x_1) = x_1,$ $f_2(x_2, ...x_m) = g \cdot h,$ $g(x_2, ...x_m) = 1 + \sum_{i=2}^{m} (x_i - G(t))^2,$ $h(f_1, g) = 1 - (\frac{f_1}{g})^{H(t)}$ $H(t) = 0.75 \cdot \sin(0.5\pi \cdot t) + 1.25$ $G(t) = \sin(0.5\pi \cdot t)$ where $m = 10, \quad x_i \in [0, 1]$
3	dMOP3	$f_1(x_r) = x_r,$ $f_2(\mathbf{x} \backslash x_r) = g \cdot h,$ $g(\mathbf{x} \backslash x_r) = 1 + \sum_{i=1}^{\mathbf{x} \backslash x_r} (x_i - G(t))^2,$ $h(f_1, g) = 1 - \sqrt{\frac{f_1}{g}}$ $H(t) = 0.75 \cdot \sin(0.5\pi \cdot t) + 1.25$ $G(t) = \sin(0.5\pi \cdot t)$ where $m = 10$, r=U(1,2,...,m), $x_i \in [0, 1]$

5.3.4 DSW Test Functions

Mehnen *et al* [192] proposed the DSW test suite to facilitate theoretical analysis in dynamic multi-objective optimization.

$$\min f_1(\mathbf{x}) = \Big(a_{11}x_1 + a_{12}x_1 - b_1 \cdot G(t)\Big)^2 + \sum_{i=2,\ldots,|\mathbf{x}|} x_i^2 \tag{5.23}$$

$$\min f_2(\mathbf{x}) = \Big(a_{21}x_1 + a_{22}x_1 - b_2 \cdot G(t) - 2\Big)^2 + \sum_{i=2,\ldots,|\mathbf{x}|} x_i^2 \tag{5.24}$$

where the type of spatial changes is determined by setting appropriate a_{11}, a_{12}, a_{21}, a_{22}, b_1, and b_2 values. Three different test functions are suggested and parameters are defined as follows,

$$\text{DSW1} : \mathbf{x} \in [-50, 50]^{n_x}, \mathrm{a}_{11} = 1, \mathrm{a}_{12} = 0, \mathrm{a}_{21} = 1, \mathrm{a}_{22} = 0, \mathrm{b}_1 = 1, \mathrm{b}_2 = 1 \tag{5.25}$$

$$\text{DSW2}: \mathbf{x} \in [-50, 50]^{n_x}, \mathrm{a}_{11} = 0, \mathrm{a}_{12} = 1, \mathrm{a}_{21} = 0, \mathrm{a}_{22} = 1, \mathrm{b}_1 = 1, \mathrm{b}_2 = 1 \tag{5.26}$$

$$\text{DSW3}: \mathbf{x} \in [-50, 50]^{n_x}, \mathrm{a}_{11} = 1, \mathrm{a}_{12} = 0, \mathrm{a}_{21} = 1, \mathrm{a}_{22} = 0, \mathrm{b}_1 = 0, \mathrm{b}_2 = 1 \tag{5.27}$$

DSW1 has a dynamic PS_t^* and the dynamic MOEA is required to track the optimal solution set with $[G(t), G(t) + 2]$. Like DSW1, DSW2 has a dynamic PS_t^*. The PS_t^* is discontinuous, which departs diametrically. If $G(t)$ is periodic, then PS_t^* will join and depart periodically. In DSW3, both PS_t^* and PF_t^* change with the solution set as the extend of the Pareto front increases with time. The main limitation of the DSW test problems is that it is not as intuitive as compared to TLK2 and the FDA test functions when it comes to the configuring of dynamic spatial features.

5.3.5 JS Test Functions

An interesting approach of aggregating objective functions of existing test problems through dynamically changing weights to form a lower dimensional dynamic problem is proposed by Jin and Sendhoff in [145]. A three-objective problem can be reformulated to form a two-objective dynamic problem in the following way,

$$\min f_1'(\mathbf{x}, t) = w(t) \cdot f_1 + \big(1 - w(t)\big) \cdot f_2 \tag{5.28}$$
$$\min f_2'(\mathbf{x}, t) = w(t) \cdot f_1 + \big(1 - w(t)\big) \cdot f_3 \tag{5.29}$$

where $0 \leq w \leq 1$ is a function of time that gives rise to the dynamic properties of the reformulated problem, f_1, f_2, and f_3 are the original objective functions. w can be defined as either a linear or non-linear function to produce different test properties. As an example, Jin and Sendhoff used the following three-objective problem:

$$\min f_1(\mathbf{x}) = x_1^2 + (x_2 - 1)^2 \tag{5.30}$$
$$\min f_2(\mathbf{x}) = x_1^2 + (x_2 + 1)^2 + 1 \tag{5.31}$$
$$\min f_3(\mathbf{x}) = (x_1 - 1)^2 + x_2^2 + 2. \tag{5.32}$$

Based on the aggregation approach, the dynamic multi-objective test function can be written as

$$\min f_1'(\mathbf{x}) = w(t)[(x_1^2 + (x_2 - 1)^2] + \tag{5.33}$$
$$(1 - w(t))[x_1^2 + (x_2 + 1)^2 + 1] \tag{5.34}$$
$$\min f_2'(\mathbf{x}) = w(t)[x_1^2 + (x_2 - 1)^2] + \tag{5.35}$$
$$(1 - w(t))[(x_1 - 1)^2 + x_2^2 + 2] \tag{5.36}$$

Simplicity and ease of construction are the main advantages of this approach.

5.4 Performance Metrics for Dynamic Multi-objective Optimization

While it is difficult to define a single quantitative measure of performance, there is a general consensus on what constitutes a good performance in dynamic optimization. Discussions on this topic in the domain of single-objective optimization can be found in [196, 281]. The MOEA solving for a static multi-objective problem is expected to adequately attain the optimization goals of 1) good convergence to the PF*, 2) diversity of the solution set, and 3) uniform distibution of the solutions. When it comes to a dynamic environment, obtaining a near-optimal, diverse, and uniformly distributed PF^A is no longer sufficient. The dynamic MOEA must also be capable of responding and adapting rapidly to the changing environment. It is essential that the dynamic algorithm discovers the new solution set quickly because each solution has a finite lifetime in a dynamic environment. The implementation of a new solution can be costly in many real-world applications, making it impractical to apply new solutions that will become obsolete too soon.

It is worth noting that the goals of achieving high speed convergence and PF^A optimality in dynamic optimization can be conflicting. Consider the situation where two different dynamic MOEAs are used to track PS$_t^*$ of a problem characterized by severe fitness landscape changes. The first algorithm (Alg 1) is designed with a greater emphasis on exploration, while the other algorithm (Alg 2) has strong local search tendencies. Alg 1 should find its way to the immediate vicinity of PS$_t^*$ rather quickly but will face problems attaining good convergence. On the other hand, Alg 2 will achieve good convergence at the expense of convergence speed. Therefore, performance metrics of dynamic MOEAs should provide an indication of 1) how effective the dynamic MOEA is in attaining the multi-objective optimization goals in the face of changing environment and 2) how fast the dyanmic MOEA is capable of tracking PS$_t^*$.

5.4.1 Illustrating Performance Using Static Performance Measures

This approach provides an indication of how well the static optimization goals are achieved before each landscape change. Therefore, researchers can take advantage of the great number of static performance metrics developed over the years and draw upon their experience in static multi-objective optimization directly in the evaluation of algorithmic performance. After the calculation of solution set quality, the results can either be tabulated [289] or illustrated as a plot of performance trend [58, 73].

Farina *et al* [73] suggested the following two metrics to measure convergence in the decision and objective space at different time instances.

$$c_x(t) = \frac{1}{np} \sum_{j=1}^{np} \min_{i=1:nh} || \frac{\mathrm{pf}_i^*(t) - \mathrm{pf}_j^A(t)}{\mathrm{R}(t) - \mathrm{U}(t)} || \tag{5.37}$$

$$c_f(t) = \frac{1}{np} \sum_{j=1}^{np} \min_{i=1:nh} ||\mathrm{ps}_i^*(t) - \mathrm{ps}_j^A(t)|| \tag{5.38}$$

where np is $|PF_t^A|$, $\mathrm{R}(t)$ is the time-dependent nadir point, and $\mathrm{U}(t)$ is the time-dependent utopia point. $\mathrm{pf}_i^*(t)$ and $\mathrm{pf}_j^A(t)$ are the i-th member of PF_t^* and PF_t^A, respectively, while $\mathrm{ps}_i^*(t)$ and $\mathrm{ps}_j^A(t)$ are the i-th member of PS_t^* and PS_t^A respectively. $c_x(t)$ is very similar to the metric of GD described in Chapter 1 except that the distance between $\mathrm{pf}_i^*(t)$ and $\mathrm{pf}_j^A(t)$ is normalized by $\mathrm{R}(t) - \mathrm{U}(t)$. As emphasized in Chapter 1, different metrics or combinations of metrics can be used as long as it gives an indication of how well the algorithms perform relative to the different optimization goals. Zeng *et al* applied the metrics of GD and spread to measure how well the MOEA performs in the aspects of convergence and diversity, respectively, while the metric of HVR is used by Deb *et al* [58].

5.4.2 Time Averaging Static Performance Measures

It is rather difficult to compare and evaluate the performance of multiple algorithms through the use of tables and performance trends. Therefore, as pointed out by Branke [26], it is desirable to have a more compact form to describe algorithmic performances. One way of achieving this is to extend the idea of offline error applied in dynamic single-objective optimization and calculate the time averages of static metrics.

Time-averaged convergence measure: Hatzakis and Wallace [117] applied the time averages of the convergence measures $c_x(t)$ and $c_f(t)$.

$$\overline{c}_x = \frac{1}{np\tau} \sum_{t=1}^{\tau} \sum_{j=1}^{np} \min_{i=1:nh} || \frac{\mathrm{pf}_i^*(t) - \mathrm{pf}_j^A(t)}{\mathrm{R}(t) - \mathrm{U}(t)} || \tag{5.39}$$

$$\overline{c}_f = \frac{1}{np\tau} \sum_{t=1}^{\tau} \sum_{j=1}^{np} \min_{i=1:nh} ||\mathrm{ps}_i^*(t) - \mathrm{ps}_j^A(t)|| \tag{5.40}$$

where τ is the number of time samples used. Apart from the number of samples, when sampling is performed is another factor to be considered. The sampling of performance metrics should be done at instances just before the next landscape change to eliminate unnecessary penalty on dynamic MOEAs employing diversity introduction schemes, such as random restart or hypermutation, in situations where change is small.

Off-line variable space distance and maximum spread: In [94], we define an extension of MS and a variable space variation of GD (VD), which can be expressed in the following form,

$$\mathrm{VD}_{offline} = \frac{1}{\tau}\sum_{t=1}^{\tau} \mathrm{VD} \cdot \mathrm{I}(t) \tag{5.41}$$

$$\mathrm{MS}_{offline} = \frac{1}{\tau}\sum_{t=1}^{\tau} \mathrm{MS} \cdot \mathrm{I}(t) \tag{5.42}$$

$$\mathrm{I}(t) = \begin{cases} 1, & \text{if } t\%\tau_t = 0 \\ 0, & \text{otherwise.} \end{cases} \tag{5.43}$$

where % is the modulus operator. Similar to the metric of GD in static environment, a low value of $\mathrm{VD}_{offline}$ is desirable and reflects good tracking capability. Likewise, a higher value of $\mathrm{MS}_{offline}$ reflects that the MOEA is capabable of evolving a diverse PF_t^A in a dynamic environment. VD measures the degree of convergence between PS^{*_t} and PS^A.

$$\mathrm{VD} = \frac{1}{n_{PS}} \cdot \left(n_{PS} \sum_{i=1}^{n_{PS}} d_i^2\right)^{\frac{1}{2}} \tag{5.44}$$

where $n_{PS} = |\mathrm{PS}_t^A|$, d_i is the Euclidean distance (in decision space) between the i-th member of PS_t^A and the nearest member of PS^{*_t}.

Convergent ratio, average density, and coverage rate: Zhang [290] suggested three metrics to measure algorithmic performance in the aspects of convergence and diversity in a dynamic landscape. The first metric is the convegence ratio CR which measures the consistency of the algorithm in finding PF_t^*. CR is given by,

$$CR = \frac{1}{K\tau}\sum_{t=1}^{\tau}\sum_{j=1}^{K-1}\frac{1}{K-j}\sum_{i=1}^{K} C(X_{t,j}, X_{t,i}) \tag{5.45}$$

where K is the number of simulation runs made by the algorithm and $C(\cdot)$ is the coverage metric. Note that coverage is used to measure the relative algorithmic performance across the different runs over time. Thus, CR value will be low when the algorithm performs consistently. However, it is clear that this metric provides no indication of how close the discovered solutions are to the true Pareto front.

To measure how well the algorithm distributes the non-dominated solutions evenly after each landscape change, an extension of the spacing metric is suggested,

$$AD = \frac{1}{\tau}\sum_{t=1}^{\tau}\frac{1}{\bar{d'}} \cdot \left(\frac{1}{n_{PF}} \cdot \sum_{i=1}^{n_{PF}} (d'_i - \bar{d'})^2\right)^{\frac{1}{2}} \tag{5.46}$$

$$\bar{d'} = \frac{1}{n_{PF}}\sum_{i=1}^{n_{PF}} d'_i$$

where $n_{PF} = |\mathrm{PF}_t^A|$, d_i is the Euclidean distance (in objective space) between the i-th member of PF_t^A and the nearest member of PF^{*t}.

The metric of average coverage (AS) is used to measure the diversity of the algorithm and defined as follows,

$$AS = \frac{1}{K\tau} \sum_{t=1}^{\tau} \sum_{i=1}^{K} \max_{1 \geq j,k \geq M} \{||f_{t,j}^A - f_{t,k}^A||, f_{t,j}^A, f_{t,k}^A \in \mathrm{PS}_t^A\} \tag{5.47}$$

where $f_{t,i}^A$ is the i-th objective in PF_t^A.

5.5 Evolutionary Dynamic Optimization Techniques

5.5.1 Design Issues

Most of the studies on evolutionary optimization in dynamic environments are restricted to the domain of single-objective problems. Comprehensive discussions on dynamic single-objective evolutionary optimization can be found in [30, 196]. Nonetheless, these works are important sources from which important lessons on dynamic optimization can be drawn upon. For instance, the design issue of diversity, in particular, has been much discussed, and applies to all problem domains. There are, of course, other design issues that are unique to multi-objective optimization.

Detection of environmental changes: The detection of environmental changes that necessitate the adaptation of solutions is an important mechanism for any EA- inappropriate choice of an indicator of environment change can lead to failure to respond to changes or even false alarms. For instance, the average population fitness deviation is a popular indicator of environment change in the single-objective literature. In Pareto-based MOEAs, the discovery of a solution that dominates a number of solutions in the previous Pareto optimal front will lead to a significant change in average population fitness. Therefore, this particular indicator will not be able to detect actual changes reliably. In this sense, the re-evaluation of past solutions to check for changes in the objective values is a more reliable approach. However, it assumes that a variation in the environment will be reflected on the entire fitness landscape. Futher note the objective functions can be computationally expensive and that noise (as studied previously) in the objective functions can also lead to false alarms.

Decision-making: While the dynamic MOEA must be capable of finding a diverse set of Pareto-optimal solutions, it should be noted that only one solution will be implemented. Since it is not practical to assume the availability of human guidance at all times, it may be necessary to incorporate an online decision-maker. The most straight forward approach is to make use of user knowledge and preferences. In the absence of such information, a plausible alternative is to find the knee solution.

Outdated elitist solutions: One potential problem of MOEAs in dynamic environment is their exploitation of non-dominated solutions. In the event of an environment change, the current solution set may not be indicative of the optimal Pareto front. Unless these outdated solutions are re-evaluated or removed, solutions that are non-dominated in the new environment may be kept out of the archive. If the new optimal solution set is not within the immediate vicinity of the previous optimal solution set, it is unlikely that MOEA is able to track any landscape changes. It is also highly possible that the reinsertion of outdated elitist solutions will misguide the subsequent optimization process. On the other hand, the exploitation of past information is desirable to improve the adaptation process.

Diversity: At this point, it is important to make the distinction between decision and objective space diversity. Objective space diversity offers decision-makers with a variety of tradeoff solutions while decision space diversity is the crucial ingredient that drives the evolutionary search process. In MOEAs, the diversity mechanisms (described in Chapter refch1) are used to maintain diversity in the objective space but objective space diversity do not necessarily mean that decision space is diverse. Typically, diversity is lost as the population converges to promising region. Unless explicit mechanisms are employed to maintain or introduce diversity in the decision space, there may not be sufficient diversity in the evolving population to adapt the solutions to changes in the environment quickly, if the algorithm is able to react at all. Different techniques proposed to promote diversity are based the following three classes.

- Diversity introduction: This approach introduces diversity upon the detection of landscape change [40, 108, 273] as illustrated in Fig. 5.1. Random restart or reinitialization is one of the simplest techniques for generating diversity. Other common techniques include hypermutation, where mutation is increased dramatically, and the variable local search, where mutation is increased gradually if no improvement is achieved. These approaches can be easily extended to MOEAs. The main drawback is that information gained is lost after the introduction of diversity.
- Diversity maintainance: This approach sought to maintain diversity throughout the run [91, 107, 195]. The effect of incorporating diversity maintainance is illustrated in Fig. 5.2. While the conventional MOEA tends to converge quickly to PS_t^*, the incorporation of a diversity maintainance scheme will ensure the presence of individuals in other regions of the search space. One of the techniques that can be easily incorporated in MOEAs is the random immigrant which is conceptually similar to the idea of random restart. However, in random immigrant, random individuals are introduced into the evolving population at fixed intervals and only a part

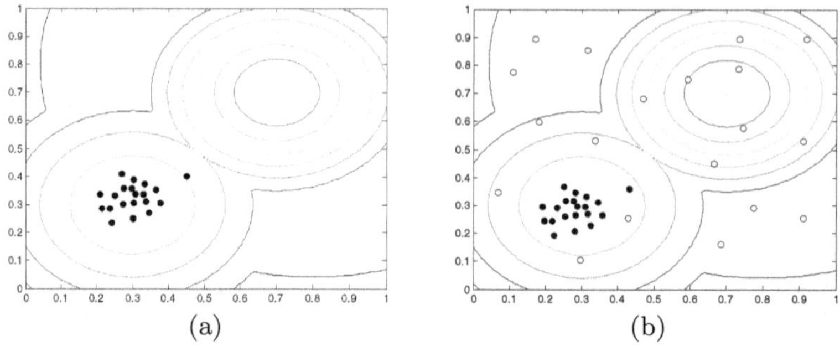

Fig. 5.1 Distribution of solutions after (a) landscape changes and (b) introduction of diversity

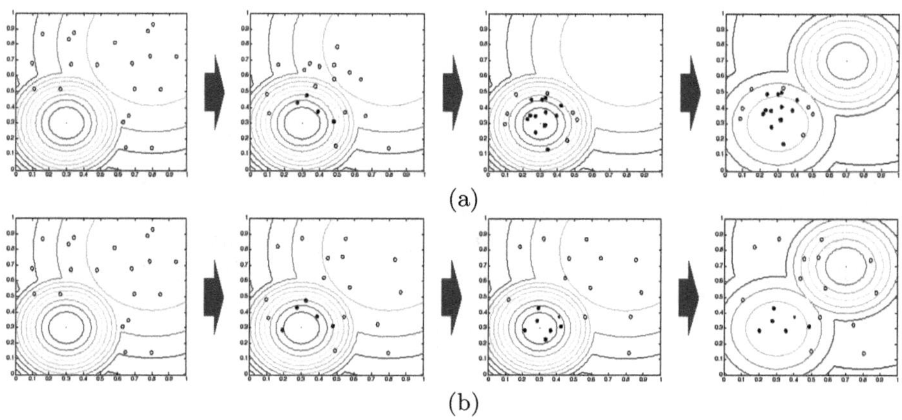

Fig. 5.2 Snap shots of the distribution of solutions at different time instances (a) without diversity maintainance and (b) with diversity maintainance

of the population is replaced. Diversity preservation techniques described in Chapter 1 can also be used, except that diversity assessment should be performed in the decision space.

- Multiple population: The basic idea of applying multiple populations is to conduct simultaneous exploration in different regions to track any change or emergence of new optimal solutions [34, 270, 282]. Typically, this approach involves a population which exploits the current optimal solution, while the other populations are encouraged to explore the search space as shown in Fig. 5.3. In some works, some form of explusion mechanism is incorporated to encourage the different populations to search different regions of the search space, allowing the other populations to locate the new PS_t^*. The population of previous PS_t^* can either be freed to explore other regions or remain in the same region in anticipation of future PS_t^* revisiting the same region again.

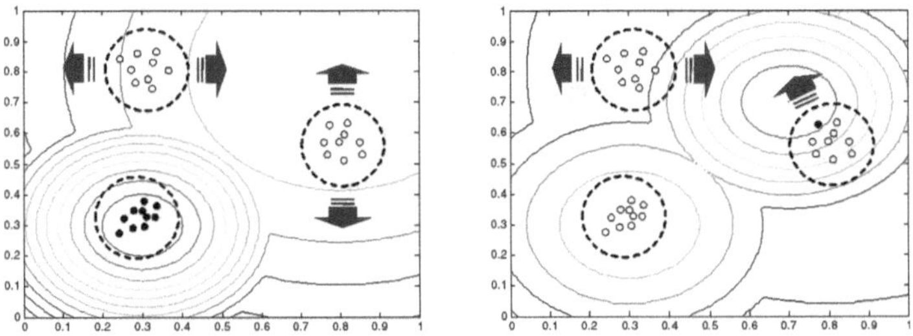

(a) (b)

Fig. 5.3 Behavior of multiple populations (a) before and (b) after a landscape change

5.5.2 Directional-Based Dynamic Evolutionary Multi-objective Optimization Algorithm

Farina *et al* [73] suggested an hybridization between evolutionary strategies (ES) and deterministic local search to improve the speed of convergence to track the dynamic PS^{*t} and PF^{*t}. This direction-based dynamic evolutionary multi-objective optimization algorithm (DB-DEMOA) involves a two-stage optimization procedure. This first stage searches for the payoff matrix, utopia, and nadir points of the PF^{*t}, while the second stage searches for a set of uniformly distributed solutions between the Utopia points. The major steps of the DB-DEMOA are shown below:

Directional-Based Dynamic Evolutionary Multi-objective Optimization Algorithm (DB-DEMOA)

Step 1: Stage 1: Search for nadir points, utopia points and payoff matrix.
Step 2: Stage 2: Search for solutions between the utopia points.
Step 3: Check for changes in environment. Restart optimization process using P_t as starting population if change is detected.

In each stage, an (1+1) ES and a local search operator are applied in an sequential manner. Once the ES is determined to have converged, the optimization process is switched to either a gradient-based algorithm (GBA) or a simplex Nelder Mead search algorithm (NMA). The convergence criterion is satisfied when solution variation in the design space between two consecutive iterations falls below a threshold.

The algorithm stops the optimization process once the set of non-dominated solutions is found by the local search process in Stage 2. Thereafter, n_ϵ solutions will be selected randomly from P_t and checked for changes in the problem periodically in the following manner:

$$\epsilon_t = \frac{1}{n_\epsilon} \sum_{j=1}^{n_\epsilon} || \frac{\mathbf{F}_{j,t}(x) - \mathbf{F}_{j,t-1}(x)}{\mathbf{R}_t - \mathbf{U}_t} || \tag{5.48}$$

where $\mathbf{R}_t$ and $\mathbf{U}_t$ are the time-dependent nadir and utopia points, respectively. Once ϵ_t falls above the values between 10^{-3} and 10^{-2}, the optimization process is restarted automatically from Stage 1. To exploit possible similarities between past and current environments, the population prior to landscape change detection is used as the starting population for Stage 1.

Stage 1: As mentioned earlier, the first stage calculates $\mathbf{R}_t$, $\mathbf{U}_t$, and payoff matrix $[\mathbf{M}]$. The utopia point is given as

$$\mathbf{U}_t = \left[\min f_{i,t}\right] = [U_{i,t}] \tag{5.49}$$

and the set of solutions associated with $\mathbf{U}_t$ is represented by the matrix $\mathbf{M}_x$.

$$\mathbf{M}_x(i,:) = \left[\mathbf{x}|f_{i,t}(x) = U_{i,t}\right]. \tag{5.50}$$

The DB-DEMOA minimizes each objective function in a sequential manner to find the utopia points associated with each objective. After which, the $[\mathbf{M}]$ is calculated from $\mathbf{U}_t$ and $\mathbf{M}_x$ in the following manner

$$[\mathbf{M}] = \begin{cases} U_{i,t}, & i = j \\ f_{j,t}(\mathbf{M}_x(i,:)), & \text{otherwise} \end{cases} \tag{5.51}$$

The nadir point is then calculated from $[\mathbf{M}]$ in the following way.

$$\mathbf{R}_{i,t} = \max \mathbf{M}(i,:) \forall\ i = 1:M. \tag{5.52}$$

Stage 2: Stage 2 attempts to find a set of n_s Pareto-optimal solutions that are uniformly distributed between the utopia points. In DB-DEMOA, the multi-objective problem is converted into a single-objective problem using the following scalar function,

$$\hat{f}^k = \frac{1}{M} \max_{i=1:M} \left[\frac{(f_i - P_i^k)}{R_{i,t} - U_{i,t}}\right], k = 1:n_s. \tag{5.53}$$

where P_i^k is the i-th objective value in the k-th chosen center. To find the n_s Pareto-optimal solutions, DB-DEMOA solves for n_s different single-objective problems, each with a different P^k. P^k is calculated iteratively as the centroid between each utopia point as well as between utopia points and successive solutions found by minimizing equation (5.53). Consider the example of a two-objective problem. P^1 will be the centroid between M_1 and M_2. After obtaining the solution by minimizing the function, P^2 and P^3 are the centroids between M_1 and S_1, and M_2 and S_1, respectively.

5.5.3 Dynamic Non-dominated Sorting Genetic Algorithm II

Deb *et al* [58] extended the NSGAII for the online optimization of a dynamic hydro-thermal power scheduling problem. To detect possible changes in the fitness landscape, the dynamic NSGAII (DNSGAII) randomly selects and re-evaluates 10% of the parent population in each generation. In the event of a landscape change, all parent solutions are re-evaluated before combining with the offspring population. The authors explored two means of introducing diversity into the evolving population, either by inserting new individuals or performing mutation on existing solutions. The major steps of this algorithm within one iteration are shown below:

Dynamic Non-Dominated Sorting Genetic Algorithm II (DNSGAII)

Step 1: Evaluate all solutions in P_t.
Step 2: Check for changes in environment.

- Select and re-evaluate 10% of individuals from A_t.
- Re-evaluate all solutions in A_t if changes are detected in the objective values.

Step 3: Combine P_t and A_t.
Step 4: Perform modified non-dominated sorting on combined population.
Step 5: Create A_{t+1}

- Perform crowding sort to truncate combined population size to $|A_t|$.
- If landscape changes detected, select ζ% of individuals from combined population.
- Replace selected individuals with random solutions/mutate selected individuals.

Step 6: Recombination process: Conduct tournament selection on A_{t+1} to form mating pool. Perform crossover and mutation to create P_{t+1}.

DNSGAII introduces diversity to part of the evolving population, allowing it to exploit information from the previous PS^{*_t}. From Step 5, it is clear that ζ% is a critical parameter since it determines how much of the previous solutions are retained. ζ% of the population are either replaced by newly created solutions or reinserting mutated variants of the selected individuals. The two schemes have their own strength and weakness. Reinitializing new solutions will allow DNSGAII to adapt faster to drastic changes in the landscape but performs poorly in environments with small changes. On the other hand, mutating existing solutions will not introduce too much diversity into the population. But this will allow NSGAII to handle small variations in the problem quickly at the expense of poor performances when faced with great changes in the environment.

Apart from tracking the dynamic PS^{*_t} and PF^{*_t}, the algorithm must be capable of making good impromptu decisions automatically once a good solution set is found. Assigning equal importance to the two objectives of the dynamic hydro-thermal power scheduling problem, Deb *et al* calculates a pseudo-weight w for each non-dominated solution found in the algorithm in the following manner:

$$w(\mathbf{F}_a) = \frac{(\overline{f}_1 - f_{a,1})/(\overline{f}_1 - \underline{f}_1)}{(\overline{f}_1 - f_{a,1})/(\overline{f}_1 - \underline{f}_1) + (\overline{f}_2 - f_{a,2})/(\overline{f}_2 - \underline{f}_2)} \tag{5.54}$$

Since equal importance is assigned to both objectives, the solution with w that is closest to 0.5 is selected and implemented.

5.5.4 Dynamic Multi-objective Evolutionary Algorithm Based on an Orthogonal Design

Zeng *et al* [289] suggested a dynamic orthogonal multi-objective evolutionary algorithm (DOMOEA) that incorporates orthogonal design methodology to improve convergence. This dynamic algorithm is similar to the orthogonal multi-objective evolutionary algorithm-II (OMOEA-II) [288] except that a periodic check is performed to detect landscape changes. The DOMOEA treats the dynamic multi-objective problem as a new problem instance after every landscape change. However, it exploits past information by using the PS_t^A prior to change as the new initial population. The major steps within an iteration is shown below:

Dynamic Multi-Objective Evolutionary Algorithm based on Orthogonal Design (DOMOEA)

Step 1: Evaluate all solutions in P_t.
Step 2: Check for changes in environment.

- Check for discrepancies.
- If landscape change is detected, re-evaluate A_t and set it as A_{t+1}. Go to Step 4.

Step 3: Selection Process

- Combine P_t and A_t.
- Perform crowding sort to truncate combined population size to $|\text{A}_t|$

Step 4: Crossover operation

- Select two random solutions from A_{t+1}.
- Perform linear crossover if random number $\leq P_c$ else perform orthogonal crossover.
- Add offspring to P_{t+1}.

The selection process of DOMOEA is very similar to the crowding-sort procedure of NSGAII. However, the clustering algorithm of SPEA is adopted instead of crowding.

In DOMOEA, diversity is maintained in the evolving population through a linear crossover operator, which generates an offspring different from its parents $\mathbf{x_a}$ and $\mathbf{x_b}$. The linear crossover first generates four random numbers $-0.5 \leq \alpha_{a,i}, \alpha_{b,i}, \beta_{a,i}, \beta_{b,i} \leq 1.5$ such that the condition $\alpha_{a,i} + \alpha_{b,i} = 1$ is satisfied. These random numbers will determine the extend in which the

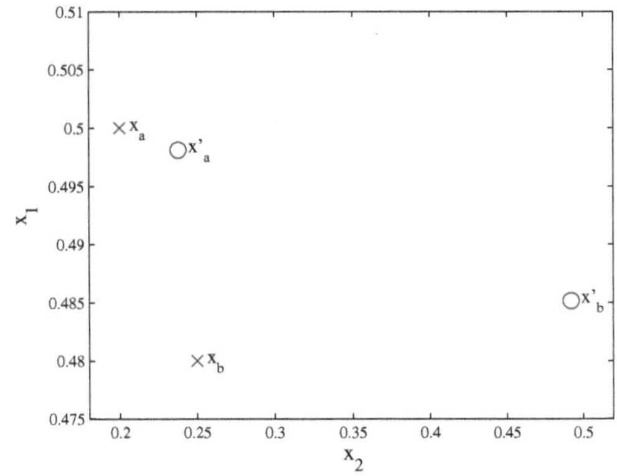

Fig. 5.4 Distribution of offspring using linear crossover. Parents and offspring are marked by (x) and (o), respectively.

Table 5.5 Orthogonal design methodology for 3 factors and 3 levels

	Factors			
Combination	x_1	x_2	x_3	Selected Objective
1	0.1	0.0	0.3	0.0760
2	0.1	0.1	0.5	0.0560
3	0.1	0.2	0.7	0.0760
4	0.15	0.0	0.5	0.0973
5	0.15	0.1	0.7	0.0772
6	0.15	0.2	0.3	0.0373
7	0.2	0.0	0.7	0.1190
8	0.2	0.1	0.5	0.0390
9	0.2	0.2	0.3	0.0590

offspring will take after a particular parent. Thereafter, the offspring are calculated as follows

$$x'_{a,i} = \alpha_{a,i}x_{a,i} + \alpha_{b,i}x_{b,i} \tag{5.55}$$

$$x'_{b,i} = \beta a, i(x_{a,i} - z_i) + \beta_{b,i}(x_{b,i} - z_i) \tag{5.56}$$

where z_i is a point in the middle of the i-th decision space. Fig. 5.4 illustrates the distribution of the offspring (o) relative to the parent (x). Notice that the offspring created by equation (5.55) tends to be located nearer to the parent. On the other hand, equation (5.56) will create offspring located far away from the parent, contributing to the diversity of the evolving population.

The orthogonal crossover is used to perform local fine-tuning and can give very good results if the region **H** bounded by the parents is linear or quadratic. The orthogonal array is first constructed and an objective is selected randomly as the optimizing goal. The orthogonal design method is then applied to identify and select the two best solutions as the offspring. Consider the instance where the solutions $\mathbf{x}_a$ =[0.1 0.2 0.3] and $\mathbf{x}_b$ =[0.2 0.0 0.7] are selected for the orthogonal crossover operation. The three-dimensional region **H** is given by $0.1 \leq x_1 \leq 0.2$, $0.0 \leq x_2 \leq 0.2$ and $0.3 \leq x_1 \leq 0.7$. In orthogonal design, the number of decision variables is also denoted as the number of factors and the number of different variation of each factor is called the level. If we are to evaluate each factor at three different levels, there will be a total of nine combinations to be tested. The nine combinations and their associated objective values are tabulated in Table. 5.5. Combination 6 provides the best results and thus, the first offspring is given as $\mathbf{x'}_a$ =[0.15 0.2 0.3]. The second offspring is determined by first calculating the mean objective attained for each level of every factor. This is calculated in the following manner:

$$\begin{aligned}
&\text{Mean objective value due to different levels of } x_1\\
\hat{f}_{x_1,Lvl1} &= \frac{1}{3}(0.0760 + 0.0560 + 0.0760) = \mathbf{0.0693}\\
\hat{f}_{x_1,Lvl2} &= \frac{1}{3}(0.0973 + 0.0772 + 0.0373) = 0.0706\\
\hat{f}_{x_1,Lvl3} &= \frac{1}{3}(0.1190 + 0.0390 + 0.0590) = 0.0723
\end{aligned}$$

$$\begin{aligned}
&\text{Mean objective value due to different levels of } x_2\\
\hat{f}_{x_2,Lvl1} &= \frac{1}{3}(0.0760 + 0.0973 + 0.1190) = 0.0974\\
\hat{f}_{x_2,Lvl2} &= \frac{1}{3}(0.0560 + 0.0772 + 0.0390) = 0.0574\\
\hat{f}_{x_2,Lvl3} &= \frac{1}{3}(0.0760 + 0.0373 + 0.0590) = 0.0574
\end{aligned}$$

$$\begin{aligned}
&\text{Mean objective value due to different levels of } x_3\\
\hat{f}_{x_3,Lvl1} &= \frac{1}{3}(0.0760 + 0.0373 + 0.0590) = \mathbf{0.0574}\\
\hat{f}_{x_3,Lvl2} &= \frac{1}{3}(0.0560 + 0.0973 + 0.0390) = 0.0641\\
\hat{f}_{x_3,Lvl3} &= \frac{1}{3}(0.0760 + 0.0772 + 0.1190) = 0.0907
\end{aligned}$$

The variable x_1 provides the best objective values at level 1, the variable x_2 provides the best objective values at either level 2 or level 3, and the variable x_3 provides the best objective values at level 1. Thus, the second offspring is given as $\mathbf{x'}_a$ =[0.1 0.1 0.3] or $\mathbf{x'}_a$ =[0.1 0.2 0.3].

5.5.5 Dynamic Queuing Multi-objective Optimizer

Hatzakis and Wallace [117] presented a dynamic queuing multi-objective optimizer (D-QMOO) which exploits past information to predict the future behavior of the dynamic multi-objective problem. An autoregressive model is employed to estimate the location of PS^*_{t+1} and the predicted individuals are used to seed the population when a change in the problem landscape is detected. Diversity is also maintained throughout the evolution. The D-QMOO is based on a fairly elaborate algorithm, the queuing multi-objective optimizer. Clustering is performed in the decision space to promote diversity and to handle the possible presence of multiple discontinuous fronts. Both clustering and ranking are only performed periodically to reduce computational cost. Genetic operators to be executed on each individual are also embedded as parameters within the chromosome as in the case of mutation strength in evolutionary strategies. Interested readers are encouraged to read [180]. The major steps of the DQMOO are shown below:

Dynamic Queuing Multi-Objective Optimizer (D-QMOO)

Step 1: Activate feed-forward prediction strategy after landscape change.
Step 2: Selection process. Only ranked solutions are allowed to participate.
Step 3: Recombination and mutation process.
Step 4: Evaluation.
Step 5: Periodic clustering process. Individuals are sorted into clusters in the decision space using Fuzzy c-means approach.
Step 6: Grouping process. Individuals within clusters are allocated to subgroups based on dominance relation. non-dominated and dominated solutions are assigned to *front* and *cruft*, respectively.
Step 7: Perform thinning process if population size exceeds limit.
Step 8: Periodic ranking process.

Feed-forward prediction strategy: No landscape change detection mechanism is incorporated because the D-QMOO assumes the availability of *a priori* knowledge on when a landscape change will be effected. The feed-forward prediction strategy refers to the incorporation of an autoregressive model to forecast the location of PS^*_{t+1} based on the accumulated time series. In the event of each problem variation, the predicted non-dominated solutions for PS^*_{t+1} will be inserted into the evolving population. When the prediction error is low, this allows the algorithm to converge upon PS^*_{t+1} faster. One time series is maintained for each objective and each time series tracks the best non-dominated solution along its associated objective. The time series is updated with the best solution to the previous landscape after each change. Since all the time series are initially empty, this strategy cannot be applied to predict the next optimum effectively in the initial stages.

Diversity maintenance: In the event that the prediction error is large and the forecasted solutions are far away from PS^*_{t+1}, the D-QMOO maintains sufficient diversity to facilitate the optimization process. Diversity is maintained in four ways. Firstly, clustering is conducted to partition the population into clusters. These clusters are evolved independently but the offspring of one cluster may inherit the genetic information of parents from other clusters.

Secondly, a thinning process is applied to ensure that a diverse set of individuals exists in the evolving population. In each generation, individuals within each cluster are further divided into two fixed-size groups, the front and the cruft, which consist of the non-dominated solutions and the dominated solutions, respectively. When the limit of the front is exceeded, solutions with the least contributions towards overall hypervolume are removed from the group. In the case of the cruft, solutions are either removed based on their crowding distance in the decision space or their age with equal probability. Least crowded and younger solutions are preferred. Finally, all duplicate solutions are removed during the recombination and ranking process.

5.5.6 Multi-objective Immune Algorithm

Zhang [292] suggested an immune algorithm approach for dynamic multi-objective optimization. The notion of applying an immune system methodology for dynamic optimization is very attractive because it has the natural capability to react to new threats and an immune memory which allows it to respond to antigens encountered before faster. For a more in-depth discussion on artificial immune systems, readers are refered to [53]. The multi-objective immune algorithm employs five different cell types, antigens (Ag), B-cells, helper T-cells (T_h), suppressor T-cells (T_s), and memory cells. The Ag models the dynamic problem, while B-cells represents the potential solution to the problem. T_h cells represents the non-dominated or archived solutions and memory cells are the collection of all previous non-dominated solutions. The major steps of the algorithm are shown below:

Multi-Objective Immune Algorithm (MOIA)

Step 1: Activate initial population scheme after landscape change.
Step 2: Calculate affinity values for B-cells in population P_t.
Step 3: Update environmental memory.
Step 4: Dynamic clonal selection to create clone population C_t.
Step 5: Generate new population:

- Perform hypermutation on C_t.
- Combine population and new solutions.
- Remove dominated and duplicate solutions to form D_t.
- Update $\eta\%$.
- Perform binary tournament selection to select $\eta\%$ of D_t.
- Create new random B-cells to replace identical individuals in population to form P_t.

Step 8: Update T_h population A_t.

- Add non-dominated solutions from P_t to A_t.
- Remove dominated solutions in A_t.
- Apply clustering if $|A_t|$ exceeds size limit.

Initial population scheme: As in the case of D-QMOO, the MOIA assumes the availability of *a priori* knowledge on when a landscape change will be effected. Upon a landscape change, the MOIA searches the environmental memory for non-dominated solutions of past environments that satisfy two conditions, 1) the solutions have the same number of objectives in the current environment and 2) the solutions exhibit similar behaviors in their previous and current environments. The MOIA first searches the memory for past solutions with the same dimensions with the current environment. If such solutions are found, the differences in objective values of these solutions before and after the landscape change are calculated in the following manner.

$$\varepsilon = \frac{1}{m_k} \sum_{j \in |m_k|} ||\mathbf{x}_{j,t} - \mathbf{x}_{j,k}|| \tag{5.57}$$

where $\mathbf{x}_{j,k}$ is non-dominated solution of the k-th environment, which satisfies the first condition, and m_k is the set of non-dominated solutions stored for the k-th environment. The set of solution satisfies the second condition if $\varepsilon < 0.5$. The first set of solutions that satisfies both conditions will be used to seed the new population. 80% of the new population will be selected from this set of solutions, while the rest are created randomly. The set of matching solution is also used to fill the A_t. In the event that no suitable solution is found, the evolving population will be reinitialized.

Calculation of affinity and activation value: After the creation of P_t, either through the immune operators or the initial population scheme, the affinity and activation values of each solution are calculated. The affinity of the solution $\mathbf{x}_a$ is measured with respect to the archived solutions

$$aff(\mathbf{x}_a) = \frac{\lambda}{1 + \exp\big(0.5 + D(\mathrm{T}_h^*, \mathbf{x}_a)E(\mathbf{x}_a, \mathrm{A}_t)\big)} + \frac{1-\lambda}{\exp(v||\mathbf{x}_a||)} \tag{5.58}$$

where T_h^* is the archived solution in A_t closest to $\mathbf{x}_a$ in the Euclidean sense and $||\cdot||$ is the Euclidean norm. The first term in equation (5.58) is a function of D and E, which measures the density of T_h^* in A_t and how well $\mathbf{x}_a$ approximates the archived solutions in A_t, respectively. D and E are calculated in the following manner,

$$D(T_h) = \frac{|\{\mathbf{x}_a^A \in \mathrm{A}_t : ||\mathbf{x}_a^A - \mathrm{T}_h^*|| < \sigma\}|}{|\mathrm{A}_t|} \tag{5.59}$$

$$E(\mathbf{F}_a, \mathrm{A}_t) = \sum_{\mathbf{F}_b^A \in \mathrm{A}_t} \langle \mathbf{F}_a, \mathbf{F}_b^A \rangle \tag{5.60}$$

where $0 < \sigma < 1$ is the density radius and $\langle \cdot, \cdot \rangle$ is the inner product between the two objective vectors. Therefore, it provides a bias towards Pareto optimal solutions that approximate rare antigens. While the first term promotes exploitation, the purpose of the second term in equation (5.58) is to promote exploration. Nonetheless, it can be easily seen that it has a bias towards solutions located near the origin of the decision space. The setting of $\lambda \in [0, 1]$ controls the balance between exploration and exploitation.

The activation level of $\mathbf{x}_a$ determines whether it will take part in the clonal process and it is calculated as follows:

$$act(\mathbf{x}_a) = aff(\mathbf{x}_a)\exp\big(-c(\mathbf{x}_a)\big) \tag{5.61}$$

$$c(\mathbf{x}_a) = \frac{|\{\mathbf{x}_b \in \mathrm{P}_t : ||\mathbf{x}_b - \mathbf{x}_a|| < \sigma\}|}{|\mathrm{P}_t|} \tag{5.62}$$

where $c(\mathbf{x}_a)$ is the density of $\mathbf{x}_a$ in P_t. From equation (5.61), we can expect that solutions located in less crowded regions with high affinity values to have high activation values. On the other hand, solutions residing in crowded areas or with low affinities will have relatively lower activation levels. By using the product of the affinity value and degree of crowding of a solution as the activation value, the MOIA ensures that none of the multi-objective optimization goals is neglected.

Update of environmental memory: After the calculation of affinity and activation values, a set of non-dominated solutions from P_t will be added to the environmental memory population E. This memory should be distinguished from the archive of non-dominated solutions A_t since it stores non-dominated solutions across different problem instantiations. To maintain a diverse set of solutions, only individuals satisfying the following criterion are updated in the memory population:

$$\min\{||\mathbf{x}_a - \mathbf{x}_b||,\ \mathbf{x}_b \in \mathrm{E}\} > \delta. \tag{5.63}$$

After all the non-dominated solutions satisfying the above condition are added to the memory, a check is performed to remove any duplicate solutions. There is a restriction on the size of the environment memory. Once this limit is reached, solutions in the most crowded region of the memory are removed using equation (5.63) with decreasing δ values until all excess solutions are removed.

Clonal selection: The selection process is based on the principle of clonal selection and it favors solutions with high degrees of affinity. The selection process starts with the calculation of the mean activation values of the solutions in P_t. After which, solutions with activation values greater than a certain threshold are selected for the clonal process, i.e $act(\mathbf{x}_a) \geq \alpha \cdot \mathrm{act}$ where act is the mean activation value of the solutions in P_t. Zhang devised a method to adapt this threshold such that the number of different solutions taking part in the clonal process is dependent on the size P_t. The selection criterion is given by

$$\alpha = \begin{cases} 0, & \text{if } |\mathrm{P}_t| \leq 20 \\ \frac{|\mathrm{P}_t|-20}{30}, & \text{if } 20 \leq |\mathrm{P}_t| \leq 50 \\ 0.6, & \text{otherwise} \end{cases} \tag{5.64}$$

Such a selection process ensures that a sufficient number of solutions will participate in the clonal process in every generation. The number of clones that each individual will produce to form C_t is based on the affinity level. It is computed as follows,

$$cl(\mathbf{x}_a) = \left\lceil (|\mathrm{A}_t| - \frac{1}{\eta aff(\mathbf{x}_a)}) \right\rceil \tag{5.65}$$

where $\eta \in [\frac{1}{2(1+aff(\mathbf{x}_a))}, \frac{1}{(1+aff(\mathbf{x}_a))}]$ is a random number. Note that of the selected solutions will be cloned a number of times and $|\mathrm{C}_t|$ is greater than $|\mathrm{A}_t|$.

Generate new population: Two different mutation operators are used to promote exploration and exploitation of the search space in MOIA. The first operator is the hypermutation operator and it is performed only on $|\mathrm{A}_t|$ solutions in C_t. This operator is defined as follows,

$$x'_{a,i} = \zeta x_{a,i} + (1-\zeta)x^A_{b,i} \tag{5.66}$$

where $\mathbf{x}^A_b \in \mathrm{A}_t$ is randomly selected from the A_t. The mutation probability of this operator is given as

$$P_{m1}(\mathbf{x}_a) = 1 - \frac{1}{M}\exp(-\frac{50}{t} - \frac{\overline{aff} - aff(\mathbf{x}_a)}{\overline{aff} - \underline{aff}}) \tag{5.67}$$

where $\overline{aff} = \max_{\mathbf{x}_a \in \mathrm{P}_t} aff(\mathbf{x}_a)$ and $\underline{aff} = \min_{\mathbf{x}_a \in \mathrm{P}_t} aff(\mathbf{x}_a)$ The other solutions in C_t are subjected to uniform mutation with the following mutation probability,

$$P_{m2}(\mathbf{x}_a) = 1 - \exp(-||\mathbf{x}_a - \mathbf{x}_b||). \tag{5.68}$$

After the mutation process, the new antibody population is combined with the non-dominated solutions from P_t to form a temporary population D_t. All duplicate and dominated solutions are then removed from D_t. Thereafter, binary tournament selection is conducted to select η% of the updated D_t to form P_t. η is also adapted along the optimization process and it is based on the number of non-dominated solutions found in each generation,

$$\eta = \begin{cases} 100, & \text{if } |\mathrm{D}| \leq 50 \\ 60 + \frac{40}{50}(100 - |\mathrm{D}|), & \text{if } |\mathrm{D}| \leq 50 \\ 50, & \text{otherwise} \end{cases} \tag{5.69}$$

To avoid over-specialization and hence losing diversity, all identical solutions are replaced by random solutions.

5.6 Conclusion

Many real-world problems involve time-dependent components. For such problems, it is unlikely that the optimal Pareto set and the Pareto front will remain invariant and the previous solution must be adapted to reflect the current requirements. Therefore, the optimization goal is not only to evolve a near optimal and diverse PF^A but also to track it as it changes with time.

In this chapter, we first presented a classification scheme for dynamic multi-objective problem and described a number of existing dynamic test functions. In order to track the time-varying Pareto optimal solution set, the MOEA must be capable of detecting the landscape change, maintaining or introducing sufficient diversity to find the new optimal solution set after each landscape variation, and handling the outdated solutions. In addition, the dynamic MOEAs must be capable of finding the new solutions quickly before the landscape change is effected again. Most algorithms presented here maintain a mechanism to re-evaluate solutions periodically to check for discrepancies as indications of problem variation. Since sufficient diversity is the key to tracking the dynamic solution set, all dynamic MOEAs incorporate some form of genetic operator to perform exploration immediately after a problem change. In addition, the directional-based dynamic evolutionary multi-objective optimization algorithm and the dynamic multi-objective evolutionary algorithm based on an orthogonal design implements local search operators that exploit similarities between the landscape prior and after change. The dynamic queuing multi-objective optimizer also applied an autoregressive model to predict the location of the next solution set to improve convergence.

Chapter 6
A Coevolutionary Paradigm for Dynamic Multi-Objective Optimization*

As pointed out in the previous chapter, it is imperative that the MOEA must be capable of attaining high convergence speeds in order to find the optimal solution set before it changes and becomes obsolete. However, high convergence speed often implies a rapid loss of diversity during the optimization process, which inevitably leads to the inability to track the dynamic Pareto front. Therefore, it is necessary to maintain or generate sufficient diversity to explore the search space when the multi-objective problem changes.

In these two regards, the notion of coevolution is very attractive. The coevolutionary paradigm, inspired by the reciprocal evolutionary change driven by the cooperative [213] or competitive interactions [222] between different species, has been extended successfully to multi-objective optimization recently [44, 134, 151, 187, 190, 251].

- On the former issue of high convergence speed, several studies [212, 272] have shown that the introduction of ecological models and coevolutionary architectures are effective methods to improve the efficacy of canonical evolutionary algorithms. As a specific instance, Tan *et al* [251] demonstrated that high convergence speeds can be achieved while maintaining a good diversity of solutions. Multi-objective coevolutionary algorithms (MOCAs) seem particularly suitable for dynamic multi-objective optimization, where the high speed of convergence can potentially be exploited for adapting quickly to the changing environment.
- On the latter issue of diversity, the works in [9, 213] demonstrated that both competitive and cooperative coevolutions have their own unique mechanisms for maintaining diversity in the species sub-population.

On the other hand, successful implementation of coevolution requires appropriate problem decomposition. In this chapter, we are concerned with the decomposition of the search space. The best way of handling problem

C.-K. Goh and K.C. Tan: Evolutionary Multi-objective Optimization, SCI 186, pp. 153–185.
springerlink.com

decomposition may not be known *a priori* and may change with time in dynamic multi-objective problem. This chapter describes a new coevolutionary paradigm [94]that incorporates both competitive and cooperative mechanisms observed in nature to solve multi-objective optimization problems and to track the Pareto front in a dynamic environment. The main idea of competitive-cooperation coevolution is to allow the decomposition process of the optimization problem to adapt and emerge rather than being hand-designed and fixed at the start of the evolutionary optimization process. In particular, each species sub-population will compete to represent a particular sub-component of the multi-objective problem, while the eventual winners will cooperate to evolve better solutions. Through this iterative process of competition and cooperation, the various sub-components are optimized by different species sub-populations based on the optimization requirements of that particular time instant, enabling the MOCA to handle both the static and dynamic multi-objective problems. A competitive-cooperation coevolutionary algorithm (COEA) for static environment is designed based on the proposed coevolutionary paradigm and subsequently extended as dynamic COEA (dCOEA) to handle dynamic multi-objective optimization problems.

6.1 Competition, Cooperation, and Competitive-Cooperation in Coevolution

Existing coevolutionary techniques can be divided into two main classes: competitive coevolution and cooperative coevolution. Regardless of the approach adopted, the design of coevolutionary algorithms for multi-objective optimization requires one to address many issues that are unique to the multi-objective problem. In this aspect, insights, such as incorporation of various elitist and diversity mechanisms obtained from the design of MOEAs, can be similarly exploited in the design of MOCAs. On the other hand, successful implementation of coevolution requires one to consider various design issues [212], such as problem decomposition, handling of parameter interactions, and credit assignment. The issues of problem decomposition and parameter interactions are often problem dependent and the approaches for solving these issues may not be known *a priori*. These factors motivated the work for an alternative coevolutionary model presented in this chapter.

This section begins with a review of both the competitive and cooperative coevolutionary algorithms for multi-objective optimization, highlighting various features and limitations of existing approaches. A competitive-cooperative coevolutionary model is then proposed, including discussions on how the different design issues in coevolutionary algorithms are addressed.

6.1.1 Competitive Coevolution

The model of competitive coevolution is often compared to predator-prey or host-parasite interactions, where preys (or hosts) implement the potential solutions to the optimization problem and the predators (or parasites) implement individual "fitness-cases". When applying this idea into optimization [5, 222], there are usually two sub-populations and an inverse fitness interaction exists between the two sub-populations. To survive, the losing sub-population adapts to counter the winning sub-population in order to become the new winner.

Although the competitive coevolution has been applied in many single-objective evolutionary algorithm studies [118, 214], this model is rarely investigated in the domain of EMOO. Laumanns *et al* [177] embodied the model of competitive coevolution in multi-objective optimization through a spatial predator-prey model. In this model, the solutions are the preys and the associated solution vectors are represented as vertices on an undirected and connected graph. There are as many predators as the number of objectives and these predators perform a random walk on the graph along their respective associated objective. The worst prey in the neighborhood of the predator, in terms of the associated objective, will be replaced by its offspring.

Lohn *et al* [187] presented a different competitive coevolutionary model which contains a population of candidate solutions and a target population with the target objective vectors. A distinct characteristic of this algorithm is the lack of any explicit diversity preservation mechanism to guide the coevolutionary optimization process. Empirical studies are conducted with well-known MOEAs, such as SPEA and NSGA, and the performance of this competitive MOCA is found to be better than the test algorithms.

There are several limitations to this coevolutionary model for numerical optimization. While competitive coevolution is a natural model for evolving objects, such as game playing programs, for which it is difficult to write an external fitness function, the need to hand-decompose the problem into antagonistic sub-components places severe limitation on its range of applicability. Adding to its complexity is the need to adapt the predator population, which is the population of target vectors in the case of [187], such that it exerts an appropriate pressure of convergence. In the context of multi-objective optimization, this pressure must be exerted to promote individuals in a direction that is normal, as well as tangential, to the tradeoff region at the same time. Intuitively, such a competitive coevolutionary approach may be sensitive to the shape of PF* in multi-objective optimization.

6.1.2 Cooperative Coevolution

Cooperative coevolution is inspired by the ecological relationship of symbiosis where different species live together in a mutually beneficial relationship. The basic idea of cooperative coevolution is to divide and conquer [213]: divide a large system into many modules, evolve the modules separately, and then combine them together to form the whole system. The cooperative coevolutionary algorithm involves a number of independently evolving species that together form complex structures for solving difficult problems. The fitness of an individual depends on its ability to collaborate with individuals from other species, which favors the development of cooperative strategies and individuals. In addition, these techniques can be implemented at two basic levels depending on the type of modules that are evolved simultaneously [152]. In the case of single-level coevolution [44, 151, 134, 190], each evolving sub-population represents a sub-component of the problem to be solved. On the other hand, a two-level coevolutionary process involves simultaneous optimization of the system and modules in separate sub-populations [9, 87].

An explicit way of implementing cooperative coevolution in optimization techniques is to split a solution vector into different sub-components and assign multiple evolving sub-populations to optimize the individual sub-components [213]. Contrary to single-objective optimization, multi-objective optimization is associated with a set of nondominated solutions, which inevitably leads to the issues of fitness assignment and representative selection. In these aspects, appropriate representatives are crucial for the search of a diverse and uniformly distributed solution set, and suitable cooperative schemes must be incorporated in order to drive the sub-populations in tandem towards the PF*.

An early attempt to integrate the cooperative model for multi-objective optimization is to decompose the problem along the decision space and each sub-population is optimized by the multi-objective genetic algorithm (MOGA) [84]. In this multi-objective cooperative coevolutionary genetic algorithm (MOCCGA) [151], each individual is evaluated twice in collaboration with either a random or the best representative from the other sub-populations and the best Pareto rank is assigned as the fitness. However, the performance of MOCCGA is limited due to the lack of elitism and the localized perception of Pareto optimality.

Maneeratana *et al* [190] later incorporated elitism in the form of a fixed-size archive to store the set of nondominated solutions and the same cooperative model is successfully extended to other MOEAs, such as Niched Pareto GA [123] and NSGA [243], with significant improvements over their canonical counterparts. Like MOCCGA, however, these MOCAs also suffer from the problem that fitness assignment conducted within a species may not be a good indicator of optimality.

Iorio and Li [134] presented a nondominated sorting cooperative coevolutionary algorithm (NSCCGA), which is essentially the coevolutionary extention of NSGAII. In NSCCGA, elite solutions are reinserted into the sub-populations and fitness assignment takes into account the set of nondominated solutions obtained via nondominated sorting. Instead of selecting nondominated individuals with the best degree of crowding, representatives are selected randomly from the best nondominated front.

Contrary to the trend of integrating the cooperative model with well-known MOEAs, Tan *et al* [251] proposed the cooperative-coevolution evolutionary algorithm (CCEA) based on a simple MOEA. Although the ranking scheme [84] of MOCCGA is adopted in CCEA, each individual is ranked against the nondominated solutions stored in the archive instead of within the sub-population. In addition, an extending operator is implemented in CCEA to improve the diversity and distribution of the PF by reinserting nondominated individuals with the best niche count into the evolving sub-population. Various representative selection scheme are also examined and good performance is observed for the scheme that retains the better solution in the cooperation between two representatives from each sub-population.

One major issue of these MOCAs is their dependence on appropriate manual decomposition of the problem into various sub-components. Since many problems exhibit parameter inter-dependencies, the decomposition of solution vector and the optimization of each sub-component independently may lead to the phenomenon of fitness landscape warping [212] and convergence to sub-optimal solutions. It should be noted that parameter interactions are usually not considered explicitly in EAs. Notable exceptions include the estimation-of-distribution algorithms (EDAs) [291] that sought to learn the inter-relation through joint probability distribution models and the covariance matrix adaptation evolution strategies (CMA-ES) [114] which has been recently extended to multi-objective optimization [135].

Iorio and Li [134] also highlighted that coevolutionary algorithms are susceptible to parameter interactions, although a higher mutation rate can often improve the algorithmic performance of rotated problems. Apparently, there is an inherent tradeoff between the fine-grain search capability and the lack of diversity due to the smaller size of sub-populations in coevolutionary algorithms. The game-theoretic approach of modeling cooperation in [240] attempts to alleviate the issue of parameter dependencies by decomposing the optimization problem into only two sub-populations. Without restricting to a single computational paradigm, an interesting approach of switching iteratively between canonical particle swarm optimization (PSO) and cooperative PSO is proposed by Van den Bergh and Engelbrecht [271] for single-objective optimization problems.

Applying a variant of the cooperative models discussed so far, Coello Coello and Sierra [44] proposed a coevolutionary MOEA (CO-MOEA) where different sub-populations cooperate to form the PF instead of a valid candidate solution. The CO-MOEA starts with a single evolving population and adaptively assigns different regions of the decision variable space to new sub-populations. This assignment process is performed by analyzing the contribution of each decision variable to the PF stored in an adaptive grid [166]. Furthermore, the sub-population size is changed in proportion to the discovery of new non-dominated solutions and any sub-population without significant contribution is eliminated. Although such an approach removes design considerations, like representative selection and parameter interactions, the CO-MOEA does not incorporate the fine-grain search capability of MOCAs with Potter and Jong's model.

6.1.3 Competitive-Cooperative Coevolution

The issue of problem decomposition poses restrictions on existing algorithmic designs and performances of both competitive and cooperative models. In retrospect, this problem should not arise in the context of coevolutionary algorithms since the role that each species plays is an emergent property in nature. On the other hand, the collaboration and competition among different species are modeled independently in coevolutionary algorithms, although these two types of interactions are rarely exclusive within an ecological system. For example, there is competition even in the veneer of seemingly perfect plant-pollinator coevolution in nature [232], where different species of bees will compete for nectar and different species of flowers will compete to attract more bees. By incorporating both elements of cooperation and competition, the proposed model represents a more holistic view of the coevolutionary forces in nature.

The proposed competitive-cooperative model involves two tightly-coupled coevolutionary processes as illustrated in Fig. 6.1. Similar to conventional cooperative coevolutionary algorithms, individuals from different species collaborate to solve the problem during the cooperative process. Each subpopulation evolves in isolation and there is no restriction on the form of representation or on the underlying EA. On the other hand, the cooperative species will also compete with other sub-populations for the right to represent the various sub-components of the problem.

The interaction between the cooperative and competitive processes may take place iteratively after each generation or at a frequency determined by the user. For the ensuing discussions, we consider that the problem at hand is decomposed along the decision variables. Each decision variable may be

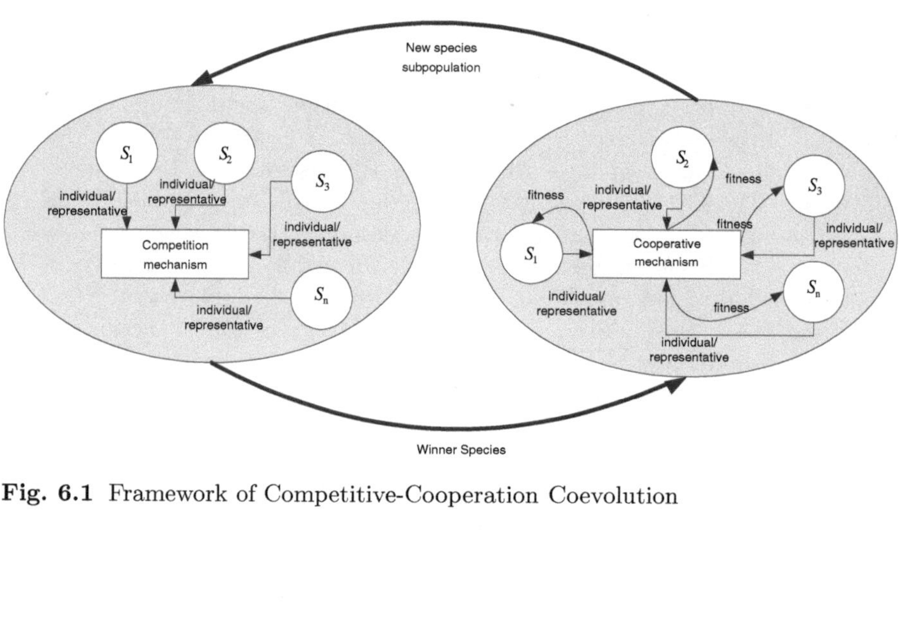

Fig. 6.1 Framework of Competitive-Cooperation Coevolution

assigned to a number of sub-populations and a sub-population may optimize for more than one decision variable.

Credit Assignment

The credit assignments for the competitive and cooperative processes are performed at the sub-population and individual levels, respectively. In the cooperative process, the different objectives are evaluated by assembling each individual with representatives of the other species to form a valid candidate solution. Accordingly, appropriate fitness assignment, such as Pareto ranking, can be performed for that particular individual. In the competitive process, the fitness of a particular species is computed by estimating how well it performs relative to its competitors in cooperating with other species to produce good solutions. For example, the species selected out of N competing sub-populations is given a higher probability for representing a particular variable in subsequent generations, while the losing species of the competition is penalized and given a lower probability.

Problem Decomposition and sub-component Inter-dependency

As mentioned earlier in the section, the issue of problem decomposition needs to be addressed for coevolutionary algorithms. The difficulty lies in the fact that information pertinent to the number or roles of sub-components is usually unknown *a priori* and many problems can only be decomposed into sub-components with complex inter-dependencies. The competitive-cooperation coevolutionary model addresses such an issue through emergent problem decomposition.

As illustrated above, the competitive process leads to a potential "arms race" among the cooperative species to improve their contributions in the associated sub-components. It should be noted that the collaboration between these two coevolutionary models can lead to the natural formation of competitive sub-populations rather than sub-components. In addition, it facilitates the interactions among different species, in possibly various roles, right at the onset of the optimization process, which benefits the discovery of inter-dependencies among the species. Therefore, the interplay of competition and cooperation provides an environment in which inter-dependent sub-components end up within similar species and reasonable problem decomposition emerges due to evolutionary pressure rather than being specified by the user.

The emergent attribute of the competitive-cooperation coevolutionary model is distinctively different from the cooperative model proposed by Potter and Jong [213]. Although the participation of a sub-population is based on its contribution made to the collaboration among species in both approaches, this feature is due to the emergence of fitter species for a particular

Table 6.1 Nomenclature

Notation	Definition
A	Archive of non-dominated solutions
A^T	Temporal archive of past non-dominated solutions
a_i	The i-th non-dominated solution of A
C_{freq}	Frequency of competition
n_x	Number of decision variables
P_i^c	Competition pool for the i-th variable
R_{size}	Number of a_i to be updated to A^T
S_i	The i-th sub-population
$s_{i,j}$	The j-th individual of S_i
$s_{i,rep}$	Representative of S_i
SC_{ratio}	Ratio of stochastic competitors in P_i^c

problem sub-component in the proposed model. One limitation of the approach in [213] is that stagnant sub-populations are simply replaced by randomly initialized sub-populations, implying that any useful information obtained previously can be discarded.

Diversity

The competitive-cooperation coevolutionary model provides a means of exploiting the complementary diversity preservation mechanism of both competitive and cooperative models. In the cooperative model, the evolution of isolated species tends to produce more diversed individuals across the different sub-populations, although this property does not necessarily extend to within each sub-population. On the other hand, a diverse sub-population is driven by the necessity to deal with different situations posed by the other sub-populations in the competitive model. Furthermore, the competitive process in competitive-cooperation coevolutionary model also allows for a more diversified search since the optimization of each sub-component is no longer restricted to one species. The competing species provides another round of optimization for each sub-component, thus increases the extent of the search and maintains an overall low computation requirement.

6.2 Applying Competitive-Cooperation Coevolution for Multi-objective Optimization

Based on the competitive-cooperation coevolutionary paradigm described in Section 6.1, this section presents a competitive-cooperation coevolutionary algorithm (COEA) for multi-objective optimization. The notations used in subsequent sections are summarized in Table 6.1. The mechanism of cooperative coevolution is described in Section 6.2.1, while the competitive element of the proposed paradigm is presented in Section 6.2.2. Finally, the implementation details of COEA are given in Section 6.2.3.

6.2.1 Cooperative Mechanism

The cooperative mechanism of the proposed COEA is extended from the model introduced by Tan *et al* [251]. By adopting this strategy, the algorithm can exploit the fine-grained search capability desirable in many applications and maintain good diversity across the sub-populations. The main steps of the cooperative mechanism are shown below:

Cooperative Coevolutionary Mechanism

Step 1: For all solutions in S_i

- Assemble complete solution with $s_{1,j}$ and representative from the other sub-populations.
- Evaluate solution
- Update A_t

Step 2: For all solutions in S_i

- Assign Pareto rank to $s_{1,j}$.
- Calculate niche count of $s_{1,j}$.

Step 3: Update representative of S_i

At the start of the optimization process, the i-th sub-population is initialized to represent the i-th variable. Concatenation between individuals in S_i and representatives from the other sub-populations is necessary to form a valid candidate solution for evaluation. As an example, consider a three-decision variable problem where sub-populations S_1, S_2, and S_3, represent the variables x_1, x_2, and x_3, respectively. When assessing the fitness of $s_{1,j}$, it will combine with the representatives of S_2 and S_3 to form a valid candidate solution.

In this approach, archive updating is conducted after the evaluation of each individual. Pareto ranking and niche count computation of individual $s_{i,j}$ are then conducted with respect to the archive. Note that only the fitness values of the individuals from S_i are updated at the i-th cycle. The Pareto rank of each individual is based on the number of archived solutions dominating it, i.e.

$$rank(s_{i,j}) = 1 + \left|\{a_k \in A | a_k \prec s_{i,j}\}\right|. \tag{6.1}$$

Similar to the ranking process, the niche count (nc) of each individual is calculated with respect to the archive of non-dominated solutions. The dynamic sharing scheme proposed in [255] is employed here to estimate the sharing radius.

The cooperative process is carried out in turn for all n_x sub-populations, where n_x is the number of decision variables. Before proceeding to the evaluation of the next sub-population, the representative of S_i denoted as $s_{i,\ rep}$ is updated to improve the speed of convergence. This updating process is based on a partial order such that ranks will be considered first and followed by niche count if there is a tie in the rank. For any two individuals $s_{i,j}$ and $s_{i,k}$, $s_{i,j}$ is selected over $s_{i,k}$ if $rank(s_{i,j}) < rank(s_{i,k})$ or $\{rank(s_{i,j}) = rank(s_{i,k})$ and $nc(s_{i,j}) < nc(s_{i,k})\}$. The rationale of selecting a non-dominated representative with the lowest niche count is to promote diversity of the solutions via the approach of cooperation among multiple sub-populations.

6.2.2 Competitive Mechanism

Given that the cooperative scheme optimizes a single variable in each sub-population, one simple approach is to allow the different sub-populations to take up the role of a particular problem sub-component in a round-robin fashion. The most competitive sub-population is then determined and the sub-component will be optimized by the winning species in the next cooperative process. Ideally, the competition is performed such that all individuals from a particular sub-population compete with all other individuals from the other sub-populations in order to determine the extent of its suitability. However, such an exhaustive approach requires extensive computational effort that is often practically infeasible. A more practical approach is thus to conduct competition with only selected individuals among a certain number of competitor sub-populations to estimate the species fitness and suitability. The competitive mechanism is illustrated in Fig. 6.2 and the main steps are shown below:

Competitive Coevolutionary Mechanism

Step 1: For all solutions in S_i

- Insert representative of sub-population representing variable i $s_{i,\ rep}$ into the competitive pool P_i^c.
- If $n_x > |S_i|$
  - Select competing sub-populations randomly.
  - Insert competitors from selected sub-population into P_i^c.
- If $n_x \leq |S_i|$
  - Insert competitors from other sub-populations into P_i^c.
  - Insert random individuals from S_i into P_i^c.

Step 2: Cooperative process.
Step 3: Determine winning sub-population S_k.
Step 4: Update $S_i = S_k$.

The competitive process to discover the most suitable sub-population is performed for each variable in an iterative manner. For the i-th variable, the

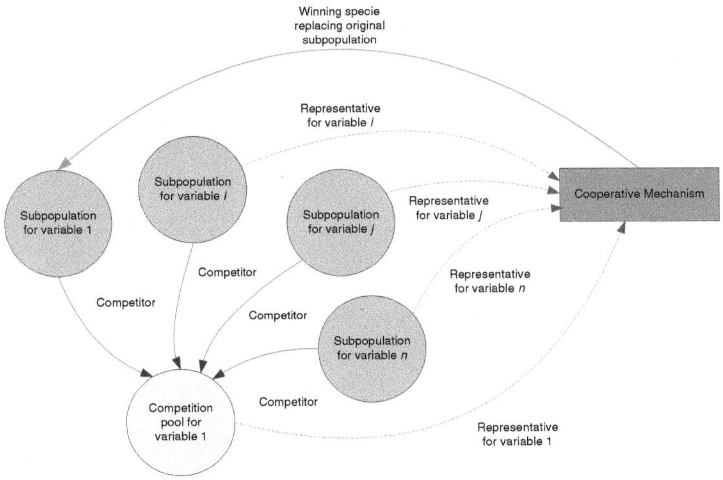

Fig. 6.2 Illustration of the competitive mechanism

representative of the associated sub-population, i.e. $s_{i,\ rep}$, is selected along with the competitors from the other sub-populations to form a competition pool. The COEA adopts a simple competitor selection scheme of choosing a random individual from each competing sub-population. Intuitively, the selection of a random competitor will enable the COEA to explore relationships among the different variables. Other types of competition schemes will be presented and analyzed in Section 6.4.3. In the case where $n_x > |S_i|$, i.e. the number of sub-populations is larger than the sub-population size, the participating sub-populations will be selected randomly before the start of the competition process. This provides an opportunity for the other sub-populations left out in this instance to participate in future competitions.

These competitors will then compete via the cooperative mechanism described earlier to determine the extent of the cooperation achieved with the representatives of the other sub-populations. In this approach, the winning species is determined by checking the originating sub-population of the representative after the representative update. At the end of the competitive process, S_i will remain unchanged if its representative wins the competition. In the case that a winner emerges from other sub-populations, S_i will be replaced by the individuals from the winning sub-population. The rationale of replacing the losing sub-population instead of associating the winning sub-population with the decision variable directly is that different variables may have close but not identical properties. Therefore, it is more appropriate to seed the losing sub-population with the desirable information and to allow it to evolve independently.

By embedding the competitive mechanism within the cooperative process, the adaptation of problem decomposition and the optimization process are conducted simultaneously. Hence, no additional computation cost is incurred from the competition. Moreover, this approach also has the advantage of allowing different sub-populations to solve a single component as a collective unit, with the competitors acting as a source of diversity.

6.2.3 Implementation

The flowchart of the proposed COEA is illustrated in Fig. 6.3. The initialization process involves the creation of n_x sub-populations of random individuals, where the i-th sub-population will represent the i-th decision variable. The individuals will then undergo the competitive-cooperation process until the stopping criterion is satisfied, which can be set based upon a fixed number of function evaluations. In this work, the number of fitness function evaluations is determined according to past experience and complexity of the test functions, which can be in multiples of the number of decision variables.

At each generation, either the cooperative or competitive mechanism is activated. In particular, the competitive mechanism is applied at a fixed frequency of $C_{freq} = 10$, otherwise the cooperative process is adopted otherwise. During the cooperative process, individuals of each sub-population are evaluated by combining them with representatives from the other sub-populations to form a complete candidate solution. As mentioned in Section 6.2.1, the archive will be updated after each evaluation. After all individuals in the sub-populations have been evaluated in the cooperative process, binary

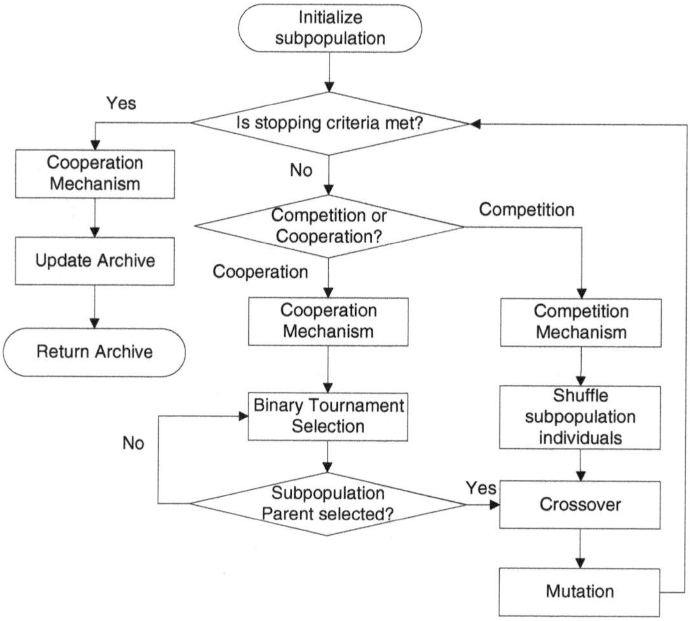

Fig. 6.3 Flowchart of COEA

tournament selection based on Pareto rank is applied to select the parents for each of the sub-populations. Here the individual with a lower niche count will be selected in the case of a tie in rank. The selected parents will then undergo the process of uniform crossover and bit-flip mutation.

During the competitive process, the archiving is also performed after each evaluation as described in Section 6.2.2. In contrast to the cooperative process, tournament selection is not employed here for the selection of parents and the sub-population individuals are shuffled randomly before undergoing crossover and mutation. It is not necessary to perform selection based on fitness measure since the replacement individuals have not been evaluated for their fitness and may not perform in an identical manner in their new rôle of optimizing another sub-component.

The algorithm applies a fixed-size archive to store non-dominated individuals along the evolution. A complete candidate solution formed by the sub-populations will be added to the archive if it is not dominated by any archived solution. Likewise, any archive member dominated by this candidate solution will be removed. When the predetermined archive size is reached, a recurrent truncation process [154] based on niche count is used to eliminate the most crowded archive member.

6.3 Adapting COEA for Dynamic Multi-objective Optimization

Besides considering the different requirements of multi-objective problems, the issues of diversity and outdated archived solutions should also be addressed before the proposed COEA is capable of dealing with environmental variations in dynamic optimization. Section 6.3.1 describes a scheme for achieving good diversity to be introduced while exploiting useful past information. Section 6.3.2 describes a simple temporal memory approach, which stores and reintroduces outdated non-dominated individuals into the archive when necessary.

6.3.1 Introducing Diversity via Stochastic Competitors

Generally, the population diversity desirable for tracking the dynamic PS_t^* in COEA can either be introduced explicitly through mechanisms such as random restart and hypermutation or be maintained by means of niching and other diversity preservation schemes. The approach of using multiple populations to explore the different regions of the search space is not applicable here since the application of sub-populations in COEA serves another purpose of optimizing a specific sub-component of the problem. Although explicit generation of diversity will allow the algorithm to react faster to

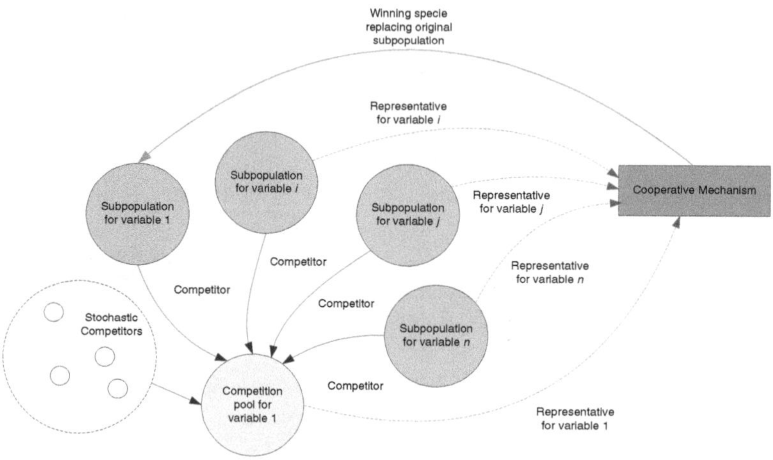

Fig. 6.4 Illustration of stochastic competitors

severe environmental changes, such an approach is unable to utilize useful past information. On the other hand, the potential for information exploitation in diversity preservation schemes is often achieved at the expense of a slower convergence. This is known as the exploration-exploitation dilemma for dynamic optimization [30].

A diversity scheme which exploits the competitive mechanism of COEA is thus implemented. In every generation, a fixed number of archived solutions are re-evaluated and the current objective values are checked against the previous values for discrepancies. If there is any environmental variation in the evolution, the competitive mechanism will be started, in addition to its fixed schedule. This strategy allows the algorithm to assess the potential of existing information within the various sub-populations for exploitation in the new problem landscape.

Furthermore, the competitive process provides a natural conduit in which the introduction of diversity into the sub-populations can be regulated. Instead of re-initialization or subjecting the entire sub-population to hypermutation, a set of stochastic competitors, which is illustrated in Fig. 6.4, are introduced together with the competitors from the other sub-populations, where the ratio between the two types of competitors is given by the parameter SC_{ratio}. The idea is to compare the potential of new regions in the search space and the past information to decide whether the sub-population should be initialized. The latin hypercube sampling is applied to generate individuals along each dimension uniformly. In the case that stochastic competitor emerges as the winner, the particular sub-population is re-initialized in the region that the winner is sampled from. Hence, diversity is introduced into the sub-populations only when it presents an advantage over the current information at hand.

6.3.2 Handling Outdated Archived Solutions

If there is any environmental change in the evolution, it is likely that the archived solutions will not remain non-dominated and these outdated archived solutions will keep out the non-dominated solutions if they are being left unchecked. Therefore, appropriate measures must be incorporated to minimize the detrimental effects of any outdated archived solution. A simple approach is to re-evaluate all the outdated solutions and to remove only the dominated solutions from the archive. Since most MOEAs are elitist in general, such an approach may mislead the optimization process with non-dominated but outdated archived solutions. Moreover, the process of re-evaluation will result in extra computational cost, which is undesirable. An alternative approach is to simply discard all the archived solutions but useful information about past PF_t cannot be exploited in the case where PS_t^* is cyclic in nature.

In order to store the potentially useful information in dCOEA, an additional external population denoted as the temporal memory is used in conjunction with the archive. In the ideal situation, the temporal memory is a repository of all the non-dominated solutions prior to any environmental variation. Due to the limited computational resources, however, decision must be made on what solutions and how the solutions are stored in the temporal memory. The main steps of the temporal archive updating mechanism are shown below:

Temporal Archive

Step 1: Select and remove the best archived solution along each dimension from A_t to temporary pool P^T.
Step 2: If $R_{size} \geq M$, randomly select and insert $M - R_{size}$ solutions from A_t to temporary pool P^T.
Step 3: If $R_{size} < M$, randomly select and remove $R_{size} - M$ solutions from temporary pool P^T.
Step 4: Add P^T to temporal archive A^T.
Step 5: If $| \mathrm{A}^T | >$ size limit, remove R_{size} oldest solutions from A^T.

To store the outdated solutions, a fixed number R_{size} of the archive is added to the temporal memory upon a landscape change. When the upper bound of the temporal memory is reached, the oldest set of R_{size} outdated solutions is removed for newer solutions. To select the R_{size} outdated solutions, the dCOEA stores the extreme solutions along each dimension in the objective space. In the case where R_{size} is greater than the number of extreme solutions, the rest of the solutions to be stored are randomly selected from the archive. On the other hand, if R_{size} is smaller than the number of extreme solutions, then R_{size} extreme solutions will be randomly selected into the temporal memory. Intuitively the value of R_{size} controls the tradeoff between the storage of information across different environmental changes and

Table 6.2 Parameter setting for different algorithms

Parameter	Settings
Populations	Population size 100 in NSGAII, SPEA2, PAES, and IMOEA; sub-population size 10 in COEA and CCEA; Archive (or secondary population) size 100.
Chromosome	Binary coding; 30 bits per decision variable.
Selection	Binary tournament selection
Crossover operator	Uniform crossover
Crossover rate	0.8
Mutation operator	Bit-flip mutation
Mutation rate	$\frac{1}{L}$ for DTLZ3 where L is the chromosome length; $\frac{1}{B}$ for FON and KUR where B is the bit size per decision variable;
Niche Radius	Dynamic sharing.

the information for a particular instance of landscape change. In particular, a smaller value of R_{size} will allow for a more diverse range of past solutions.

After the R_{size} outdated archived individuals have been added to the temporal memory, all archived solutions will be discarded. Subsequently, the temporal memory will be re-evaluated and the archive updating is conducted on this external population. The computational cost incurred by this re-evaluation process is necessary so as to exploit any useful information about the current PS_t^*. To address the concern that solutions updated into the archive through this approach may misguide the optimization process, no archived solution will be re-inserted back to the sub-populations in the generation immediately after the environmental change.

6.4 Static Environment Empirical Study

This section starts with a comparative study between COEA and MOEAs that are representative of the state-of-the-arts in Section 6.4.1. This section concludes with further investigations to gain better insights to the dynamics of competitive-cooperation evolution in Section 6.4.2 and Section 6.4.3.

6.4.1 Comparative Study of COEA

In order to examine the effectiveness of COEA, a comparative study including COEA, CCEA [251], SPEA2 [298], and NSGAII [61] is carried out based upon FON, KUR, and DTLZ3. The simulations are implemented in C++ on an Intel Pentium 4 2.8 GHz personal computer. Thirty independent runs are performed for each of the test functions to obtain the statistical information, such as consistency and robustness of the algorithms. The various parameter

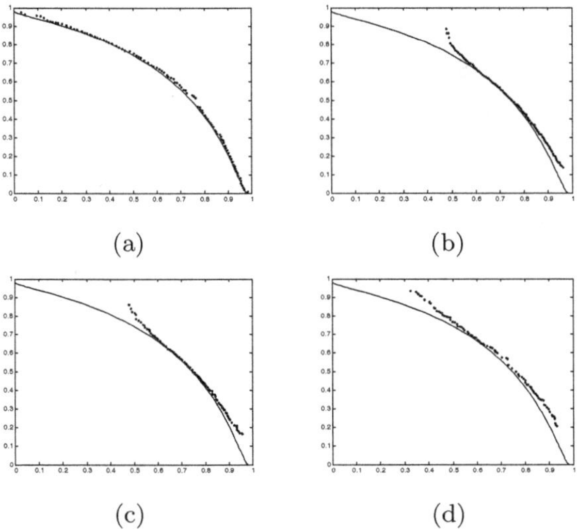

Fig. 6.5 The evolved Pareto front from (a) COEA, (b) CCEA, (c) NSGAII, and (d) SPEA2 for FON

settings are listed in Table 6.2. All the algorithms here are implemented using the same binary coding scheme, tournament selection, uniform crossover, and bit flip mutation.

FON

The FON challenges the algorithms' ability to find and maintain the entire tradeoff curve uniformly. Since the tradeoff curve is non-convex and non-linear in FON, it is difficult to maintain a stable evolving population for this problem. A stopping criterion of 20,000 evaluations is used here. The PFs obtained from the different algorithms using the same random seed are shown in Fig. 6.5(a)-(d), while the distributions of the different performance metrics are represented by box plots in Fig. 6.6(a)-(d). The advantage of the proposed competitive-cooperation model in handling parameter interactions is shown in Fig. 6.6 and by comparing the evolved PF in Fig. 6.5(a) and Fig. 6.5(b).

KUR

The KUR is characterized by PF^* that is non-convex and disconnected, which contains three distinct and disconnected regions on the final trade-off. The decision variables corresponding to the final tradeoff for KUR are difficult to find since they are disconnected in the decision variable space. Like FON, there are high interactions between the decision variables, which pose a

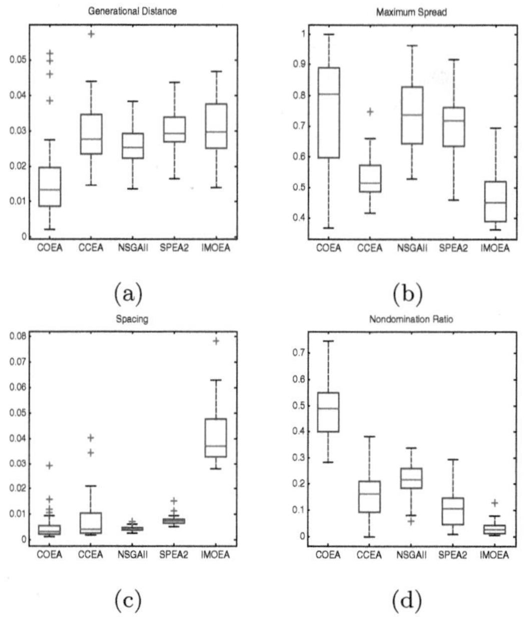

Fig. 6.6 Performance metric of (a) GD, (b) MS, (c) S, and (d) NR for FON

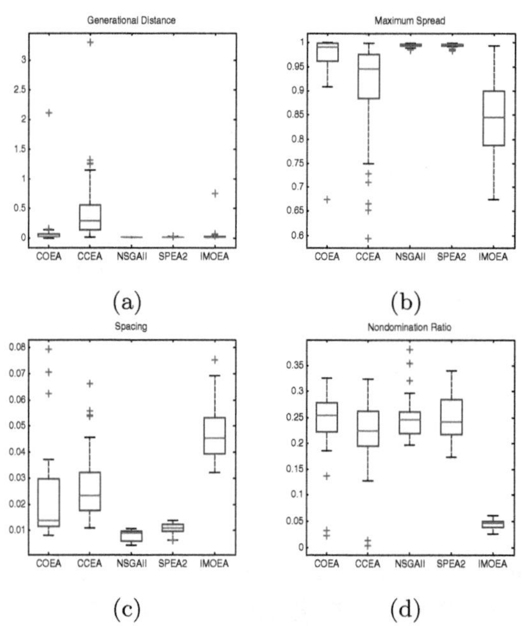

Fig. 6.7 Performance metric of (a) GD, (b) MS, (c) S, and (d) NR for KUR

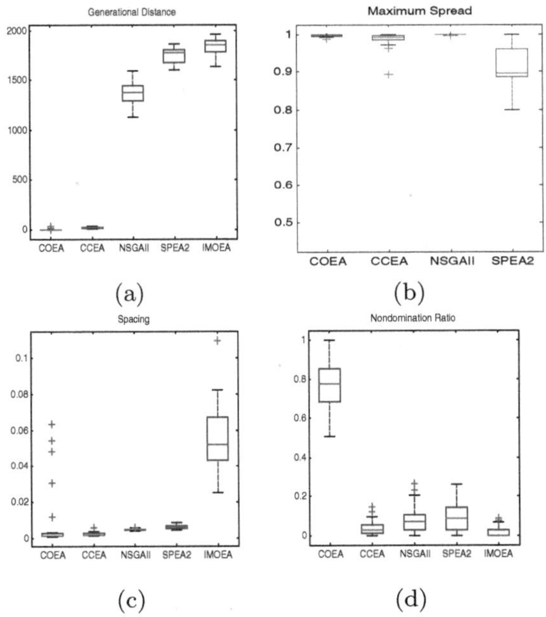

Fig. 6.8 Performance metric of (a) GD, (b) MS, (c) S, and (d) NR for DTLZ3

challenge to MOCAs. A stopping criterion of 30,000 evaluations is used for this problem. The distributions of the different performance metrics are represented by box plots in Fig. 6.7(a)-(d). The main difficulty stemming from the high parameter interactions in this problem is the finding of all the four disconnected regions of PF. Although CCEA is capable of evolving a PF that is close to PF*, it can be observed from Fig. 6.7(b) and Fig. 6.7(c) that it faces difficulty in finding a diverse PF. As shown in the metric of MS, the competitive-cooperation paradigm allows COEA to evolve a more diverse solution set as compared to the CCEA.

DTLZ3

DTLZ3 is used to challenge the MOEA's capability to produce adequate pressure in driving individuals towards the high-dimensional PF*. Moreover, the DTLZ3 is also characterized by the challenge of multi-modality. A stopping criterion of 28,000 evaluations is used for this problem. The distributions of the different performance metrics for DTLZ3 are shown in Fig. 6.8(a)-(d). It can be observed that although SPEA2 and NSGAII are unable to find good solutions near the PF*, they manage to evolve a good spread of solutions. On the other hand, the COEA is seen to scale well with increasing objectives and to produce competitive performance for GD, S, and MS. The metric of NR also shows that the COEA outperforms CCEA as given in Fig. 6.8(d).

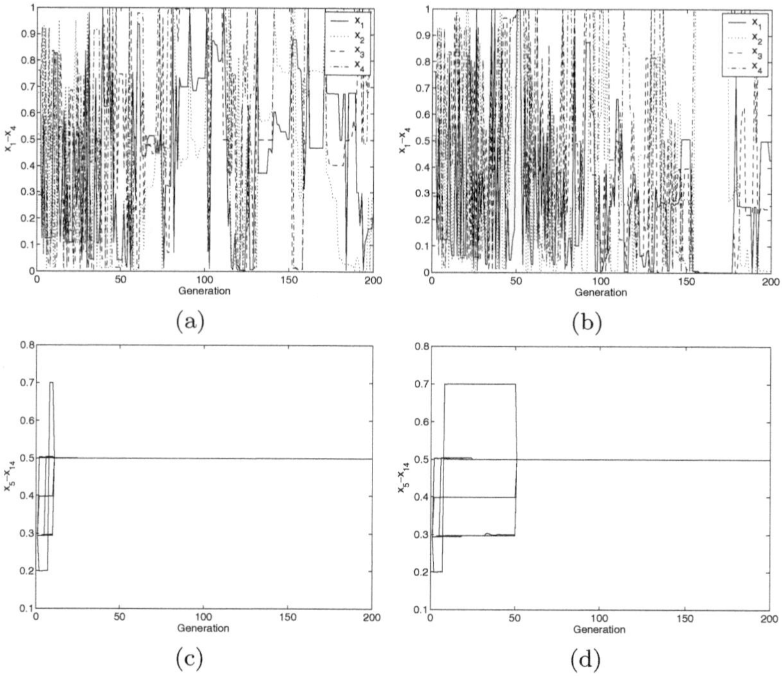

Fig. 6.9 Evolutionary trend of variables (a) $x_1 - x_4$ for $C_{freq} = 10$, (b) $x_1 - x_4$ for $C_{freq} = 50$, (c) $x_5 - x_{14}$ for $C_{freq} = 10$ and (d) $x_5 - x_{14}$ for $C_{freq} = 50$. $x_5 - x_{14}$ are all represented by straight lines.

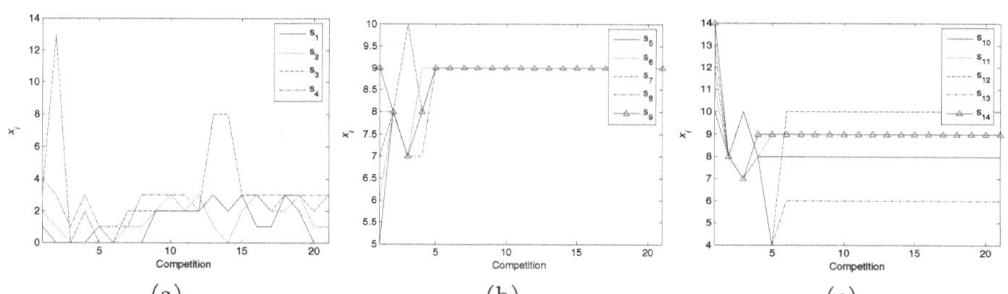

Fig. 6.10 Trace of the winning sub-population

6.4.2 *Effects of the Competitive Mechanism*

In this section, experiments are conducted at C_{freq}={1, 5, 10, 30, 50, inf} to study the effects and dynamics of incorporating both competitive and cooperative processes in a common framework based on the benchmark problems of FON, KUR, and DTLZ3. As mentioned earlier, the FON and KUR have severe parameter interactions, which are useful to examine the performance improvement of the competitive mechanism in COEA. The DTLZ3

Table 6.3 Performance of COEA for FON with different C_{freq}. The best results are highlighted in bold.

		1	5	10	30	50	Inf
GD	1st quartile	0.0080	**0.0050**	0.0086	0.0107	0.0119	0.0235
	Median	0.0116	**0.0075**	0.0133	0.0157	0.0207	0.0276
	3rd quartile	0.0171	**0.0090**	0.0198	0.0217	0.0243	0.0347
MS	1st quartile	**0.9492**	0.5394	0.5991	0.6313	0.6121	0.4857
	Median	**0.9741**	0.8916	0.8036	0.7510	0.6882	0.5159
	3rd quartile	**0.9975**	0.9466	0.8891	0.8547	0.7280	0.5732

Table 6.4 Performance of COEA for KUR with different C_{freq}. The best results are highlighted in bold.

		1	5	10	30	50	Inf
GD	1st quartile	0.0349	**0.0256**	0.0329	0.0370	0.0521	0.1414
	Median	0.0425	**0.0365**	0.0376	0.0864	0.2946	0.2941
	3rd quartile	**0.0499**	0.0549	0.0807	0.3078	0.4924	0.5592
MS	1st quartile	**0.9995**	0.9822	0.9608	0.9458	0.9214	0.8841
	Median	**0.9998**	0.9939	0.9902	0.9678	0.9610	0.9461
	3rd quartile	**1.0000**	0.9988	0.9987	0.9906	0.9730	0.9752

Table 6.5 Performance of COEA for DTLZ3 with different C_{freq}. The best results are highlighted in bold.

		1	5	10	30	50	Inf
GD	1st quartile	28.6021	0.0000	0.0000	0.0000	0.0000	15.0409
	Median	58.4115	0.0039	0.0009	0.0000	0.0000	18.4015
	3rd quartile	100.8232	0.0252	**0.0248**	0.0271	0.1414	23.4576
MS	1st quartile	0.6744	**0.9990**	0.9972	0.9950	0.9958	0.9860
	Median	0.7575	**0.9998**	0.9990	0.9987	0.9979	0.9933
	3rd quartile	0.8702	**1.0000**	0.9998	0.9996	0.9995	0.9986

is included here since most algorithms are unable to deal with this problem effectively as observed in the previous section.

The performances of COEA with C_{freq}={1, 5, 10, 30, 50, inf} for FON, KUR and DTLZ3 are summarized in Table 6.3, Table 6.4, and Table 6.5, respectively. Note that no competition takes place when C_{freq}=inf, which effectively reduces the competitive-cooperative paradigm to a conventional cooperative model. From the tables, it can be observed that COEA gives better performances for the three benchmark problems at lower settings of

C_{freq} and the performances deteroriate when the competitive mechanism is absent. By comparing the results over different C_{freq}, it is clear that a larger C_{freq} allows the COEA to adapt faster to the problem requirements and to evolve a more diverse and near optimal PF. On the other hand, it can be seen that the improvement for MS is attained at the expense of GD for FON and KUR. In the case of DTLZ3, the algorithmic performance deteriorates sharply at $C_{freq} = 1$ since constant competition may restrict the time necessary for the sub-populations to adapt to the decision variables. Nonetheless, it can be observed that the incorporation of competitive mechanism with reasonable C_{freq} can result in significant improvement of convergence and diversity for the problems of FON, KUR, and DTLZ3.

Fig. 6.9 shows the evolutionary trend of the best solution for each variable in DTLZ3 with $C_{freq} = 10$ and $C_{freq} = 50$. In order to evolve a near-optimal, diverse and uniformly distributed PF, the algorithm needs to maintain a wide range of values for x_1-x_4, while finding the optimal value of 0.5 for x_5-x_{14}. For both settings, it can be seen that x_1-x_4 oscillate continuously along the evolution process in order to span the entire range of feasible values. Likewise, x_5-x_{14} are able to converge to the optimal value of 0.5 as shown in Fig. 6.9(c)-(d). By comparing Fig. 6.9(a) and Fig. 6.9(b), it can be seen that the COEA with $C_{freq} = 10$ converges to the optimal value of 0.5 at the 10th generation. However, the COEA with $C_{freq} = 50$ only converges to the optimal value at the 50th generation. It is also observed that the convergence of the algorithm coincides with each competition process in the evolution.

To analyze the influence of the competitive mechanism on the emergent decomposition process, the winning sub-population for each round of the competition is shown in Fig. 6.10. To facilitate the introduction of diversity for variables x_1-x_{14}, it is observed that S_1-S_3 emerge as the most suitable sub-populations and each takes over the role of optimizing a variable within x_1-x_4 in an almost iterative manner. For the variables x_5-x_{14}, it is observed that S_8 took over the rest of the sub-populations at the first competition. Although subsequent winners include S_4, S_7, S_8, S_9, and S_{10}, S_9 is shown to be the dominant sub-population for these variables. It can also be observed from the sub-population distribution that individuals of S_1-S_3 are distributed throughout the search space, while individuals of S_4-S_{14} are concentrated around the value of 0.5.

6.4.3 Effects of Different Competition Schemes

In this section, three different competition models are incorporated in COEA and their effectiveness for multi-objective optimization are investigated. These models are as follows,

- Random: Before the start of each competition process, an individual is selected randomly from each competing sub-population as the participant. This set of competitors will remain fixed during the entire course of the competition for that particular sub-component. This scheme is implemented in the COEA adopted for the comparative study in the previous section.

- Elitist: Before the start of each competition process, each competing subpopulation selects the best individual for its associated sub-components as the participant. This set of competitors will remain fixed during the whole course of the competition for that particular sub-component. This scheme is expected to perform well when the different sub-components have similar properties.
- Hybrid: Before the start of each competition process, each competing subpopulation randomly selects either the best individual or a random individual as the participant. This set of competitors will remain fixed during the entire course of the competition for that particular sub-component. This model represents a tradeoff between the random and the elitist scheme.

The experiments are conducted for COEA having different competition schemes with $C_{freq} = 10$. The results of 30 independent runs for the problems of FON, KUR, and DTLZ3 are summarized in Table 6.6, Table 6.7, and Table 6.8, respectively. It can be seen that the elitist scheme is capable of evolving the PF with a good convergence for all the three problems. It also gives the best performance in the metric of GD for DTLZ3. This result is expected since the optimal values for variables x_5-x_{14} are identical and the elitist scheme is able to exploit this relationship quickly. On the other hand, it is observed that the random scheme and the hybrid scheme demonstrate better performances when parameter interactions are present. The limitation of high selection pressure introduced by the elitist scheme is also evident from the relatively poor performance in the metric of MS for all problems. Although the random scheme demonstrates the best performance for KUR where the PS* is discontinuous in the decision space, it produces relatively poor convergence results for FON and DTLZ3. It can also be seen that the hybrid scheme provides competitive results in all cases, and gives the best performance in the metric of MS for FON and DTLZ3.

Table 6.6 Performance of COEA for FON with different competition models. The best results are highlighted in bold.

		Random	Elitist	Hybrid
	1st quartile	0.0086	**0.0069**	0.0071
GD	Median	0.0133	**0.0083**	0.0102
	3rd quartile	0.0198	0.0190	**0.0158**
	1st quartile	0.5991	**0.671203**	0.6646
MS	Median	0.8036	0.7565	**0.8288**
	3rd quartile	0.8891	0.8796	**0.9125**

Table 6.7 Performance of COEA for KUR with different competition models. The best results are highlighted in bold.

		Random	Elitist	Hybrid
GD	1st quartile	0.0329	**0.0264**	0.0306
	Median	**0.0376**	0.0400	0.0537
	3rd quartile	**0.0807**	0.1056	0.0918
MS	1st quartile	**0.9608**	0.8244	0.9491
	Median	**0.9902**	0.9691	0.9868
	3rd quartile	**0.9986**	0.9948	0.9955

Table 6.8 Performance of COEA for DTLZ3 with different competition models. The best results are highlighted in bold.

		Random	Elitist	Hybrid
GD	1st quartile	0.0000	0.0000	0.0000
	Median	**0.0009**	0.0010	0.0033
	3rd quartile	0.0248	**0.0125**	0.0172
MS	1st quartile	**0.9972**	0.9956	0.9956
	Median	0.9990	0.9979	**0.9995**
	3rd quartile	0.9998	0.9992	**0.9999**

The elitist scheme is the greediest method which may restrict the exploration of possible relationships among the variables. This explains the reason that it performs well for problems with low variable interactions but provides relatively poor results for problems with high parameter interactions. In contrast, the random scheme is the least greedy approach which allows it to consider the different parameter relationships for maintaining diverse solutions in the evolution. Hence, it performs well for problems with high parameter interactions but the random nature of the competitor selection also makes it incapable of exploiting the characteristic that the optimal solutions for FON and DTLZ3 lie in the same region. Nonetheless, it is also this property that allows the random scheme to evolve a more diverse PF as compared to the elitist scheme. On the other hand, the hybrid scheme demonstrates features of both the random and elitist schemes, thus allowing it to attain competitive results that are at least comparable to the other two schemes. Although the three competition schemes behave differently for the different problems, the proposed coevolutionary model produces better performance as compared to conventional approaches. Note that these three schemes are only examples of how different competition models can be applied and other variants can also be considered if so desired.

6.5 Dynamic Environment Empirical Study

6.5.1 Comparative Study

To examine the performance of dCOEA, two different dynamic MOEAs based on a basic MOEA and the CCEA, respectively, are adopted in this study. The basic MOEA is similar to the model presented in Chapter 2. The algorithm employs a fixed-size population and an archive to store non-dominated solutions along the evolution. The archive is updated at each cycle, i.e. a candidate solution will be added to the archive if it is not dominated by any member in the archive. Likewise, any archive member dominated by this solution will be removed from the archive. When the predetermined archive size is reached, a recurrent truncation process based on niche count is used to eliminate the most crowded archive member. Elitism is implemented by selecting individuals to a mating pool through binary tournament selection of the combined archive and evolving population. The selection criterion is based on Pareto rank and niche count is used in the event of a tie. In both the dynamic MOEA (dMOEA) and dynamic CCEA (dCCEA), a fixed number of archived solutions are re-evaluated in every generation. In the case where a change in the landscape is detected, the temporal memory described previously will be applied and random restart is incorporated to generate diversity within the evolving population.

The parameter settings for the different algorithms are tabulated in Table 6.9. Thirty independent simulation runs with randomly generated initial populations are performed for each of the test problems. The experiments are conducted at different severity levels of $n_T = \{1, 10, 20\}$ and different frequencies of $\tau_T = \{5, 10, 25\}$ so as to study the impact of dynamics in uncertain environments. Since each generation involves 100 function

Table 6.9 Parameter setting for different algorithms

Parameter	Settings
Populations	Population size 100 in dMOEA; sub-population size 10 in dCOEA and dCCEA; Archive (or secondary population) size 100.
Chromosome	Binary coding; 30 bits per decision variable.
Selection	Binary tournament selection
Crossover operator	Uniform crossover
Crossover rate	0.8
Mutation operator	Bit-flip mutation
Mutation rate	$\frac{1}{L}$ for FDA1, dMOP1, dMOP2 and dMOP3;
Niche Radius	Dynamic sharing.
Evaluation number	20,000

Table 6.10 Performance of MOEA, dCCEA, and dCOEA for FDA1 at different settings of τ_T and n_T. The best results are highlighted in bold only if it is statistically different based on the KS test.

(τ_t, n_T)		$VD_{offline}$ MOEA	dCCEA	dCOEA	$MS_{offline}$ MOEA	dCCEA	dCOEA
(5,10)	1st quartile	0.666	0.243	**0.107**	0.789	0.829	**0.939**
	Median	0.683	0.255	**0.110**	0.801	0.834	**0.944**
	3rd quartile	0.695	0.264	**0.113**	0.801	0.841	**0.953**
(10,10)	1st quartile	0.489	0.154	**0.034**	0.870	0.863	**0.963**
	Median	0.508	0.163	**0.038**	0.878	0.873	**0.970**
	3rd quartile	0.521	0.167	**0.039**	0.890	0.882	**0.977**
(25,10)	1st quartile	0.485	0.080	**0.001**	0.876	0.926	**0.979**
	Median	0.528	0.091	**0.002**	0.894	0.939	**0.985**
	3rd quartile	0.583	0.102	**0.003**	0.914	0.947	**0.989**
(10,1)	1st quartile	1.008	0.135	**0.020**	0.535	0.857	**0.973**
	Median	1.031	0.149	**0.022**	0.585	0.866	**0.981**
	3rd quartile	1.064	0.156	**0.025**	0.599	0.883	**0.984**
(10,20)	1st quartile	0.542	0.152	**0.039**	0.847	0.858	**0.970**
	Median	0.584	0.162	**0.042**	0.868	0.875	**0.975**
	3rd quartile	0.606	0.171	**0.044**	0.881	0.888	**0.979**

evaluations, the setting of $\tau_T = 5$ implies a change of the landscape in every 500 evaluations. In this study, SC_{ratio} and R_{size} are set as 0.5 and 5, respectively.

FDA1

The FDA1 challenges the dynamic MOEAs' ability to track and converge towards the PF_t^* with every landscape change. One interesting characteristic of this problem is that the distribution and diversity of the solutions along the PF_t are not affected by the landscape change. The simulation results for $VD_{offline}$ and $MS_{offline}$ with various settings of τ_T and n_T are summarized in Table 6.10. In general, the coevolutionary paradigm is shown to be more appropriate than canonical MOEAs in handling dynamic landscapes. As can be seen, the dCOEA outperforms dCCEA in both the aspect of tracking and finding a diverse solution set. Table 6.10 also shows a better convergence and diversity performance of dMOEA, dCCEA and dCOEA for larger value of τ_T or less frequent landscape changes. Although the dMOEA gives a better convergence for larger value of n_T or less severe landscape changes, it is observed that better results of dCCEA and dCOEA can be obtained with a more severe landscape change.

Table 6.11 Performance of MOEA, dCCEA, and dCOEA for dMOP1 different settings of τ_T and n_T. The best results are highlighted in bold only if it is statistically different based on the KS test.

		$VD_{offline}$			$MS_{offline}$		
(τ_t, n_T)		MOEA	dCCEA	dCOEA	MOEA	dCCEA	dCOEA
(5,10)	1st quartile	0.114	0.230	**0.005**	0.891	0.825	**0.977**
	Median	0.128	0.242	**0.007**	0.911	0.838	**0.983**
	3rd quartile	0.137	0.252	**0.008**	0.933	0.846	**0.989**
(10,10)	1st quartile	0.103	0.111	**0.002**	0.916	0.880	**0.988**
	Median	0.114	0.121	**0.003**	0.916	0.880	**0.988**
	3rd quartile	0.131	0.132	**0.004**	0.935	0.888	**0.994**
(25,10)	1st quartile	0.065	0.023	**0.001**	0.916	0.931	**0.989**
	Median	0.077	0.026	**0.00**	0.940	0.948	**0.991**
	3rd quartile	0.093	0.030	**0.001**	0.962	0.962	**0.996**
(10,1)	1st quartile	0.106	0.120	**0.002**	0.891	0.870	**0.986**
	Median	0.116	0.126	**0.003**	0.914	0.877	**0.990**
	3rd quartile	0.128	0.137	**0.004**	0.934	0.893	**0.992**
(10,20)	1st quartile	0.101	0.115	**0.002**	0.904	0.871	**0.982**
	Median	0.117	0.123	**0.003**	0.921	0.881	**0.988**
	3rd quartile	0.130	0.133	**0.003**	0.939	0.890	**0.993**

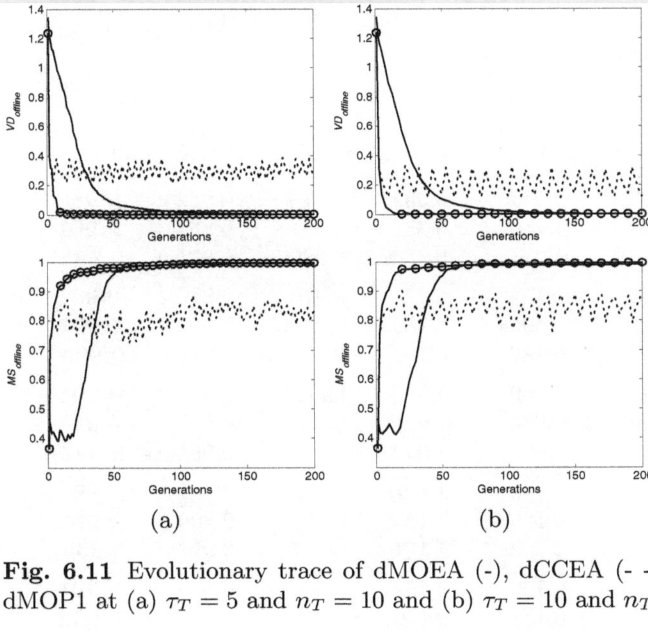

Fig. 6.11 Evolutionary trace of dMOEA (-), dCCEA (- -), and dCOEA (o) for dMOP1 at (a) $\tau_T = 5$ and $n_T = 10$ and (b) $\tau_T = 10$ and $n_T = 10$.

dMOP1

Unlike FDA1, the convexity of dMOP1 changes with time while the location of PS* remains fixed. The dMOP1 challenges the dynamic MOEA's ability to maintain a diverse PF_t^* with every landscape change. The simulation results for $VD_{offline}$ and $MS_{offline}$ with various settings of τ_T and n_T are summarized in Table 6.11. Similar to the problem of FDA1, the dCOEA outperforms dMOEA and dCCEA in both tracking and finding a diverse solution set. It is also observed that the dMOEA produces better results than dCCEA for the settings of $\tau_T = 5$ and $\tau_T = 10$. The evolutionary trace of $VD_{offline}$ and $MS_{offline}$ for these settings are shown in Fig. 6.11. While the dCOEA and dCCEA behave similarly in the initial generations before the first landscape change, it can be seen that the dCCEA is greatly affected by the change in the shape of PF. On the other hand, the dMOEA is capable of finding the PS_t^* and a diverse PF_t despite having a slower convergence. Based on previous studies in dynamic single-objective optimization, diversity schemes, like random restart, tend to perform poorly in situation where the change is small. As compared to the problem of FDA1, the severity of change has less impact on the metric of $VD_{offline}$ for these three algorithms. This is due to the incorporation of temporal memory that allows the algorithm to rediscover the PS* quickly, even though random restart is utilized in the dMOEA and dCCEA.

Table 6.12 Performance of MOEA, dCCEA, and dCOEAS for dMOP2 at different settings of τ_T and n_T. The best results are highlighted in bold only if it is statistically different based on the KS test.

(τ_t, n_T)		$VD_{offline}$ MOEA	$VD_{offline}$ dCCEA	$VD_{offline}$ dCOEA	$MS_{offline}$ MOEA	$MS_{offline}$ dCCEA	$MS_{offline}$ dCOEA
(5,10)	1st quartile	0.642	**0.285**	0.352	0.973	0.852	**0.988**
	Median	0.666	**0.291**	0.372	0.981	0.861	**0.991**
	3rd quartile	0.680	**0.300**	0.384	0.986	0.871	**0.994**
(10,10)	1st quartile	0.495	0.159	0.173	0.976	0.886	**0.991**
	Median	0.517	0.169xx	0.180	0.980	0.902	**0.993**
	3rd quartile	0.535	0.187	0.192	0.987	0.915	**0.996**
(25,10)	1st quartile	0.462	0.069	**0.059**	0.9817	0.949	**0.991**
	Median	0.514	0.075	**0.063**	0.989	0.958	**0.994**
	3rd quartile	0.557	0.093	**0.071**	0.993	0.964	**0.997**
(10,1)	1st quartile	1.137	0.176	**0.140**	0.965	0.881	**0.991**
	Median	1.166	0.186	**0.152**	0.978	0.899	**0.996**
	3rd quartile	1.188	0.202	**0.176**	0.985	0.912	**0.998**
(10,20)	1st quartile	0.466	0.166	0.162	0.966	0.889	**0.991**
	Median	0.487	0.177	0.170	0.979	0.899	**0.992**
	3rd quartile	0.519	0.185	0.184	0.986	0.916	**0.996**

dMOP2

The convexity and PS_t^* of dMOP2 change with time, which challenge the dynamic MOEA's ability to track the PS_t^* and to maintain a diverse PF_t^* with every landscape change. The simulation results for $VD_{offline}$ and $MS_{offline}$ with various settings of τ_T and n_T are summarized in Table 6.12. In contrast to the previous two problems, the dCOEA is outperformed by dCCEA for $VD_{offline}$ when (τ_T, n_T) is set as (5,10) and (10,10). Since random restart is applied in dCOEA, further investigations in the next section demonstrate that a lower SC_{ratio} allows the dCOEA to give a better performance. On the other hand, the dCOEA outperforms both dMOEA and dCCEA in tracking and maintaining better diversity for other parameter settings. By comparing the values of $MS_{offline}$ in Table 6.10, Table 6.11, and Table 6.11, it can be observed that dCCEA is unable to find a diverse PF_t when the shape of PF_t^* is dynamic.

dMOP3

Although the dMOP3 has similar characteristics as FDA1, the variable that determines the spread of the solution set in dMOP3 is not fixed and changes with time. The dynamic MOEA thus faces an additional challenge in tracking

Table 6.13 Performance of MOEA, dCCEA, and dCOEAS for dMOP3 at different settings of τ_T and n_T. The best results are highlighted in bold only if it is statistically different based on the KS test.

(τ_t, n_T)		$VD_{offline}$ MOEA	dCCEA	dCOEA	$MS_{offline}$ MOEA	dCCEA	dCOEA
(5,10)	1st quartile	0.679	0.226	**0.083**	0.619	0.824	**0.906**
	Median	0.701	0.2398	**0.087**	0.637	0.835	**0.913**
	3rd quartile	0.727	0.249	**0.09**	0.658	0.841	**0.927**
(10,10)	1st quartile	0.460	0.140	**0.013**	0.802	0.856	**0.943**
	Median	0.482	0.149	**0.017**	0.822	0.867	**0.957**
	3rd quartile	0.507	0.162	**0.021**	0.843	0.880	**0.965**
(25,10)	1st quartile	0.424	0.068	**0.001**	0.903	0.927	**0.976**
	Median	0.467	0.078	**0.002**	0.914	0.9338	**0.983**
	3rd quartile	0.515	0.096	**0.003**	0.927	0.949	**0.987**
(10,1)	1st quartile	1.055	0.129	**0.011**	0.505	0.861	**0.977**
	Median	1.087	0.138	**0.014**	0.539	0.873	**0.981**
	3rd quartile	1.108	0.15	**0.018**	0.565	0.886	**0.987**
(10,20)	1st quartile	0.477	0.138	**0.019**	0.837	0.855	**0.946**
	Median	0.505	0.147	**0.022**	0.857	0.865	**0.954**
	3rd quartile	0.538	0.155	**0.025**	0.866	0.883	**0.966**

the diverse PF_t^* for this problem. The simulation results for $\mathrm{VD}_{offline}$ and $\mathrm{MS}_{offline}$ with various settings of τ_T and n_T are summarized in Table 6.13. By comparing the results in Table 6.10 and Table 6.11, it can be seen that the three algorithms give poorer performances for $\mathrm{MS}_{offline}$ with $\tau_T = 5$ and $\tau_T = 10$. Nonetheless, similar to the problems of FDA1 and dMOP2, it is observed that the dCOEA outperforms dMOEA and dCCEA in both tracking and finding a diverse solution set for the different settings of τ_T and n_T.

6.5.2 Effects of Stochastic Competitors

The SC_{ratio} determines the degree of diversity introduced in the proposed dCOEA after every landscape change for good tracking performance. The relationships between $SC_{ratio} = \{0.3, 0.5, 0.7, 1\}$ and various settings of n_T and τ_T for FDA1 are shown in Fig. 6.12. Note that no stochastic competitor is introduced when $SC_{ratio} = 1$. These relationships are similarly investigated for the problems of dMOP1, dMOP2, and dMOP3. The trends for dMOP1 are illustrated in Fig. 6.13, while the plots for dMOP2 and dMOP3 are omitted here since similar traits with FDA1 have been exhibited.

It can be seen from the metric of $\mathrm{MS}_{offline}$ in Fig. 6.12 and Fig. 6.13 that the diversity of the evolved PF_t improves generally with the introduction of stochastic competitors. On the other hand, Fig. 6.13 shows that the tracking performance deteriorates with increasing diversity for dMOP1. This is because the location of PS_t^* remains unchanged for this problem and it is improbable that the new set of non-dominated solutions introduced by the stochastic competitors will be better or even comparable to the archived solutions before any landscape change. Nonetheless, it is clear that stochastic competitors play an important role in the tracking of dynamic PS_t^* for the problems of FDA1, dMOP2, and dMOP3.

It can also be observed from Fig. 6.12 that the dCOEA produces the best results when $n_T = 1$ for FDA1. In contrast, Fig. 6.13 shows that the dCOEA gives the worst results for dMOP1 for the same setting of $n_T = 1$. This observation is similar to past findings from dynamic single-objective optimization that a higher degree of diversity is required with severe changes in the environment.

6.5.3 Effects of Temporal Memory

The R_{size} determines the extent in which information about past PS_t^* is stored. A larger R_{size} results in a higher degree of information exploitation at the expense of a more diverse repertoire of past PS_t^*. On the other hand, limited information regarding past PS_t^* is avaliable when R_{size} is small. The relationships between $R_{size} = \{0, 5, 10, 20\}$ and various settings of n_T and τ_T for FDA1 are shown in Fig. 6.14. Note that no memory is kept when

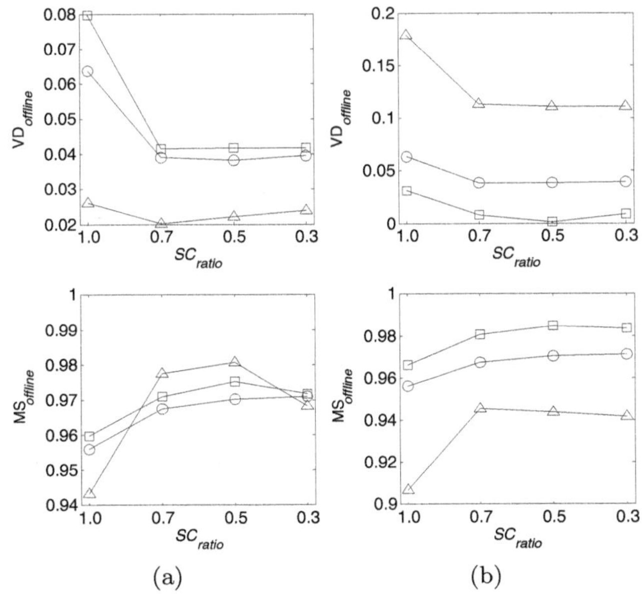

Fig. 6.12 Performance metrics of $\mathrm{VD}_{offline}$ (top) and $\mathrm{MS}_{offline}$ (bottom) for FDA1 over different settings of SC_{ratio} (a) at n_t=1 (△), n_t=10 (○), and n_t=20 (□) and (b) τ_T=5 (△), τ_T=10 (○), and τ_T=25 (□)

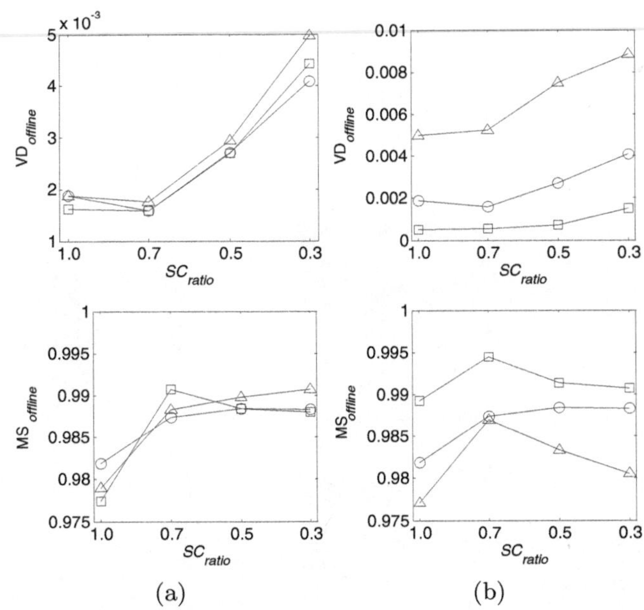

Fig. 6.13 Performance metrics of $\mathrm{VD}_{offline}$ (top) and $\mathrm{MS}_{offline}$ (bottom) for dMOP1 over different settings of SC_{ratio} (a) at n_t=1 (△), n_t=10 (○), and n_t=20 (□) and (b) τ_T=5 (△), τ_T=10 (○), and τ_T=25 (□)

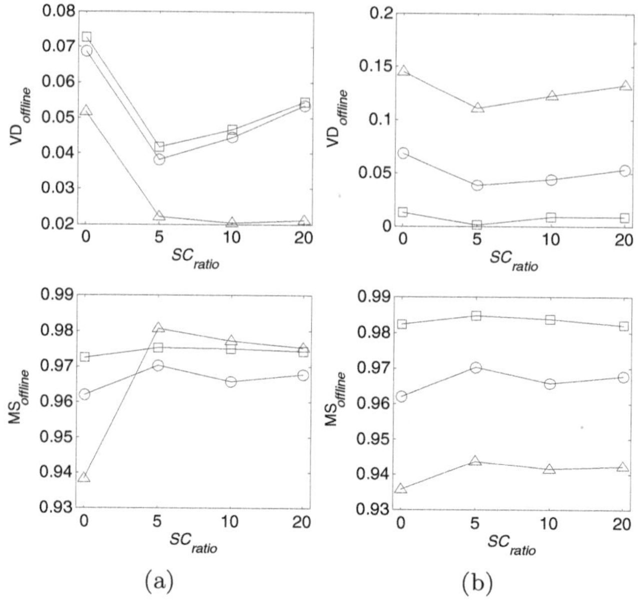

Fig. 6.14 Performance metrics of $\text{VD}_{offline}$ (top) and $\text{MS}_{offline}$ (bottom) for FDA1 over different settings of R_{size} at (a) n_t=1 (△), n_t=10 (○), and n_t=20 (□) and (b) τ_T=5 (△), τ_T=10 (○), and τ_T=25 (□)

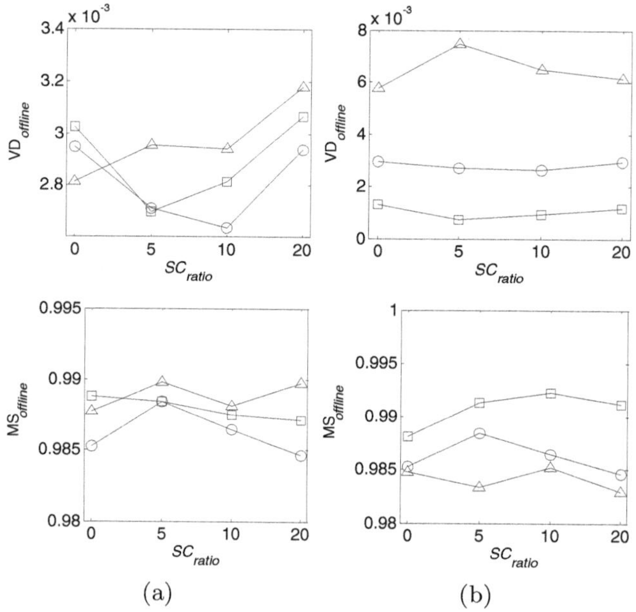

Fig. 6.15 Performance metrics of $\text{VD}_{offline}$ (top) and $\text{MS}_{offline}$ (bottom) for dMOP1 over different settings of R_{size} at (a) n_t=1 (△), n_t=10 (○), and n_t=20 (□) and (b) τ_T=5 (△), τ_T=10 (○), and τ_T=25 (□)

$R_{size} = 0$. These relationships are similarly investigated for dMOP1, dMOP2, and dMOP3. The trends for dMOP1 are shown in Fig. 6.15, while the plots for dMOP2 and dMOP3 are excluded here since similar traits have been observed.

Similar to the findings in Section 6.5.2, better tracking performances are observed for higher τ_T and at $n_T = 1$ in FDA1, dMOP2, and dMOP3 for the different settings of R_{size}. Fig. 6.15 illustrates that the incorporation of temporal memory tends to improve convergence according to the metric of $\text{VD}_{offline}$. The only exception occurs at the setting of $n_T = 1$ and $\tau_T = 5$ for dMOP1. The tradeoff between the exploration and exploitation of information is also evident from the figures with increasing R_{size}. For instance, when the repetition of similar PS_t^* is very frequent for $n_T = 1$, a large R_{size} can be used to mine information from past PS_t^* since the number of different PS_t^* that needs to be represented in the memory is small and vice versa.

6.6 Conclusion

This chapter presented a new coevolutionary paradigm that incorporates both competitive and cooperative mechanisms observed in nature to solve multi-objective optimization problems and to track the Pareto front in a dynamic environment. The proposed competitive-cooperation coevolution is capable of overcoming the limitations of conventional coevolutionary models by allowing the decomposition process of the optimization problem to emerge based on problem requirements as well as exploiting the high speed of convergence to allow the algorithm to adapt quickly to the changing environment. Based on this coevolutionary model, a competitive-cooperation coevolutionary algorithm (COEA) is proposed for multi-objective optimization. Subsequently, this algorithm is extended as a dynamic COEA (dCOEA) and incorporated the features of stochastic competitors, which allows the algorithm to track the changing solution set, and temporal memory, which allows the algorithm to exploit past information. Extensive studies upon three benchmark problems demonstrate that COEA is capable of evolving near-optimal, diverse, and uniformly distribution Pareto fronts even for problems with severe parameter interactions. The parameter settings and working dynamics of the competitive mechanism as well as different competitive schemes are also examined, illustrating the robustness and importance of both competitive and cooperative elements in a common framework. Likewise, extensive studies are performed to investigate the performances of dCOEA over different settings of change severity and change frequency. Simulation result shows that dCOEA is capable of tracking the different environmental changes effectively and efficiently. In addition, the contributions and parameter settings of the diversity scheme and the temporal memory are also analyzed over various problem settings.

Part III
Evolving Robust Solution Sets

Chapter 7
Robust Evolutionary Multi-objective Optimization*

Branke [30] considered robust optimization as a special case of dynamic optimization, where solutions cannot be adapted fast enough to keep in pace with environmental changes. In such cases, it would be desirable to find solutions that perform reasonably well within some range of change. In fact, many real-world applications involve the simultaneous optimization of several competing objectives and are susceptible to decision or environmental parameter variations, which result in large or unacceptable performance variations. Robust optimization of multi-objective problems is the third and final type of uncertainty considered in this work and it involves the optimization of a set of Pareto optimal solutions that remains satisfactory in face of parametric variations.

This chapter addresses the issue of robust multi-objective optimization by presenting a robust continuous multi-objective test suite with features of noise-induced solution space, fitness landscape and decision space variation. In addition, the vehicle routing problem with stochastic demand (VRPSD) is presented a practical example of robust combinatorial multi-objective optimization problems.

7.1 Robust Multi-objective Optimization Problems

Apart from the multi-objective optimization goals of evolving a set of near optimal, diverse, and uniformly distributed Pareto solution set, robust multi-objective optimization sought to find a set of tradeoff solutions that is robust or reliable [59]. A robust solution can be defined as a solution that provides satisfactory performance in face of parametric variations, i.e it is insensitive to small variations in design and/or environment variables.

C.-K. Goh and K.C. Tan: Evolutionary Multi-objective Optimization, SCI 186, pp. 189–211.
springerlink.com

Definition 7.1. Robust Pareto Optimal Front: The robust Pareto optimal front is the set of objective vectors that are non-dominated with respect to the optimization objectives, the adopted robust measure, and the applied noise intensity.

Definition 7.2. Robust Pareto Optimal Set: The robust Pareto optimal set is the set of solutions whose corresponding objective vectors are non-dominated with respect to the optimization objectives, the adopted robust measure, and the applied noise intensity.

In order to reduce the consequences of uncertainties on optimality and practicality of the solution set, factors, such as decision variable variation, environmental variation, and modeling uncertainty, have to be considered explicitly. Therefore, the minimization multi-objective problem is redefined as the following.

$$\begin{aligned} \min \mathbf{f}(\mathbf{x}, \boldsymbol{\delta}_\mathbf{x}, \boldsymbol{\delta}_\mathbf{e}) = \{&f_1(\mathbf{x}, \boldsymbol{\delta}_\mathbf{x}, \boldsymbol{\delta}_\mathbf{e}), f_2(\mathbf{x}, \boldsymbol{\delta}_\mathbf{x}, \boldsymbol{\delta}_\mathbf{e}), \\ &..., f_M(\mathbf{x}, \boldsymbol{\delta}_\mathbf{x}, \boldsymbol{\delta}_\mathbf{e})\} \end{aligned} \tag{7.1}$$

$$\text{s.t. } \mathbf{g}(\mathbf{x}, \boldsymbol{\delta}_\mathbf{x}, \boldsymbol{\delta}_\mathbf{e}) \geq 0, \mathbf{h}(\mathbf{x}, \boldsymbol{\delta}_\mathbf{x}, \boldsymbol{\delta}_\mathbf{e}) = 0$$

where δ_x and δ_e represent, respectively, the uncertainties associated with $\mathbf{x}$ and environmental conditions. Both forms of uncertainties may be treated equivalently. Different noise models, such as normal, Cauchy, and uniform distributions, have been considered in the literature.

In order to avoid any confusion in the subsequent discussions, it will be instructive to make a distinction between the notations used for deterministic multi-objective and robust multi-objective optimization. The terms PF* and PS* refer to the desired Pareto front and solution set in the general sense, without representing any specific case. The optimal Pareto front and the corresponding Pareto solution set of a particular deterministic multi-objective problem will be denoted as PF^*_{det} and PS^*_{det}, respectively. Note that PF^*_{det} may not be known *a priori* and it is fixed for any particular multi-objective problem. The final set of non-dominated solutions evolved by MOEA will be termed as PF^A_{det}.

In the case of robust multi-objective optimization, the optimal robust Pareto front and solution set are dependent on the noise model and the robust measure. Therefore, the notation should reflect the noise model and the robust measure used. In this chapter, the optimal robust Pareto front and optimal solution set are denoted as $\text{PF}^*_{rm,\delta}$ and $\text{PS}^*_{rm,\delta}$, respectively. The terms rm and δ refers to the robust measure and noise model in consideration, respectively. Accordingly, $\text{PF}^A_{rm,\delta}$ refers to the final set of non-dominated solutions evolved by robust MOEA based on the robust measure rm and noise model δ.

7.2 Robust Measures

There are several possible notions of robustness and many different robust measures have been applied in the literature. The most popular and

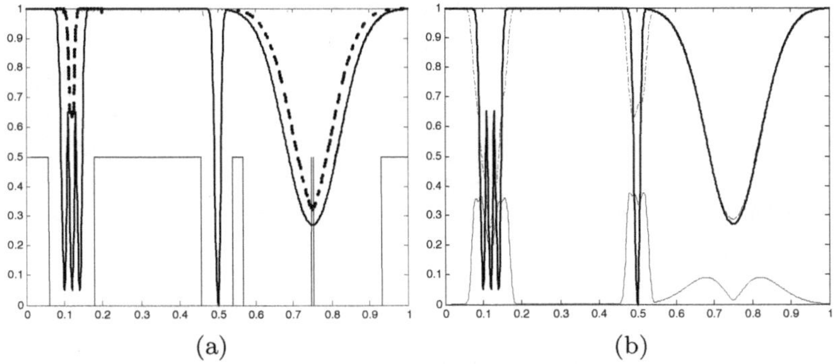

Fig. 7.1 Illustration of the different robust measures, (a) worst case (– –) and constrained (—) and (b) effective (-·-·) and standard deviation (—), with respect to the deterministic landscape (—)

straight-forward measure is the optimization of the expected performance. Since it is usually hard to compute the integration of the solution over the possible disturbances analytically, Monte Carlo integration, i.e. $E(f_i) = \frac{1}{N} \cdot \sum_{i=1}^{N} f_i(\mathbf{x} + \boldsymbol{\delta}_\mathbf{i})$, is usually applied to obtain an estimate of the expected performance. Solutions that are optimized based on expected fitness are known as effective solutions. Hence, for multi-objective optimization, the resulting Pareto front is known as the effective Pareto front ($\mathrm{PF}^A_{\mathrm{eff},\delta}$). Other measures include the optimization based on the worst case scenario [205], as a constraint to be satisfied [56], and using various forms of variances.

Each of these robust measures reflects the different aspects of robustness and Fig. 7.1 illustrates the behavior of the different robust measures for an arbitrary function of varying sensitivities in the search space. The various plots are generated by sampling the values of x with an uniform distribution of $[-0.025, 0.025]$. If the model is known with absolute certainty and the solution can be implemented exactly, then the global optimum represented by the deterministic solution at $x = 0.5$ is the ideal solution. However, if variable x is stochastic, then the solutions presented by the other approaches would be more viable and the location of the optimum is also different. In particular, it can be noted that the expected fitness approach will favor the solution at $x = 0.11$, while the approaches based on variance and worst case scenario will favor the solution at $x = 0.75$. On the other hand, the constrained approach indicates the feasible solutions which satisfy the predefined robustness criterion. This chapter is focused on the optimization of the expected performance.

7.3 Robust Optimization Problems

This section presents a set of guidelines for the construction of robust multi-objective test problems. Based on the existing literature on robust optimization, Section 7.4.1 reviews the different categorization of robust problems and presents a classification scheme applicable to multi-objective optimization. Desirable properties of robust multi-objective problems are highlighted and some existing test problems are analyzed empirically in Section 7.4.2. Subsequently, the robust landscape generator and detailed construction guidelines are presented in Section 7.5. Finally, a vehicle routing problem with stochastic demand (VRPSD) test suite is proposed as an example of a real-world representation of combinatorial robust multi-objective problems in Section 7.6.

7.4 Robust Continuous Multi-objective Test Problem Design

7.4.1 Robust Multi-objective Problem Categorization

Robust optimization is very similar to noisy optimization and the two are often considered in the same context. However, there are significant differences between these two forms of uncertainties. For noisy optimization, uncertainty is *inherent* to the objective functions and it tends to mislead the optimization process, resulting in convergence to sub-optimal solutions. In the case of robust optimization, noise is *incorporated* into the objective functions to guide the optimization process to regions that are less sensitive to parametric variations.

Different categorization of robust problems have been considered in the literature. Based on the source of uncertainty, Jin and Branke [144] states that robust optimization can be considered from the perspective of solution sensitivity to decision variable variations or environmental variable variations. Decision variable variations stem from the fact that deviations from design specifications are inevitable in manufacturing. On the other hand, environmental variable variations arise from variations in operational or environmental conditions. Instances of environmental variable variations include temperature changes in circuit design [264], speed changes in aerodynamic shape and turbine blade design, and machine breakdowns in machine scheduling.

An alternative classification of single-objective problems based on the relationships between the efficient and effective fitness landscapes is presented recently in [207]. Paenke *et al* proposed four categories: 1) identical optimum where efficient and robust optima are identical, 2) neighborhood optimum where efficient and robust optima are located on the same peak or trough, 3) local-global flip where one of the local optima corresponds to the robust optimum, and 4) max-min flip where the global maximum corresponds to the robust optimum.

Deb and Gupta [56] considered a similar classification that is specific to the context of multi-objective optimization: 1) the global efficient front is robust, 2) a part of the global efficient front is not robust, 3) the robust front is represented by a local efficient front, and 4) the robust front is represented by both the global and local efficient fronts.

Robust multi-objective problems are certainly much more complex than single-objective problems as both PF^* and PS^* are susceptible to change due to uncertainties. Recent studies [22, 236] have shown that some problems have the interesting property of demonstrating fitness topological changes in the presence of noise. To be precise, topological variation strictly refers to the introduction of new problem features to the deterministic problem under the influence of noise. According to Huband *et al* [168], features refer to problem characteristics that define problem difficulty. For the two classification schemes described above, problems of the first category are typically considered to be less interesting as compared to problems of the other classes. On the other hand, it is possible that noise-induced features can actually result in a more challenging optimization problem even if the location of the optimum remains the same. For instance, Beyer and Sendhoff [22] observed the phenomenon of noise-induced multi-modality whereby a uni-modal problem transforms into a multi-modal problem when noise is applied. Therefore, it will certainly be more interesting to classify robust multi-objective problems according to the aspects of change under the influence of noise, i.e how the fitness landscape behaves in the face of uncertainties.

Most benchmark problems in the literature are commonly characterized by the emergence of a local optimum as the most robust solution in the presence of noise, signifying a change in the location of the optimum and in the context of multi-objective problem, a change in PS^*. Moreover, as mentioned above, it is distinctly possible that new problem features are introduced. As noted by Deb and Gupta [56], the PF^* is also susceptible to changes.

For the classification of robust multi-objective problems, this paper defines a three-bit binary number where the bits, in decreasing significance, represent the presence of PF^*, PS^*, and noise-induced features, respectively. Therefore, there is a total of eight classes. At one extreme, we have a class 0 problem representing a problem that does not change under the influence of noise. On the other end, we have a class 7 problem which exhibits all three types of changes. As a specific instance, a multi-objective problem that demonstrates feature and PS^* changes is a class 3 problem under this classification.

The above classification will be useful in the investigation of the various problem characteristics' impacts on evolutionary multi-objective optimization as well as identifying the suitability of the different robust handling techniques. Other aspects of robust multi-objective problem that are worth considering include the effects of the different robust measures on the landscape transformation and the degree of change with increasing noise levels. As shown in Fig. 7.1, the various robust measures result in different transformations. For the latter case, the change in landscape properties, such as the height of each peak, may change gradually with noise or there may be a sudden change in landscape feature once a certain noise threshold is reached.

7.4.2 Empirical Analysis of Existing Benchmark Features

Several desirable properties of deterministic benchmarks and test suites have been suggested in the EA literature. In addition to these guidelines, the following issues should be considered in the development of robust benchmark problems in the context of multi-objective optimization:

- Robust multi-objective problems are essentially multi-objective problems and guidelines for the construction of multi-objective benchmark problems established in the existing literature should be taken into account;
- The robust multi-objective test problems should not have any bias towards $PS^*_{rm,\delta}$;
- Some test problems should contain noise-induced features that pose more difficulty to the optimization algorithm;
- The benchmark problem component, which determines how the problem behaves in the presence of noise, should be scalable;
- Some test problems should contain possible tradeoffs between robustness and the different objectives.

In general, any test function should be simple enough to allow for analysis of algorithmic behavior but, at the same time, complex enough to allow conjectures to the real-world. However, a quick survey of past works will reveal the lack of problem characteristics beyond the basic landscape featuring contrastive sharp and broad peaks or troughs in the evaluation of uncertainty-handling techniques. In particular, some robust single-objective test functions may be too simplistic for proper algorithmic evaluation with the apparent lack of difficulties that may hinder the selection of robust multi-objective solutions. Furthermore, some robust benchmarks are distinctly multi-modal in nature and it may be difficult to ascertain whether the robust solution found is the consequence of premature convergence or the effectiveness of the particular robust optimization technique.

Therefore, empirical investigations are conducted in this section to analyze the behavior of four existing benchmark problems found in the literature. Three of the problems studied are extended from single-objective benchmark problems in [26, 31, 207] using the ZDT framework [299], which allows the easy incorporation of problem characteristics that hinder MOEA progress to the Pareto front. The fourth is a robust multi-objective problem proposed in [56]. These test functions are characterized by multiple dips of varying widths and depths in the fitness landscape and the global minimum is much narrower as compared the local minima. Intuitively, it is more sensitive to noise and one of the local PS will emerge as $PS^*_{rm,\delta}$ when noise is applied. All four benchmark problems are class 2 test functions since only PS^* changes. The definitions of these extended benchmarks are summarized in Table 7.1. To examine the scalability of these problems, experiments are conducted for $N = \{2,5,10\}$.

Table 7.1 Definition of robust Test Problems

Problem	Source	Definition
rMOP1	f_1 [31]	$f_1(x_1) = x_1$, $f_2(x_2, ...x_{n_{x,r}}) = g \cdot h$, $g(x_2, ...x_{\|n_{x,r}\|}) = 1 + \sum_{i=2}^{n_{x,r}} (2 - sin\sqrt{\|40x_i\|} - \frac{20-\|x_i\|}{20})$, $h(f_1) = 1 - \sqrt{f_1}$ where $x_1 \in [0, 1]$, $-20 \leq x_i < 20$, $\forall i = 2, ..., n_{x,r}$
rMOP2	f_1 [26]	$f_1(x_1) = x_1$, $f_2(x_2, ...x_{n_{x,r}}) = g \cdot h$, $g(x_2, ...x_{n_{x,r}}) = 1 + \sum_{i=2}^{n_{x,r}} G(x_i)$, $\mathrm{G}(x_i) = \begin{cases} 0.2 + (x_i + 1)^2 + 0.8 \cdot \|\sin(6.283x_i)\|, & \text{if } -2 \leq x_i < 0 \\ 0.6411 - 0.6 \cdot 2^{-8\cdot\|x_i-1\|} + 0.8 \cdot \|sin(6.283x_i)\|, & \text{if } 0 \leq x_i < 2 \end{cases}$ $h(f_1) = 1 - \sqrt{f_1}$ where $x_1 \in [0, 1]$, $-2 \leq x_i < 2$, $\forall i = 2, ..., n_{x,r}$
rMOP3	TP5 [207]	$f_1(x_1) = x_1$, $f_2(x_2, ...x_{n_{x,r}}) = g \cdot h$, $g(x_2, ...x_{n_{x,r}}) = 1 + \sum_{i=2}^{n_{x,r}} G(x_i)$, $\mathrm{G}(x_i) = \begin{cases} 0.6 - 0.5 \exp(-0.5 \cdot \frac{(x_i-0.4)^2}{0.05^2}), & \text{if } x_i < 04693 \\ 0.6 - 0.6 \exp(-0.5 \cdot \frac{(x_i-0.5)^2}{0.02^2}), & \text{if } x_i \leq 0.4693`0.5304 \\ 0.6 - 0.5 \exp(-0.5 \cdot \frac{(x_i-0.6)^2}{0.05^2}), & \text{if otherwise} \end{cases}$ $h(f_1) = 1 - \sqrt{f_1}$ where $x_i \in [0, 1]$, $\forall i = 2, ..., n_{x,r}$
rMOP4	TP3 [56]	$f_1(x_1) = x_1$, $f_2(x_2, ...x_m) = h \cdot (g + S)$, $g(x_{\|n_{x,r}\|+2}, ...x_m) = \sum_{i=\|n_{x,r}\|+2}^{m} 50x_i^2$, $S(f_1) = 1 - \sqrt{f_1}$ $h(x_2, ..., x_{\|n_{x,r}\|+1}) = 2 - 0.8 \exp(\sum_{i=2}^{\|n_{x,r}\|+1} (\frac{x_i-0.35}{0.25})^2) - \exp(\sum_{i=2}^{\|n_{x,r}\|+1} (\frac{x_i-0.85}{0.03})^2)$ where $x_1 \in [0, 1]$, $-20 \leq x_i < 20$, $\forall i = 2, ..., m$

In the simulation studies, two state-of-the-art MOEAs, the non-dominated sorting genetic algorithm II (NSGAII) and the strength Pareto evolutionary algorithm 2 (SPEA2), are applied to determine the difficulty of finding PF*. Both algorithms are implemented using the same binary coding scheme of 15 bits per variable, binary tournament selection, uniform crossover, and bit flip mutation. The simulations are implemented in C++ on an Intel Pentium 4 2.8 GHz computer and 30 independent runs are performed for each of the test functions in order to obtain statistical information, such as consistency and robustness of the algorithms. The simulation results with respect to the metrics of ratio of convergence, spacing (S), and maximum spread (MS) are shown in Table 7.2. The ratio of convergence is based on the average number of non-dominated solutions in each run that are located in the vicinity of

Table 7.2 Empirical Results of NSGAII and SPEA2 for the different robust multi-objective test functions

		NSGAII			SPEA2		
		Ratio	Space	Maximum Spread	Ratio	Space	Maximum Spread
rMOP1	2-D	0.8867	0.5411	1.0	0.7280	0.6582	1.0
	5-D	0.0	0.5431	1.0	0.0	0.6783	1.0
	10-D	0.0	0.5626	0.9999	0.0	0.6773	0.9999
rMOP2	2-D	0.9947	0.5314	1.0	0.9890	0.6362	1.0
	5-D	0.9883	0.5369	1.0	0.9840	0.6278	1.0
	10-D	0.9850	0.5781	0.9999	0.9807	0.6849	0.9999
rMOP3	2-D	0.9853	0.5332	1.0	0.9833	0.6354	1.0
	5-D	0.7193	0.4965	1.0	0.6203	0.6484	1.0
	10-D	0.0	0.5039	0.9999	0.0	0.6268	1.0
rMOP4	2-D	0.5250	0.5066	0.9999	0.4243	0.6500	0.9999
	5-D	0.0920	0.5012	0.9997	0.0	0.6442	0.9999
	10-D	0.0	04900	0.9997	0.0	0.6174	0.9998

PS^*_{det}. A solution is considered to be in the vicinity of PS^*_{det} if it has a Euclidean distance of less than 0.05 difference from the nearest point in PS^*_{det}.

From the simulation results, it is observed that NSGAII and SPEA2 generally perform similarly for the set of benchmark problems. It is evident from the metrics of S and MS in Table 7.2 that both algorithms are capable of consistent performance in the aspects of solution distribution and diversity. This is due to the manner in which the test problems are constructed, where the distribution and diversity of the solution set is optimized only through the h function. With the exception of rMOP2, NSGAII and SPEA2 are unable to locate PS^*_{det} consistently and this situation worsens with increasing solution space dimensionality. This fact can also be observed from Fig. 7.2 which illustrates the tradeoffs evolved by NSGAII and SPEA2 for rMOP3 at $N = \{2,10\}$. As mentioned before, all four problems are characterized by at least one local Pareto set that becomes more desirable than PS^*_{det} when noise is introduced. Nonetheless, the failure to find PS^*_{det} clearly indicates that the two algorithms can converge to more robust regions readily even without the incorporation of any robust handling mechanism. This is because PS^*_{det} is located at a much narrower region and hence, harder to locate as compared to the broader dips corresponding to the more robust solutions. This implies that rMOP1, rMOP3, and rMOP4 are not suitable for the evaluation of robust MOEA techniques since it is not possible to ascertain whether robust solutions are found due to the adopted robust measures or due to the algorithms' failure to locate PS^*_{det} in the first place.

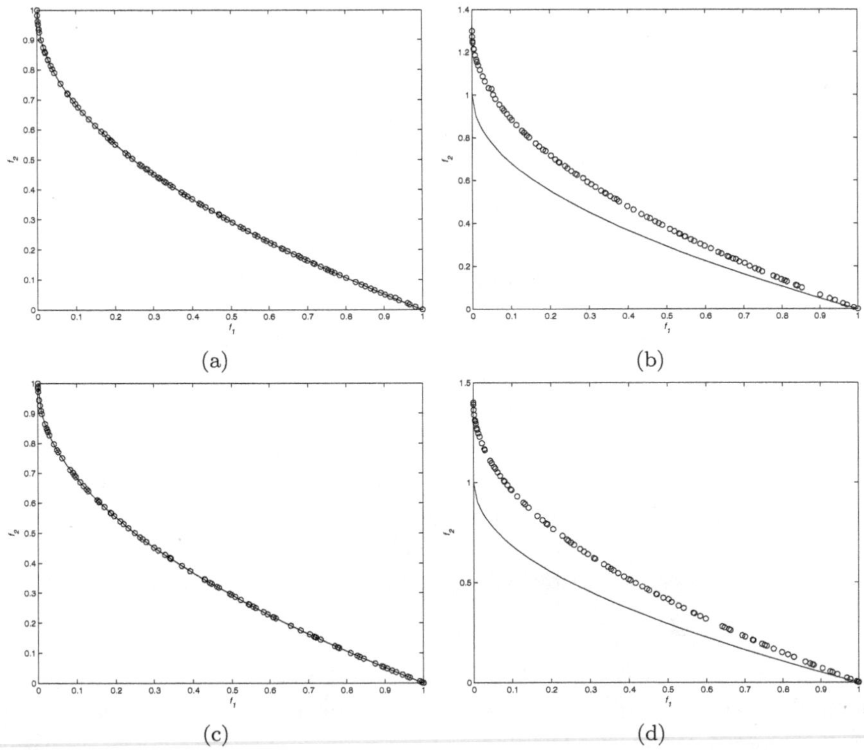

Fig. 7.2 The tradeoffs evolved by NSGAII at (a) $N = 2$ and (b) $N = 10$ and SPEA2 for rMOP3 at (c) $N = 2$ and (d) $N = 10$. The PF^*_{det} is represented by (-) while the evolved solutions are represented by (∘).

On the other hand, NSGAII and SPEA2 are able to find the PS^*_{det} of rMOP2 consistently. This is because the basin of attraction and areas under the curve of the robust and sensitive peaks are designed to be the same to prevent any initial bias. This implies that the basin of attraction is an important feature to be considered in the design of robust multi-objective test functions.

7.5 Robust Continuous Multi-objective Test Problem Design

The fundamental component of the robust multi-objective test functions proposed in this work is a Gaussian landscape generator that introduces various parametric sensitivities to the deterministic fitness landscape. It generates a set of $n_{x,r}$-dimensional minima throughout the fitness landscape and is given by:

$$b(\mathbf{x_r}) = 1 - \frac{1}{|x_r|} \sum_{i \in x_r} \max_{j \in J} \Big\{ h_{ij} \cdot \exp\big[(\frac{x_i - \mu_{ij} \cdot E_{ij}(\sigma, s_{ij})}{w})^2 \big] \Big\} \tag{7.2}$$

$$E_{ij}(\sigma, s_{ij}) = 1 + s_{ij} \cdot U(-\sigma, \sigma) \tag{7.3}$$

where J is the number of basis functions, d_j, μ_{ij}, and w denote the amplitude, location, and the width of the basis functions respectively. E_{ij} controls how the environmental variable behaves with noise σ and the degree of sensitivity s_{ij}. Intuitively, the robustness of a particular basin will depend on the associated E_{ij} function, while the amplitude will determine the optimality of the solution. From Eqn 7.2 and 7.3, it can be noted that test functions designed using this landscape generator are different from most previous works in two aspects:

- Any solution space or objective space transformation is a consequent of *environmental* variation. Although environmental parameter variation is rarely considered in the literature, it is definitely more flexible compared to decision parameter variation when it comes to the design of different possible scenarios.
- As observed from the simulation studies conducted in the previous section, it is important for the basins of attraction of the various troughs to be very similar. This ensures that there is no initialization bias towards any particular region of the search space.

The max function has been used successfully in previous works [86, 96] to combine the different Gaussian components and it ensures that the landscape feature at any one point is determined and influenced only by the dominant basin. Without the overlapping influences from the other basis functions, each basis function can be considered independently and this facilitates the design and analysis of the robust test function. In particular, it is possible to define explicitly the locations and depths of the different basins to create different test functions with specific characteristics. For the purpose of evaluating algorithmic performance, it is necessary to know the relative degree of robustness for each minimum. The theoretical values for each basis function can be easily worked out to be:

$$B_j = d_j \cdot \Big(\frac{w\sqrt{\pi}}{2 s_{ij} \sigma} \cdot erf\big(\frac{s_{ij}\sigma}{w}\big) \Big) \tag{7.4}$$

One desirable property of this test generator is that it provides a means to extend existing multi-objective test problems to robust multi-objective test functions without changing the original problem characteristics. The rationale is to allow researchers to investigate the impact of robust optimization on test functions with different characteristics such as deception, multi-modality, and discontinuities. As a specific instance, consider the i-th objective function of an arbitrary multi-objective benchmark problem. The corresponding objective function of the extended robust multi-objective test function can be written as:

$$f_i'(\mathbf{x}) = f_i(\mathbf{x_d}) + b(\mathbf{x}_r) \tag{7.5}$$

where $\mathbf{x}_d$ represents the subset of decision variables associated with the original problem, while $\mathbf{x}_r$ represents the subset of decision variables of the robust component of the problem.

In this chapter, the robust multi-objective test problems are built upon the ZDT framework, which has been applied earlier in Section 7.4.2 to extend the robust single-objective problems. The flexibility of this framework has also been demonstrated by the development of a suite of dynamic multi-objective problems by Farina *et al* in [73]. The guidelines for the construction of the deterministic ZDT test functions are formally described by the following

$$\begin{aligned} \min f_1(\mathbf{x_{d1}}) &= x_1 \\ \min f_2(\mathbf{x_{d2}}) &= g(\mathbf{x_{d2}}) \cdot h(f_1, g) \end{aligned} \tag{7.6}$$

where $\mathbf{x_{d1}}, \mathbf{x_{d2}} \in \mathbf{x}$ and the g and h functions control the problem difficulty and the shape of the Pareto front, respectively. For the ensuring discussions, we assume that the particular ZDT problem to be extended has the following functional form,

$$\begin{aligned} g(\mathbf{x_{d2}}) &= 1 + \textstyle\sum_{x \in x_{d2}} x_i \\ h(f_1, g) &= 1 - (\tfrac{f_1}{g})^{\alpha}. \end{aligned} \tag{7.7}$$

7.5.1 Basic Landscape Generation

Noise-induced changes to the PF*, PS*, and fitness landscape can be introduced by incorporating b into either the h and/or g functions to construct different classes of robust test problems. A straight-forward approach of introducing robust features into the problem is to change g in the form of $g(\mathbf{x}) = 1 + b(\mathbf{x_r})$, with h and f_1 unchanged. $\mathbf{x_r}$ is also a subset of $\mathbf{x}$. Let us consider a two-dimensional landscape generated by

$$\begin{aligned} b(\mathbf{x_r}) = 1 - \tfrac{1}{|x_r|} \textstyle\sum_{i \in \mathbf{x_r}} \max\Big\{ & 0.8 \exp\big[(\tfrac{x_i - 0.25 E_{i1}(\sigma, s)}{0.1})^2\big] \\ & , \exp\big[(\tfrac{x_i - 0.75 E_{i2}(\sigma, s)}{0.1})^2\big] \Big\}. \end{aligned} \tag{7.8}$$

The problem landscape presented by b at $\sigma = \{0, 0.15\}$ and the resulting Pareto fronts are shown in Fig. 7.3. The minimum located at (0.75,0.75) is the global minimum in a deterministic setting and failure to converge to this point will result in a dominated solution. With $s = 1$, notice that the effect of noise on the basin at (0.75,0.75) is actually three times more than that on the basin at (0.25,0.25). Thus, when noise is incorporated into the problem, the local minimum at (0.25,0.25) constitutes to the $\text{PF}^*_{rm,n}$ as it is more robust

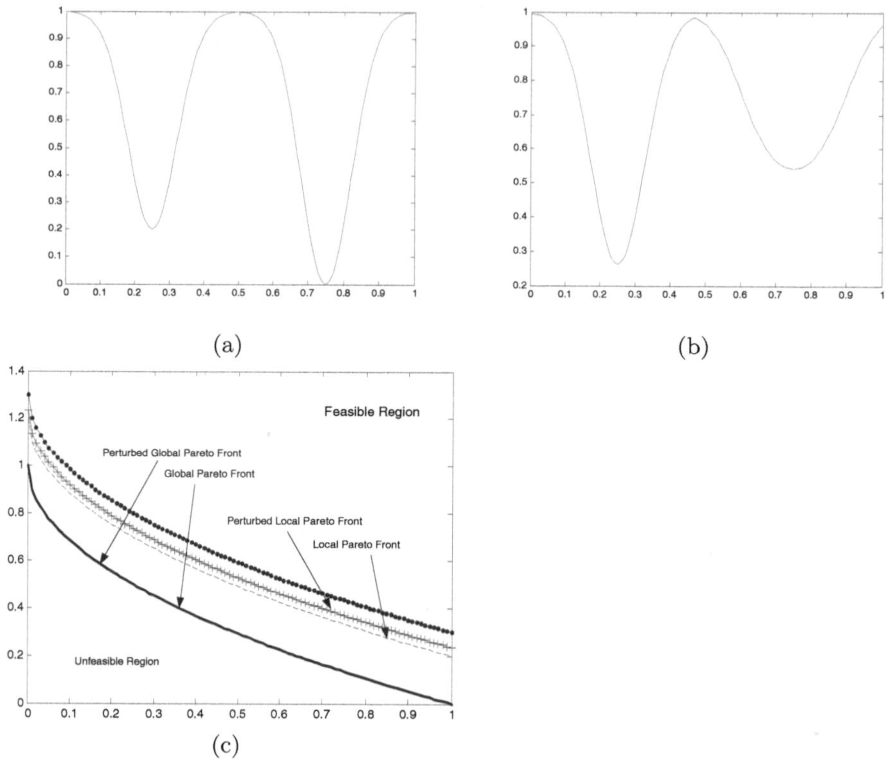

Fig. 7.3 An example of a landscape with two basins with $s = 1$ at (a) $\sigma = 0.0$ and (b) $\sigma = 0.15$. The minima at 0.75 is optimal under a deterministic setting while the minima at 0.25 emerges as the global robust minima at $\sigma = 0.15$. The corresponding Pareto fronts of the resulting problem in (c) shows the relationship between the two minima.

and its performance is less affected by noise. Since the only change induced by noise is the location of PS*, this is a simple instance of a class 2 robust problem.

On the other hand, if the g and b functions are combined such that

$$g(\mathbf{x}) = \big(1 + \sum_{x \in x_{d2}} x_i\big) + b(\mathbf{x_r}), \tag{7.9}$$

the resulting problem is also a class 2 problem. However, such a formulation allows the analysis of the effects of robust optimization on the original problem. Thus, the robust MOEA must be capable of finding the global robust minimum associated with b as well as dealing with the difficulties posed by the deterministic problem in order to find $\mathrm{PF}^*_{rm,n}$. It is also possible to redefine f_1 as $f_1(\mathbf{x}) = x_1 + b(\mathbf{x_r})$ to construct a class 2 problem with similar properties.

Instead of hand-designing b, the landscape generator can be parameterized to generate arbitrary landscapes by specifying key geometric properties such as the number of basins, the noise level at which investigations are conducted, and the locations of the deterministic optimum and robust minimum. The geometric properties of the other $J - 2$ basins must be selected such that their effective minimum computed by Eqn. 7.4 are worse than the pre-defined global robust optimum by some pre-defined ratio. A basin j is considered to be more robust than basin k if the following criterion is met: $s_{ij} \cdot \mu_{ij} < s_{ij} \cdot \mu_{ik}$. In addition, the sensitivity of each basin to noise should adhere to the condition of $s_{1j} \cdot \mu_{1j} = s_{2j} \cdot \mu_{2j} = ... =_{ij} \cdot \mu_{ij}$ for the landscape to behave properly.

Accordingly, the general form of the landscape generator can be written as

$$\langle RLS : [\boldsymbol{\mu}_r, h_r], [\boldsymbol{\mu}_g, h_g], \sigma, J, w, \beta \rangle \tag{7.10}$$

where $\boldsymbol{\mu}_r$ and h_r, $\boldsymbol{\mu}_g$ and h_g specify the locations and depths of the global robust and deterministic minima, respectively, while β is the factor at which the next best robust optimum is worse compared to the global robust optimum. An example of a two-D landscape generated using the specification of

$$\langle [(0.25, 0.25), 0.8], [(0.75, 0.75), 1.0], 0.1, 40, 0.1, 0.1 \rangle$$

is shown in Fig 7.4. Note that the landscape illustrated in Fig. 7.3 can be generated by specifying $J = 2$.

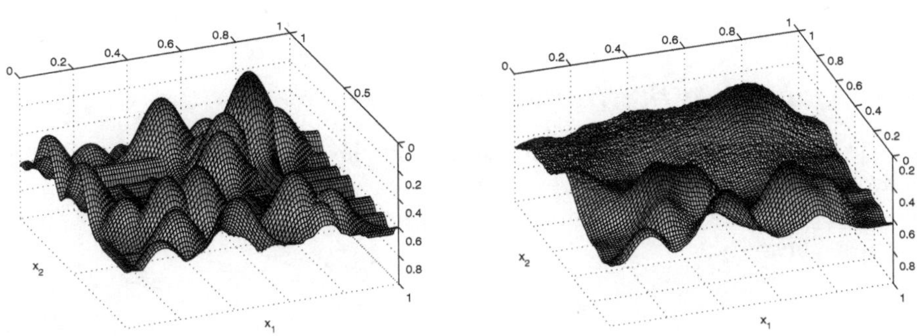

(a) (b)

Fig. 7.4 An example of an arbitrary two-D landscape with $J = 40$ at (a) $\sigma = 0.0$ and (b) $\sigma = 0.15$. The minimum at (0.75,0.75) is optimal under a deterministic setting while the minimum at (0.25,0.25) emerges as the global robust minimum at $\sigma = 0.15$.

7.5.2 Changing the Decision Space

When combined with g in the ways described above, the b function gives rise to the element of noise-induced changes to the PS* and results in class 2 test problems. Features of noise-induced search space variation can be easily incorporated into the problem by changing g in the following form,

$$g(\mathbf{x}) = 1 + (\sum_{x \in x_{d2}} x_i)^{\beta - b(\mathbf{x_r})}, \tag{7.11}$$

which forces the distribution of the solutions to change. Notice that g is now a function of b. In this particular instance, it is possible to apply equation 7.8 as the b function but finding its minimum will have no direct contribution to solution optimality. Interestingly, finding the optimal value for b will improve the distribution of the solutions near PS* and hence, simplify the problem somewhat. Thus, the resulting problem can be considered as a class 1 test problem.

More complex fitness topology variations can be induced by making h a function of g instead. In particular, consider the scenario where we define b such that the width, i.e. the size of the basin of attraction, of the selected minimum is a function of g and replace the g function by

$$g'(\mathbf{x}) = g(\mathbf{x_{d2}}) + b(\mathbf{x_r}, g). \tag{7.12}$$

The corresponding problem depends on the characteristics of the b function; it is a class 1 test problem if $J = 1$ and class 3 test problem if $J \geq 1$ and deterministic and robust optima are different. In any case, the robust MOEA must be able to deal with the features that arise due to noise in order to find $PF^*_{rm,n}$.

7.5.3 Changing the Solution Space

Since the shape of the PF^*_{det} is determined by the h function in the ZDT framework, $PF^*_{rm,n}$ can be easily controlled by combining the b and h in some appropriate ways. The simplest way to introduce changes in PF* is to control its convexity:

$$h(f_1, g, \mathbf{x_r}) = 1 - (\frac{f_1}{g})^{\alpha + b(\mathbf{x_r})}. \tag{7.13}$$

If the g function is unchanged and b defines a single basin, only the convexity of the PF* is affected by noise, while PS* remains the same. Thus, the resulting problem is a class 4 test problem and the robust MOEA must be capable of distributing the solutions along PF* with varying noise-induced convexity. However, if b is characterized by multiple basins as illustrated in Fig. 7.3 or Fig. 7.4, both the PF* and PS* will change and the corresponding problem becomes a class 6 test problem instead.

It is also possible to redefine the h function as,

$$h(f_1, g, b) = b(\mathbf{x_r}, f_1) - \sqrt{(\frac{f_1}{g})}. \tag{7.14}$$

where b is now a function of f_1 as well. One interesting implication of such a formulation, particularly if the sensitivities of the relevant basins increase with f_1, is the resulting tradeoff between the robustness and optimality of f_2. Therefore, a part of the PF* will become dominated in the presence of noise and hence, only part of x_1 makes up the $\mathrm{PF}^*_{rm,n}$. Intuitively, the corresponding problem is a class 6 test problem.

7.5.4 Example of a Robust Multi-objective Test Suite

Having described the possible modifications to extend the ZDT test problems, we are now in the position to suggest a suite of five robust multi-objective test problems that satisfy the requirements described in Section 7.4.2. Although not all seven classes of problems are represented, these problems embody the most challenging aspects of robust multi-objective optimization that have been described previously. Nonetheless, interested readers are encouraged to construct more interesting problems based on the guidelines made in the previous sections. At this point, it is worth mentioning that the proposed b function can also be employed as a non-optimizable component of the problem and as a noise-sensitive environment variable instead, i.e. $b(R)$. The definitions and characteristics of the test suite are summarized in Table 7.3 and Table 7.4, respectively.

GTCO1 utilizes the effects of equation 7.12 to bring about a change from unimodal at $\sigma = 0.0$ to multi-modal fitness landscape at $\sigma = 0.2$ as shown in Fig. 7.5. The PS^*_{det} and $\mathrm{PS}^*_{rm,n}$ are the same at all noise levels and correspond to $x_i \in \mathbf{x_{d1}} = 0$ and $x_i \in \mathbf{x_r} = 0$. The problem becomes increasingly multi-modal with increasing σ and this is an instance where the problem becomes more challenging and the robust MOEA will face difficulties finding PF* due to the landscape change. The settings of $|\mathbf{x_{d2}}| = 10$, $|\mathbf{x_r}| = 5$, and $\sigma = 0.2$ are recommended for GTCO1.

GTCO2 is an instantiation of the two-minima scenario considered in Section 7.5.1. This problem is similar to the problem of rMOP4 in the sense that the deterministic global and local minima are switched when noise is increased beyond a threshold as shown in Fig. 7.6. However, as mentioned before, the basins of attraction for both minima are the same, eliminating initialization bias. The PS^*_{det} corresponds to $\mathbf{x_r} = 0.75$, while $\mathrm{PS}^*_{rm,n}$ corresponds to $\mathbf{x_r} = 0.25$. The settings of $|\mathbf{x_r}| = 10$ and $\sigma = 0.2$ are recommended for GTCO2.

GTCO3 represents a combination of GTCO2 and the effects of equation 7.11 to induce both fitness landscape and PS* changes in the presence of noise. Noise-induced changes to the decision space are similar to GTCO2

Table 7.3 Definitions of the GTCO test suite

Problem	Type	Definition
GTCO1	Class 1	$f_1(\mathbf{x}_{D1}) = x_1$
		$g(\mathbf{x}) = 1 + \sum_{i=2}^{\lvert \mathbf{x}_{D2} \rvert} \{(x_{D2,i} - 0.4)^2 + b(\mathbf{x}_R, x_{D2,i})\}$
		$h(f_1, g) = 1 - \sqrt{\frac{f_1}{g}}$
		$b(\mathbf{x}_R, x_i) = 1 - \frac{1}{\lvert \mathbf{x}_R \rvert} \sum_{j \in \mathbf{x}_R} \exp\left[\left(\frac{x_{R,j}^5 - E(\sigma,s)}{W(x_i)}\right)^2\right]$
		$W(x_i) = 0.1 + 0.1\cos(20(x_i - 0.4)\pi) \cdot (1 - \lvert x_i - 0.4 \rvert)^5$
		$E(\sigma, s) = U(-\sigma, \sigma), \quad \mathbf{x}_{D1}, \mathbf{x}_{D2}, \mathbf{x}_R \in [0, 1]$
GTCO2	Class 2	$f_1(\mathbf{x}_{D1}) = x_1$
		$g(\mathbf{x}) = 1 + b(\mathbf{x}_R)$
		$h(f_1, g) = 1 - \sqrt{\frac{f_1}{g}}$
		$b(\mathbf{x}_R) = 1 - \frac{1}{\lvert \mathbf{x}_R \rvert} \sum_{i \in \mathbf{x}_R} \max\left\{0.8\exp\left[\left(\frac{x_{R,i} - 0.25E(\sigma,s)}{0.1}\right)^2\right], \exp\left[\left(\frac{x_{R,i} - 0.75E(\sigma,s)}{0.1}\right)^2\right]\right\}$
		$E(\sigma, s) = 1 + U(-\sigma, \sigma), \quad \mathbf{x}_{D1}, \mathbf{x}_R \in [0, 1]$
GTCO3	Class 3	$f_1(\mathbf{x}_{D1}) = x_1$
		$g(\mathbf{x}) = 1 + 10\left(\sum_{i=2}^{\lvert \mathbf{x}_{D2} \rvert} \frac{x_{D2,i}}{\lvert \mathbf{x}_{D2} \rvert - 1}\right)^{1.25 - b_1(\mathbf{x}_{R1})} + b_2(\mathbf{x}_{R2})$
		$h(f_1, g) = 1 - \sqrt{\frac{f_1}{g}}$
		$b_1(\mathbf{x}_{R1}) = 1 - \frac{1}{\lvert \mathbf{x}_{R1} \rvert} \sum_{i \in \mathbf{x}_R} \exp\left[\left(\frac{x_{R1,i}^5 - E_1(\sigma,s)}{0.05}\right)^2\right]$
		$b_2(\mathbf{x}_{R2}) = 1 - \frac{1}{\lvert \mathbf{x}_{R2} \rvert} \sum_{i \in \mathbf{x}_{R2}} \max\left\{0.8\exp\left[\left(\frac{x_{R2,i} - 0.25E_2(\sigma,s)}{0.1}\right)^2\right], \exp\left[\left(\frac{x_{R2,i} - 0.75E_2(\sigma,s)}{0.1}\right)^2\right]\right\}$
		$E_1(\sigma, s) = U(-\sigma, \sigma),$
		$E_2(\sigma, s) = 1 + U(-\sigma, \sigma), \quad \mathbf{x}_{D1}, \mathbf{x}_{D2}, \mathbf{x}_{R1}, \mathbf{x}_{R2} \in [0, 1]$
GTCO4	Class 6	$f_1(\mathbf{x}_{D1}) = x_1$
		$g(\mathbf{x}_{D2}) = 1 + 10\sum_{i=2}^{\lvert \mathbf{x}_{D2} \rvert} x_{D2,i}$
		$h(f_1, g, \mathbf{x}_R) = 1 - (\frac{f_1}{g})^{\alpha}$
		$\alpha = 0.5 + b(\mathbf{x}_R)$
		$b(\mathbf{x}_R) = 1 - \frac{1}{\lvert \mathbf{x}_R \rvert} \sum_{i \in \mathbf{x}_R} \max\left\{0.8\exp\left[\left(\frac{x_{R,i} - 0.25E(\sigma,s)}{0.05}\right)^2\right], \exp\left[\left(\frac{x_{textR,i} - 0.75E_{ij}(\sigma,s)}{0.05}\right)^2\right]\right\}$
		$E(\sigma, s) = 1 + U(-\sigma, \sigma), \quad \mathbf{x}_{D1}, \mathbf{x}_{D2}, \mathbf{x}_R \in [0, 1]$
GTCO5	Class 7	$f_1(\mathbf{x}_{D1}) = x_1$
		$g(\mathbf{x}) = 1 + \sum_{i=2}^{\lvert \mathbf{x}_{D2} \rvert} \{x_{D2,i} + 5b_1(\mathbf{x}_{R1})\}$
		$h(f_1, g) = 1 + 2b_2(\mathbf{x}_{R2}, f_1) - \sqrt{\frac{f_1}{g}}$
		$b_1(\mathbf{x}_{R1}, x_i) = 1 - \frac{1}{\lvert \mathbf{x}_{R1} \rvert} \sum_{i \in \mathbf{x}_{R1}} \exp\left[\left(\frac{x_{R1,i} - E(\sigma,s)}{W_1(x_i)}\right)^2\right]$
		$b_2(\mathbf{x}_{R2}, f_1) = 1 - \frac{1}{\lvert \mathbf{x}_{R2} \rvert} \sum_{i \in \mathbf{x}_{R2}} \exp\left[\left(\frac{x_{R2,i} - E(\sigma,s)}{W_2(f_1)}\right)^2\right]$
		$W_1(x_i) = \begin{cases} 0.2, & \text{if } x_i < 0.05 \\ 0.1x_i + 0.05, & \text{otherwise} \end{cases}$
		$W_2(f_1) = 0.2 \cdot (1.1 - \sqrt{f_1})$
		$E(\sigma, s) = U(-\sigma, \sigma), \quad \mathbf{x}_{D1}, \mathbf{x}_{D2}, \mathbf{x}_{R1}, \mathbf{x}_{R2} \in [0, 1]$

Table 7.4 Characteristics of the GTCO test suite

Problem	PF* change	PS* change	Feature change
GTCO1	-	-	Increasing multi-modality
GTCO2	-	Global to local	-
GTCO3	-	Global to local	Decreasing density of solutions near PF*
GTCO4	Increasing non-convexity	Global to local	-
GTCO5	Decreasing coverage of PF*	Range of x_1 reduces	Linear to deceptive

except that the density of the Pareto optimal solutions is now adversely affected by noise. The behaviors of the solution space at $\sigma = 0.0$ and $\sigma = 0.2$ are shown in Fig. 7.7, where it can be noted that the bias away from the PS* will be attenuated with increasing σ values. The density of Pareto optimal solutions is at its highest and hence, easiest to find when $\mathbf{x_{r2}} = 0.0$. The settings of $|\mathbf{x_{d2}}| = 10$, $|\mathbf{x_{r1}}| = |\mathbf{x_{r2}}| = 5$ and $\sigma = 0.2$ are recommended for this problem.

Noise-induced changes in PS* and PF* in GTCO4 are achieved through the implementation of equation 7.13. Once again, the b function governed by equation 7.8 is applied to generate PS* changes and the corresponding Pareto fronts at different σ levels are shown in Fig. 7.9. At low levels of σ, PS* corresponds to $\mathbf{x_r} = 0.75$ and the PF* becomes increasingly non-convex with noise. At sufficiently high σ levels, the PS* corresponds to $\mathbf{x_r} = 0.25$. Note that non-convexity is one of the problems that posed considerable difficulties to early multi-objective algorithms. Therefore, a robust MOEA has to be capable of distributing the discovered solutions uniformly along the Pareto front for the various degrees of convexity. The settings of $|\mathbf{x_{d2}}| = |\mathbf{x_r}| = 10$ and $\sigma = 0.2$ are recommended for this problem.

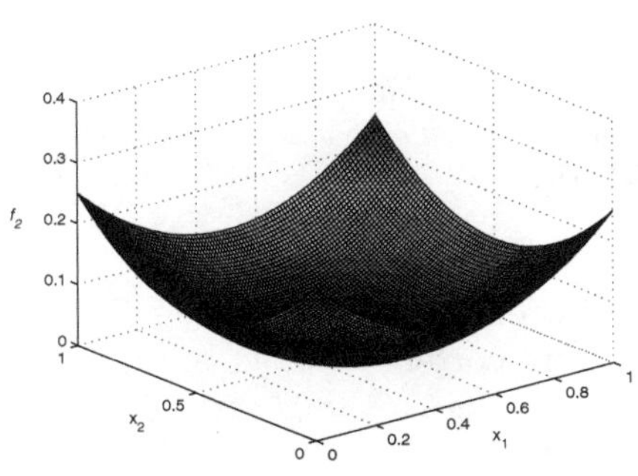

(a)

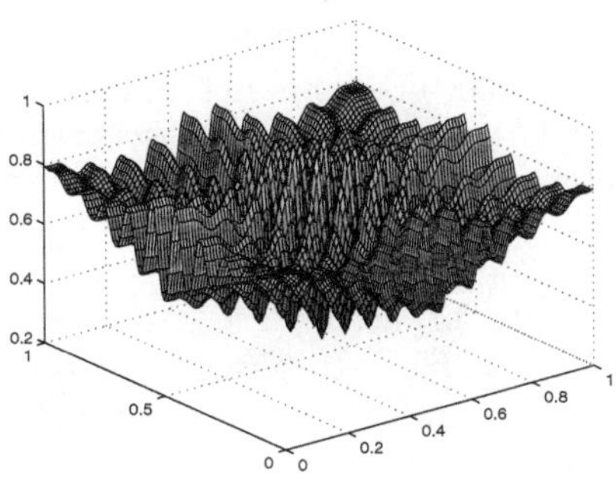

(b)

Fig. 7.5 Fitness landscape of GTCO1 with $|x_r| = 2$ at (a) $\sigma = 0.0$ and (b) $\sigma = 0.2$. GTCO1 is unimodal under a deterministic setting and becomes increasingly multimodal as noise is increased.

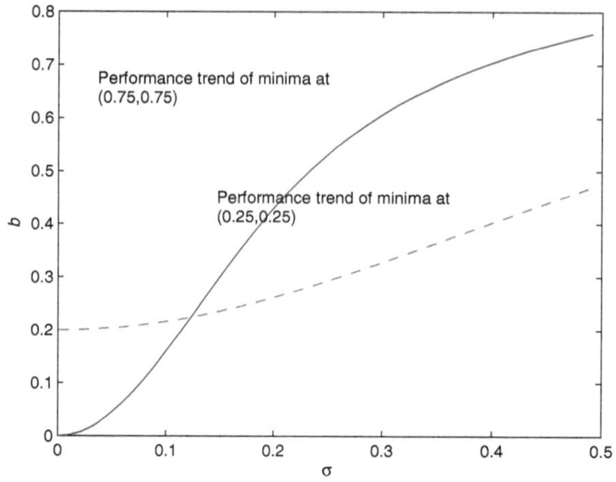

Fig. 7.6 Performance variation of the two minima with increasing σ for GTCO2

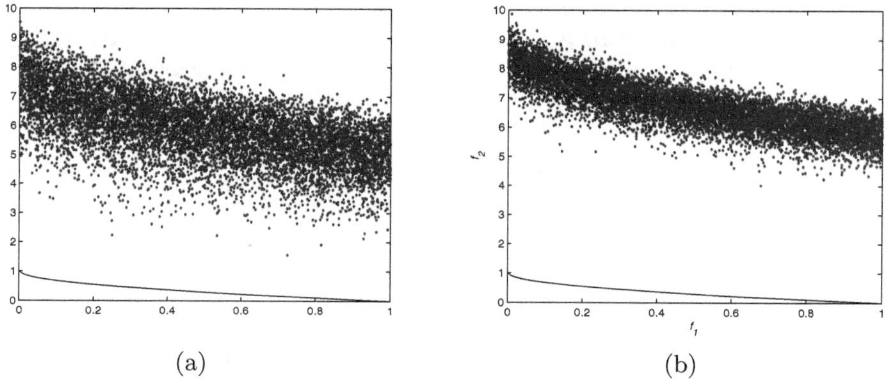

Fig. 7.7 10000 random solutions for GTCO3 at (a) $\sigma = 0.0$ and (b) $\sigma = 0.2$. The density of the solutions near the Pareto front is adversely affected in the presence of noise and deteriorates with increasing uncertainties.

GTCO5 is based on equation 7.14 which introduces noise-induced PS* and PF* changes. Robustness of the solutions are correlated to f_1 and this presents a conflict with the optimality of f_2. Considering the effects of this tradeoff alone, the remaining region of PF* becomes increasingly smaller with noise as illustrated in Fig. 7.9(a). Fitness topological changes are based on the principle adopted in GTCO1. However, the associated b function gives rise to a deceptive landscape in this instance as shown in Fig. 7.9(b). The only decision variable associated with PS* that varies with σ is x_1, while the other remains at $\mathbf{x_r} = 0.0$ and $\mathbf{x_{d2}} = 0.0$. The settings of $|\mathbf{x_{d2}}| = |\mathbf{x_r}| = 10$ and $\sigma = 0.08$ are recommended for this problem.

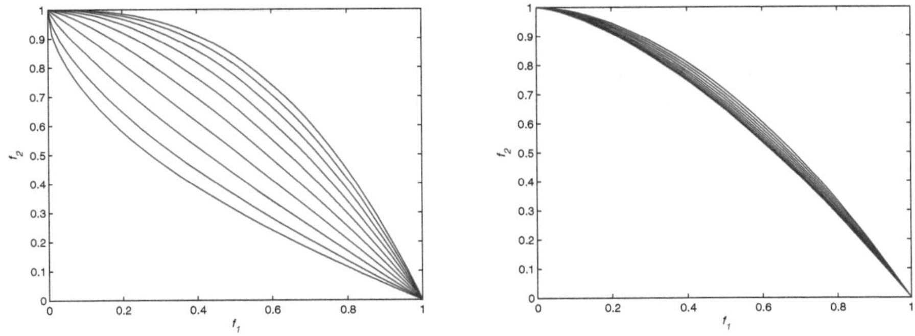

(a) (b)

Fig. 7.8 The resulting Pareto front of GTCO4 at (a) $\mathbf{x_r} = 0.75$ and (b) $\mathbf{x_r} = 0.75$ for $\sigma = [0.01, 0.1]$

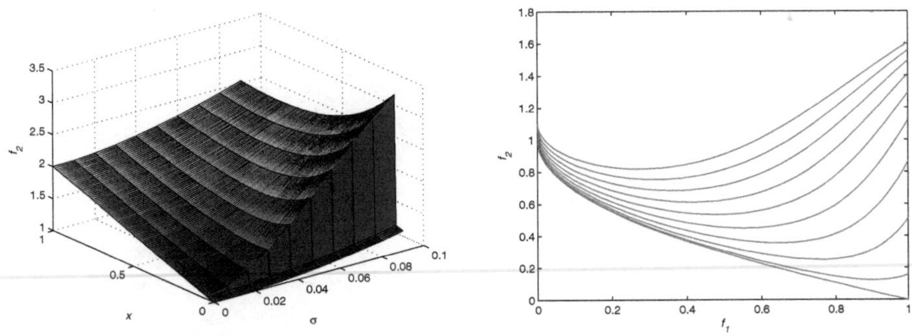

(a) (b)

Fig. 7.9 Effects of (a) decision space variation and (b) solution space variation across different σ values for GTCO5

7.6 Vehicle Routing Problem with Stochastic Demand

This section presents the vehicle routing problem with stochastic demand (VRPSD) as a practical example of robust multi-objective optimization problems. The VRPSD is a variant of the classical vehicle routing problem, where customers' demands are stochastic and all other parameters are known *a priori*. The demands are treated as random variables whose distributions are known and the actual demand of each customer is revealed only when a vehicle arrives at the customer's location. This combinatorial optimization problem appears in the delivery of home heating oil [66], trash collection, sludge disposal [172], beer and soft drinks distribution, the provision of bank automates with cash, and the collection of cash from bank branches [169]. The VRPSD has been shown to be naturally multi-objective [249] and involves not only generating minimal cost solutions but also robust solutions whose expected costs are good approximation of the actual costs of implementation. Due to the stochastic nature of the customers' demands, the cost of a particular solution cannot be known with certainty and various robust measures, such as the expected cost, have to be employed.

7.6.1 Problem Features

The VRPSD is characterized by many factors, which can influence the behavior of multi-objective routing and scheduling algorithms. A simple example of the routing plan with $|V| = 6$ is illustrated in Fig. 7.10. The routing schedule S is given as $\mathbf{R}=\{\mathbf{r}_1, \mathbf{r}_2\}$ where $\mathbf{r}_1$ is represented by $\mathbf{r}_1= \langle v_1, v_6, v_2, v_3\rangle$ and $\mathbf{r}_2$ is represented by $\mathbf{r}_2= \langle v_5, v_4\rangle$. The depot is omitted since all vehicles must depart and return to the depot. It can be seen that the number of routes $|S|$ is equal to the number of vehicles (C_v) used in the plan. The condition of $\bigcup_i^{|\mathbf{R}|} \mathbf{r}_i = V$, i.e. all customers are routed, must be satisfied.

As a robust multi-objective test function, we suggest the following parameterization of the problem landscape,

$$\langle \text{VRPSD} : \mathbf{F}, |V|, \mathbf{L_c}, L_d, \mathbf{D}, \mathbf{t_s} \rangle$$

- *Topology of customers,* $\mathbf{L_c}$ If it is possible to obtain real-world information about the problem, then the actual geographical distribution of the customers can be easily used to construct the problem. Otherwise, the locations of the customers can be generated randomly based on some reasonable probability distributions. The spatial distribution of customers can be categorized into three main classes [242]: 1) Type-R problems where all the customers are remotely located, 2) Type-C problems where the customers are grouped into clusters, and 3) Type-RC problems which is a mixture of remote and clustered customers. The effects that the three customer topologies have on robust multi-objective optimization will be revealed in Section 8.2.
- *Customer demand distribution,* $\mathbf{D}$ The demand distribution determines the extent to which the robust problem deviates from the deterministic one. A uniform distribution assumes a fixed range of demands and the problem can be solved conservatively by optimizing on worst-case demands. On the other hand, there is an outside chance for the occurrence of outlier demands if a normal distribution is adopted, which results in a more challenging problem. One common approach of generating the normal demand distribution model is to use the original demand quantity from some existing deterministic VRP datasets as the mean demand and generating the standard deviation of the demand distribution of each customer randomly such that it falls between zero and one-third of the mean demand of the customer [65, 249].
- *Location of depot,* L_d The location of the depot has a significant impact on the optimization process. For instance, a depot that is located at the centre

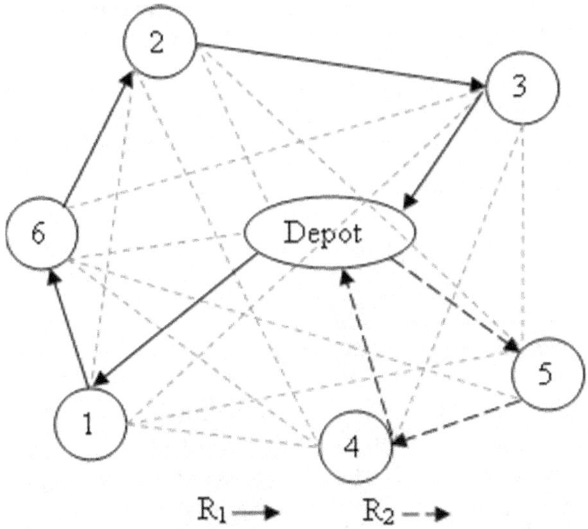

Fig. 7.10 Graphical representation of a simple vehicle routing problem

of a map would give the depot better proximity from all the customers and allow shorter trips back to the depot for restocking in the event of route failures. The other extreme case would be to locate the depot at a corner of the map. This would make recourse actions more expensive and emphasize the importance of robust solutions to the stochastic problem.

- *Service duration,* $\mathbf{t_s}$ A service time $\mathbf{t_s}$ is associated with each customer and the depot, which will be incurred each time the vehicle arrives at the customer or returns to the depot for restocking. A convenient way of specifying the service duration is to set the service times of all customers to be the same [249]. However, it is quite unlikely that this would be the case in reality. It would be more appropriate if service times were given physical meanings such as the time required for loading and unloading of cargo or the time required for clearing the customs. In either case, the service time could be proportional to the amount of cargo to be loaded at the depot and unloaded at the customer. Problems with longer service times would also amplify the need for robust solutions that minimize the occurrences of route failures since multiple trips to the depot for restocking would be more costly.
- *Number of customers,*$|V|$ V is the set of all customers. It is clear that the problem gets more difficult as the number of customers increases. Typical problem sizes that have been adopted in conventional vehicle routing problem with time windows (VRPTW) test problems range from 100 [242] to 1000 [122] customers and this serves as a good guide for VRPSD.
- *Optimization criteria,* **F** As in all real-world optimization problems, it is desirable to minimize overall operational cost which includes factors such as travel distance (C_d), the number of vehicles involved (C_v) and monetary

cost such as driver remuneration (C_m). Thus, the multi-objective VRPSD is to find the routing schedule S such that it:

$$\min \mathbf{F}(\mathbf{R}) = \{C_d(\mathbf{R}), C_v(\mathbf{R}), C_m(\mathbf{R})\}. \tag{7.15}$$

7.6.2 Problem Formulation

One common assumption made in this problem model is the homogeneity of all vehicles in the fleet and each one has a capacity limit which acts as a hard constraint. In the case of a route failure, i.e. a vehicle finds that a customer's demand cannot be satisfied upon reaching the customer, a recourse policy is employed to maintain the feasibility of solutions [89, 90, 170, 171, 260, 261]. The recourse policy requires the vehicle to unload all remaining goods at the particular customer, return to the depot to restock before going back to complete the service and/or continue with the scheduled route. These recourse actions will of course incur additional transportation cost, in terms of the travel distance and time for the to and fro trips to the depot. Additional service times will also be incurred when a vehicle visits a customer more than once or returns to the depot for restocking. As a matter of practicality, note that each customer can only be serviced by one vehicle.

Distance cost, C_d: The distance cost of route $\mathbf{r}_i$ is given by

$$C_d(\mathbf{r}_i) = \sum_{j=1}^{|\mathbf{r}_j|-1} \left(d(r_{i,j}, r_{i,j+1})\right) + d(v_0, r_{i,1}) + d(r_{i,|\mathbf{r}_j|}, v_o) + Q_d(\mathbf{r}_i). \tag{7.16}$$

where $d(r_{i,j}, r_{i,j+1})$ is the travel distance between the j-th and the $j+1$-th customer in the route. Q_d is the additional transportation cost incurred for the recourse policy.

Monetary cost, C_m: The travel time incurred for route $\mathbf{r}_i$ is given by

$$C_t(\mathbf{r}_i) = \sum_{j=1}^{|\mathbf{r}_j|-1} \left(T(r_{i,j}, r_{i,j+1})\right) + T(v_0, r_{i,1}) + T(r_{i,|\mathbf{r}_j|}, v_o) + Q_t(\mathbf{r}_i). \tag{7.17}$$

where $T(r_{i,j}, r_{i,j+1})$ is the travel time between the j-th and the $j+1$-th customer in the route. Q_t is the additional transportation time incurred for the recourse policy. It also includes the additional service times incurred when a vehicle visits a customer more than once or returns to the depot for restocking due to route failures.

The time constraint is such that $C_t(\mathbf{r}_i)$ should not exceed a given bound. This is a soft constraint and B is calculated as the time for a vehicle to travel diagonally across the map from one corner to the other and back. This time is assumed to be eight hours, equivalent to a driver's workday. Remuneration is such that drivers are paid \$10 for each of the first eight hours of work and \$20 for every additional hour of work subsequently. This is done to penalize exceedingly long routes which may not be feasible to implement in the context of the real-world. Therefore, the driver remuneration can be calculated as,

$$C_m(\mathbf{r}_i) = \begin{cases} 80 \cdot \frac{C_t(\mathbf{r}_i)}{B}, & \text{if } C_t(\mathbf{r}_i \leq B \\ 80 + 160 \cdot \frac{C_t(\mathbf{r}_i) - B}{B}, & \text{otherwise.} \end{cases} \tag{7.18}$$

Number of Vehicles, C_v: The number of vehicles is simply the number of routes in the schedule.

7.7 Conclusion

Apart from the need to satisfy several competing objectives, many real-world applications are also sensitive to decision or environmental parameter variations which result in large or unacceptable performance variations. Although the application of evolutionary multi-objective optimization to real-world problems is gaining popularity from researchers in different fields, there is a distinct lack in studies investigating the issues of robust optimization in the literature. This chapter examines the suitability of existing robust test problems for multi-objective optimization and presents a set of guidelines for the construction of robust multi-objective test problems. The fundamental component of the robust test problems is a Gaussian landscape generator that facilitates the specification of robust optimization-specific features such as noise-induced solution space, fitness landscape, and decision space variation. This generator is developed with the purpose of generating noise-sensitive landscapes in conjuction with existing multi-objective test problems and due to its independent nature, it can be used to generate robust single-objective test problems as well. Subsequently, a robust continuous multi-objective test suite is built upon the ZDT framework. Additionally, the vehicle routing problem with stochastic demand (VRPSD) is presented as a practical example of robust combinatorial multi-objective optimization problems.

Chapter 8
Evolving Robust Solutions in Multi-Objective Optimization

As described in Chapter 7, variations in design variables or the environment may affect solution quality and design performance adversely and robust optimization considers the effects explicitly and seeks to minimize the consequences without eliminating efficiency. Many different approaches, including Taguchi orthogonal arrays, response surface methodology, probabilistic design analysis, have been applied for robust optimization. In operational research, robust optimization is considered as a modeling methodology where robust problems are reformulated into the form of linear, conic quadratic, and semi-definite programming problems. Nonetheless, assumptions or approximations are often made during problem reformulation to ensure computational tractability, resulting in more uncertainties in the problem model. In addition, it does not allow for the incorporation of any domain knowledge to achieve better performance. On the other hand, evolutionary optimization techniques do not have such limitations, making it appropriate for robust optimization.

Despite the importance and practicality of robust multi-objective problems, they are rarely considered in the literature until recently (56, 111). In particular, Deb and Gupta (56) considered the optimization of the mean objective values as well as the formulation of the robust multi-objective problem as a constrained problem. In contrast to single-objective optimization, it is also essential to obtain a well-distributed and diverse solution set in multi-objective optimization. Although the general concepts of existing evolutionary robust single-objective techniques can be easily extended for robust multi-objective optimization, the robust optimization process is complicated by the need to consider explicitly several issues that are unique to multi-objective optimization. Therefore, the challenge in the design of any robust MOEA is to deal with issues such as the notion of Pareto-optimality, elitism, and balance between exploration and exploitation with the additional consideration of robustness.

C.-K. Goh and K.C. Tan: Evolutionary Multi-objective Optimization, SCI 186, pp. 213–227.
springerlink.com

8.1 Evolutionary Robust Optimization Techniques

While MOEAs have been demonstrated to be capable of discovering good tradeoff solutions for various multi-objective problems, it is necessary to ensure that these solutions are implementable in practice. Conventional MOEAs, without the necessary mechanisms to identify robust solutions, are not capable of finding $PS^*_{rm,\delta}$ unless it coincides with PS^*_{det}. Therefore, the selection of an appropriate robust measure is of utmost importance in robust optimization.

It can be noted that studies on evolutionary robust optimization are mainly conducted in the domain of robust single-objective optimization. Nevertheless, the robust measures and uncertainty handling mechanisms adopted in these works are generally applicable for robust multi-objective optimization; subsequent discussions are largely based on these studies and on its suitability in the context of robust MOEA. Specific issues such as diversity preservation and fitness assignment must be considered in robust MOEA design.

Based on the state-of-the-art, EAs for robust optimization can be classified into single-objective and multi-objective approaches depending on how the various measures are incorporated into the EA.

1. The *single-objective* approach optimizes the selected robust measures in place of the original objectives.
2. The *multi-objective* approach considers the selected robust measures as additional objectives to be optimized.

As noted by Jin and Branke (144), the former is the more popular approach. This is perhaps because of its ease of implementation, whereas there is a need to consider the implications brought about by the increase in problem dimensionality for the latter.

8.1.1 Single-Objective Approach

Since it is usually difficult to compute the various robust measures analytically, this approach is also characterized by the stochastic evaluation of the adopted robust measure to account for uncertainties, i.e. these measures are usually estimated over a number of randomly sampled perturbations. The optimization of the expected objective values estimated from the mean of the sampled points is also known as *explicit averaging* and has been applied successfully for robust multi-objective optimization (56). In the same work, the effects of sample size and noise level on $PF^A_{\text{eff},\delta}$ are investigated. On the other hand, Hughes (127) introduced a probabilistic approach for Pareto ranking scheme to account for the presence of uncertainties.

Although simple to implement, stochastic evaluation is computationally intensive since additional solution evaluations are required. It can be expected that this situation would be exacerbated by the presence of multiple objectives in multi-objective problems. Therefore, suitable methods for reducing the number of evaluations will be required to the lower total computational cost. To this end, the Latin hypercube sampling is applied by (30, 56) to get a more efficient fitness estimate. Other methods in the robust single-objective optimization literature that are appropriate for reducing computational cost of robust MOEAs include:

- Allocation of more computational resource for the evaluation of Pareto optimal solutions,
- Use of similar individuals evaluated in the past to estimate the expected fitness,
- Adaptation of computational resource allocation for evaluation through the evolutionary process, and
- Use of approximate models in place of the original objective functions.

A viable option for the efficient optimization of expected objective values is the method of *implicit averaging* (267, 268) where each solution is perturbed once before evaluation. This approach is based on the concept that solutions are implicitly averaged over a set of perturbed samples as the MOEA tends to revisit promising regions of the search space. Tsutsui and Ghosh also showed, by means of the Schema theorem, that an EA with infinite population size working on perturbed evaluations has the same effects as one working on the effective fitness.

8.1.2 Multi-objective Approach

The multi-objective approach involves both deterministic and stochastic evaluation of the various objectives and robust measures, respectively. Therefore, computational cost is also an important issue as in the case of the single-objective approach. When the multi-objective approach is applied to solve robust single-objective problems, it allows us to consider the tradeoff between robustness and solution quality. On the other hand, when it is applied to solve robust multi-objective problems, this approach poses an interesting decision-making exercise particularly if there are also tradeoffs between the robustness of the different objectives. Furthermore, the scaling capability of the MOEA becomes extremely important, for example, a five-objective problem can become a ten-objective problem if the effective performance of each objective is considered explicitly.

At present, there are two variants of the multi-objective approach for robust MOEA. The first approach optimizes the selected robust measures on top of the existing deterministic objective functions and sought to discover the inherent tradeoff between optimality and robustness. This is also known as the multi-objective approach (146) and various combinations of different measures, such as expected fitness and variance-based measures, have been applied in the evolutionary robust single-objective optimization literature.

In (218), Ray utilized three objectives, the nominal objective value, the effective objective value, and the standard deviation, to evolve designs that remain feasible under decision variable variations.

More recently, Goh and Tan (96) extended this method to evolve the tradeoff between Pareto optimality and the robust measure of worst case performance for constrained multi-objective problems. In order to improve computational efficiency, the mechanisms of Tabu-restriction and μGA are implemented for the computation of the worst case performance and constraint violation. Apart from the nominal objective values, Li *et al* (181) applied the concept of the worst case sensitivity region (WCSR) and sought to maximize the radius of this region. The WCSR approximates the worst violation that a particular design can absorb before it violates some predetermined threshold on performance variation.

In (182), Lim *et al* also presented a single-objective/multi-objective inverse evolutionary optimization methodology for robust design. In contrast to conventional forward robust optimization, the inverse approach avoids making assumptions about the uncertainty when insufficient field data exists for estimating its structure. Apart from the objectives of nominal fitness and robustness, Lim *et al* considered the possible benefits as the uncertainty prevails by introducing an opportunity criterion in the inverse search scheme as the third objective.

The second variant is proposed by Deb and Gupta (56) as a more practical approach to the single-objective method and treats the selected robust measures as hard constraints. Adopting a similar approach in an independent work, Gunawan and Azarm (110) considered a designer specified WCSR radius as a constraint to be satisfied. Therefore, in both works, the goal is to evolve the best PF^{A}_{det} that satisfies the tolerable bounds on performance deviation.

8.1.3 Robust Multi-Objective Optimization Evolutionary Algorithm

In order to explore the tradeoffs between robustness and Pareto optimality for constrained MO problems, we suggested a robust multi-objective evolutionary algorithm (RMOEA) which considers robustness as an independent optimization criterion and implements the features of micro-genetic algorithm (μGA), Tabu restriction, and archival re-evaluation in (96). The main steps of RMOEA within each iteration are shown below.

Robust Multi-Objective Evolutionary Algorithm (RMOEA)

Step 1: Evaluate all solutions in P_t and update tabu list.
Step 2: Perform constrained Pareto ranking.
Step 3: Update archive:

- Add non-dominated solutions in P_t to A_t.
- Remove dominated solutions from updated A_t.
- Remove most crowded solutions based on niche count if $|\mathrm{A}_t|$ exceeds size limit.

Step 4: Combine updated A_t and P_t. Perform binary tournament selection to form mating pool.
Step 5: Recombination process.
Step 6: Apply AMO operator.

The evolutionary process starts with the initialization of the initial population of candidate solutions. The initialization process can either be random or created by means of Design of Experiment (DOE) techniques. After which, all individuals are evaluated based on the respective objective functions, the robust measure as well as the constraint violation. In particular, this chapter utilizes a measure that is calculated based on the worst case situation and will be described in the next section.

The evaluation process involves the selection of neighboring points for the assessment of each candidate solution. In order to minimize the effects of stochasticity associated with the random and DOE creation of neighboring points, RMOEA applies a μGA to search for the worst solution within a specified bound defined by the noise model considered. In addition, Tabu restriction is applied to guide the search and reduce computational load. After the evaluation process, archiving is performed to store the non-dominated solutions found along the evolutionary process. Since robust optimization is inherently uncertain, archived solutions are re-evaluated at periodic intervals.

After which, the selection of individuals into the mating pool is conducted. The selection process is elitist in nature and is conducted in a two-stage process: 1) the archive and evolving populations are combined and 2) binary tournament selection of this combined population is conducted to fill up the mating pool. In order to promote the exploration and exploitation of robustness and optimality, the selection criterion adopted in each tournament is randomly based on either a constrained Pareto ranking scheme or the robust measure. In the event of a tie, i.e. same rank or same robust measure, the individual with the lower niche count (102) will be selected. Note that the mechanism of niche sharing is used in the tournament selection and diversity maintenance in the archive. After the selection process, the individuals will undergo the process of uniform crossover and adaptive mutation operator (AMO) (252).

Multi-Objective Robust Measure: The RMOEA implements a multi-objective approach, which allows the algorithm to explore the tradeoffs between robustness and Pareto-optimality. The worst case measure is adopted and the robust measure for the $i-th$ objective can be written as,

$$f_i'(\mathbf{x}) = \frac{\max_{\mathbf{x} \in \mathbf{X}^{n_x}} \left(f_i(\mathbf{x}')\right) - f_i(\mathbf{x})}{f_i(\mathbf{x})} \tag{8.1}$$

where $\mathbf{x}' \in [\mathbf{x}' - lb, \mathbf{x}' + up]$. From equation (8.1), it can be noted that the measure reflects the degree of variation resulting from the worst objective value.

Constrained Pareto Ranking: In RMOEA, the feasibility of each individual for each constraint is represented by the worst case violation in its neighborhood in order to evolve solutions that remain valid under perturbations. An individual $\mathbf{F}_a$ constraint-dominates the individual $\mathbf{F}_b$ if,

1. $\mathbf{F}_a$ is feasible and $\mathbf{F}_b$ is infeasible.
2. Both solutions are infeasible and $\mathbf{F}_a$ dominates $\mathbf{F}_b$ in terms of the different constraint violations.
3. Both solutions are infeasible, incomparable in terms of the different constraint violations and $\mathbf{F}_a$ dominates $\mathbf{F}_b$ in terms of the objective values.
4. Both solutions are feasible and $\mathbf{F}_a$ dominates $\mathbf{F}_b$ in terms of the objective values.

Actual ranking is based on the scheme presented by Fonseca and Fleming (84), which assigns the same smallest rank for all non-dominated individuals, while the dominated individuals are ranked according to how many individuals in the population dominating them.

Tabu Restriction: Tabu restriction is used in conjunction with the evaluation process to reduce the number of evaluations required for solutions that are similar to those that have been assessed to be infeasible. In addition, it avoids the classification of a previously known infeasible solution as feasible solution due to the uncertainties involved in the evaluation process. During the evaluation process of each individual, each of the neighboring points will be examined against the Tabu list. Similar points will not be evaluated and will inherit the attributes of the matching Tabu solution in the Tabu list. At the same time, the Tabu list will be updated with new solutions that violate the constraints. Similar to the archive, the most crowded member will be removed when the maximum size is reached.

μGA Evaluation of Worst Case Performance: The RMOEA incorporates a μGA to perform a directed search for each individual's worst case performance and constraint violation within the neighborhood of the particular individual. The μGA begins with the creation of POP_SIZE_{GA} neighboring points about the individual under evaluation. The Tabu restriction-assisted assessment presented in the previous section is implemented for the evaluation of these individuals. Since the μGA sought to discover the worst case performance, the μGA is actually optimizing a MO problem that maximizes 1) the worst case objective in equation (8.1) and 2) the worst constraint violation. At every generation, the worst performance found will be updated and

the best individual will be stored. Elitism is applied by reinserting the best individual into the mating pool. Binary tournament selection is then conducted on the evolving population to fill up the mating pool. After which, the mating pool will undergo uniform crossover and AMO.

8.2 Empirical Analysis

8.2.1 Fitness Inheritance for Robust Optimization

Since the analytical effective objective values are often unavailable, Monte Carlo integration is applied to estimate the expected fitness values. However, this explicit averaging approach requires a large number of evaluations, which is a computationally expensive task. Therefore, the method of fitness inheritance is employed here to reduce the number of evaluations.

The key design issues in implementing fitness inheritance are when and how to inherit fitness from the parents. In this work, we adopt a similar strategy applied by Bui *et al* (33), in which the offspring inherits the mean fitness of the two parents if some similarity criteria are satisfied. After the genetic processes of crossover and mutation, the distance between the offspring and the parents is calculated. If the distance is smaller than a threshold T along all dimensions in the decision space, the offspring will inherit the mean fitness from the parents, otherwise the offspring will be evaluated for H samples to compute the expected objective values. T is set as 0.1 in the normalized decision space. Fitness inheritance is implemented in both NSGAII and SPEA2 for subsequent empirical studies.

8.2.2 Evaluating GTCO Test Suite

In this section, simulation studies are conducted to analyze the performances of NSGAII and SPEA2 on the proposed GTCO test suite. As before, both algorithms are implemented using the same binary coding scheme of 15 bits per variable, binary tournament selection, uniform crossover, and bit flip mutation. Thirty independent runs of 500 generations are performed for each of the test problems.

Empirical results

The performance trends of NSGAII and SPEA2 over the number of samples H={1, 10, 20, 40} and σ={0.0, 0.05, 0.1, 0.2} for the different problems are shown in this section. For each problem, we will show how the algorithms' abilities to find a diverse and near optimal PF^A are affected by the introduction of noise.

GTCO1 is a class 1 problem where multi-modality is introduced into the fitness landscape with noise while $PS^*_{det}=PS^*_{\text{eff},\sigma}$. From the metric of VD

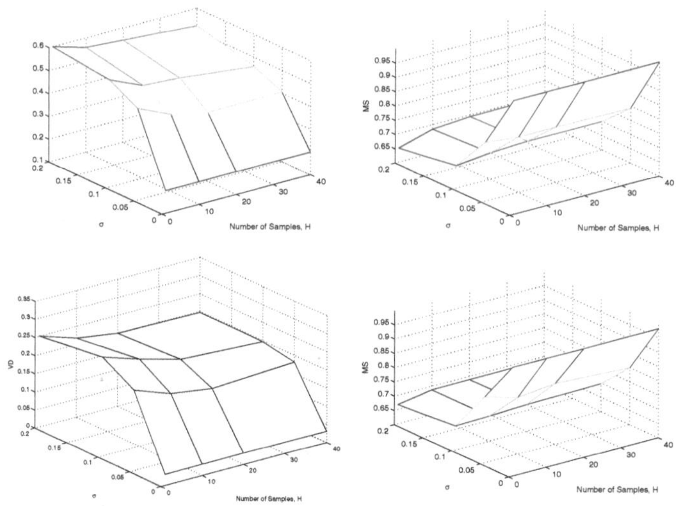

(a) (b)

Fig. 8.1 GTCO1 Performance trend of NSGAII (first row) and SPEA2 (second row) over H={1, 10, 20, 40} and σ={0.0, 0.05, 0.1, 0.2} for (a) VD and (b) MS

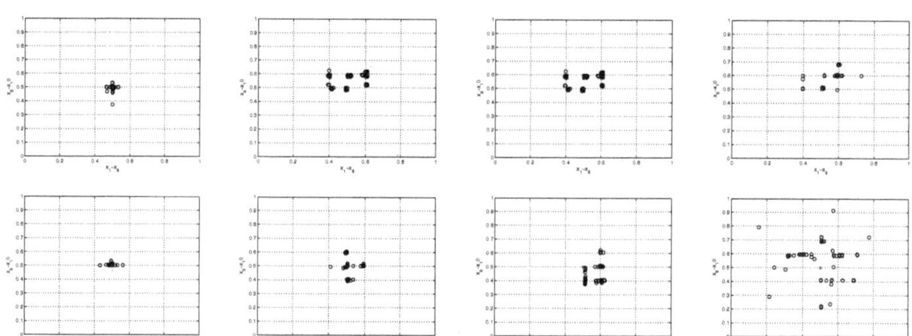

(a) (b) (c) (d)

Fig. 8.2 The evolved solutions of NSGAII (first row) and SPEA2 (second row) with number of samples H=1 for GTCO1 along $\mathbf{x}_{D2}$ with number of samples (a) $\sigma = 0$, (b) $\sigma = 0.05$, (c) $\sigma = 0.1$, and (d) $\sigma = 0.2$. The PS* is represented by (x) while the evolved solutions are represented by (o).

in Fig. 8.1(a) as well as the distribution of the solutions in Fig. 8.2(a)-(d), NSGAII and SPEA2 encounter no difficulty finding PS^*_{det} but both algorithms face increasing difficulties with increasing σ. The diversity of $\text{PF}^A_{\text{eff},\sigma}$ is also affected. This clearly demonstrates how robust optimization can be more difficult in the face of noise-induced landscape features.

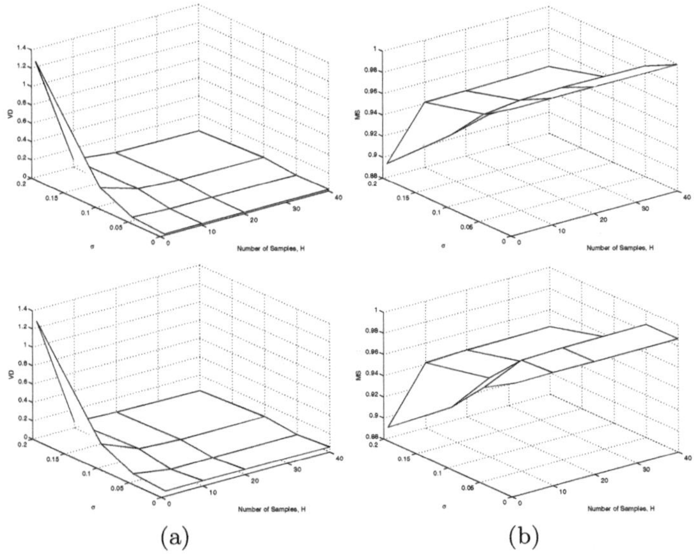

Fig. 8.3 GTCO2 Performance trend of NSGAII (first row) and SPEA2 (second row) over H={1, 10, 20, 40} and σ={0.0, 0.05, 0.1, 0.2} for (a) VD and (b) MS

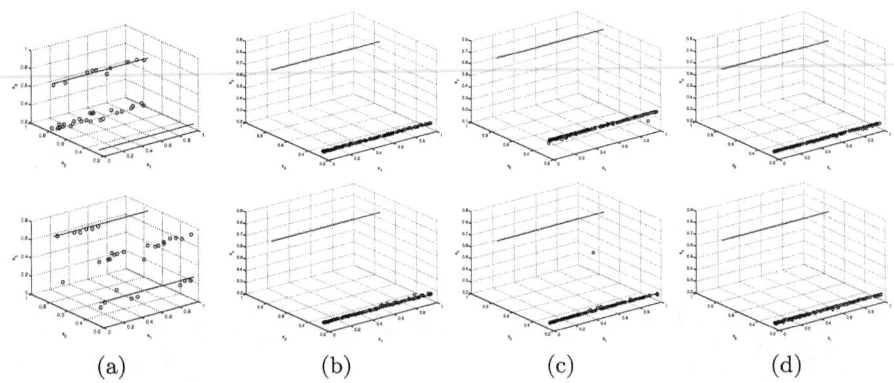

Fig. 8.4 The evolved solutions of NSGAII (first row) and SPEA2 (second row) at $\sigma = 0.2$ for GTCO2 as seen in the decision space with number of samples (a) H=1, (b) H=5, (c) H=10, and (d) H=20. The PS* is represented by (-) at (x_2, x_3)=(0.25,0.25) while the evolved solutions are represented by ($\circ$). The solutions have to be distributed along x_1 in order to find the entire span of PF*.

Fig. 8.3 shows the performance trends of SPEA2 and NSGAII for GTCO2. GTCO2 is a class 2 problem which exhibits a change in PS* once σ exceeds a certain threshold. The variations of $PS^A_{\text{eff},\sigma}$ for both algorithms can be seen in Fig. 8.4. Interestingly, the effects of implicit averaging can be observed in Fig. 8.4(a) and both algorithms are capable of finding $PS^*_{\text{eff},\sigma}$ as well as a

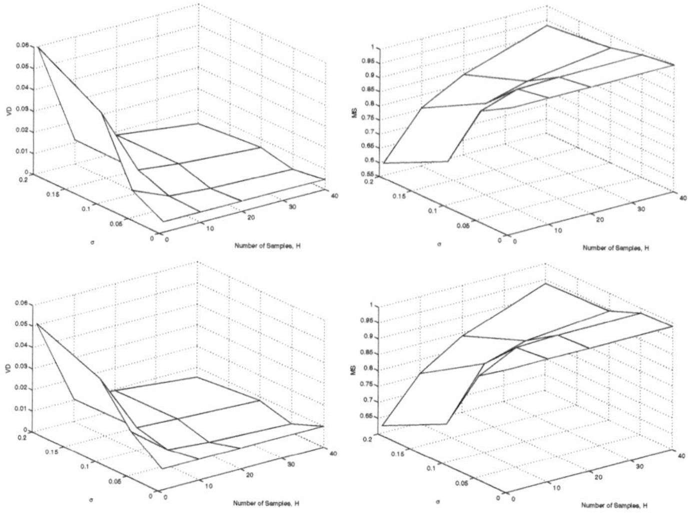

(a) (b)

Fig. 8.5 GTCO3 Performance trend of NSGAII (first row) and SPEA2 (second row) over H={1, 10, 20, 40} and σ={0.0, 0.05, 0.1, 0.2} for (a) VD and (b) MS

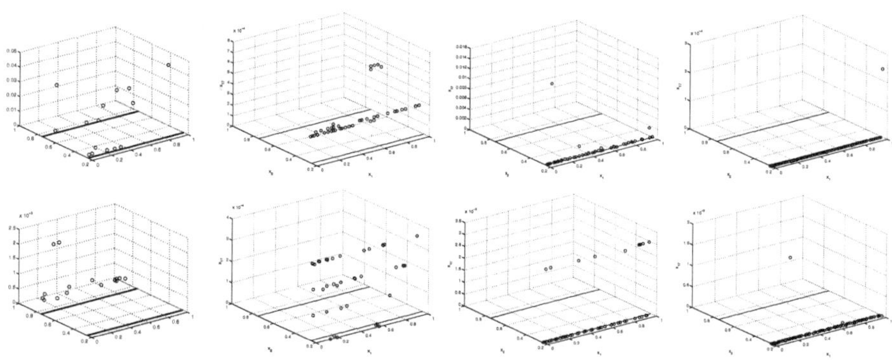

(a) (b) (c) (d)

Fig. 8.6 The evolved solutions of NSGAII (first row) and SPEA2 (second row) at $\sigma = 0.2$ for GTCO3 as seen in the decision space with number of samples (a) H=1, (b) H=5, (c) H=10, and (d) H=20. The PS* is represented by (-) at (x_2, x_{17})=(0.25,0) while the evolved solutions are represented by ($\circ$). The solutions have to be distributed along x_1 in order to find the entire span of PF*.

diverse $\mathrm{PF}^A_{\mathrm{eff},\sigma}$ given a sufficient number of samples. While both algorithms exhibit similar performances, it can be noted from Fig. 8.3(b) and Fig. 8.4(d) (along the x_1 axis) that the diversity maintenance mechanism of SPEA2 is more susceptible to noise.

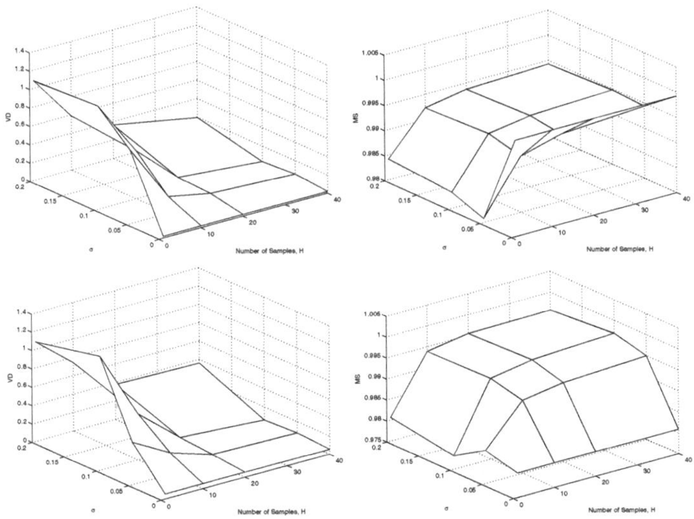

(a) (b)

Fig. 8.7 GTCO4 Performance trend of NSGAII (first row) and SPEA2 (second row) over H={1, 10, 20, 40} and σ={0.0, 0.05, 0.1, 0.2} for (a) VD and (b) MS

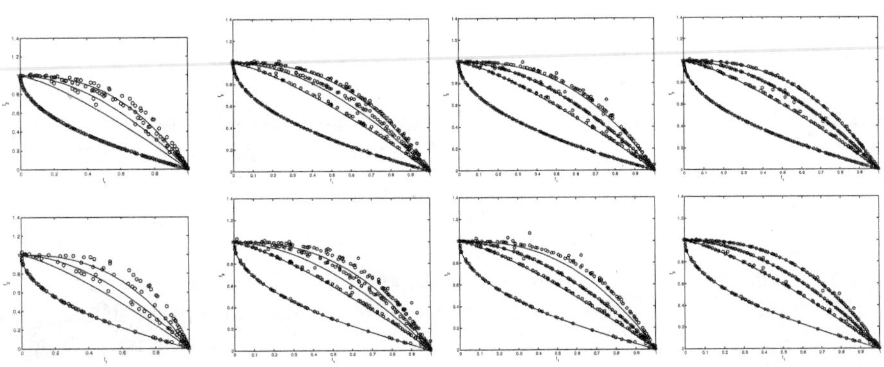

(a) (b) (c) (d)

Fig. 8.8 The PF^A of NSGAII (first row) and SPEA2 (second row) at various $\sigma = 0.2$ values for GTCO4 as seen in the decision space with number of samples (a) H=1, (b) H=5, (c) H=10, and (d) H=20. The PF^* is represented by (-) while the evolved solutions are represented by (∘).

GTCO3 is a class 3 problem which exhibits noise-induced landscape and PS^* changes. The performance trends of NSGAII and SPEA2 are shown in Fig. 8.5. As in the problem of GTCO2, NSGAII and SPEA2 are capable of finding $PS^*_{\text{eff},\sigma}$ as well as a diverse $PF^A_{\text{eff},\sigma}$ given a sufficient number of samples. Nonetheless, even though GTCO2 and GTCO3 undergo the same

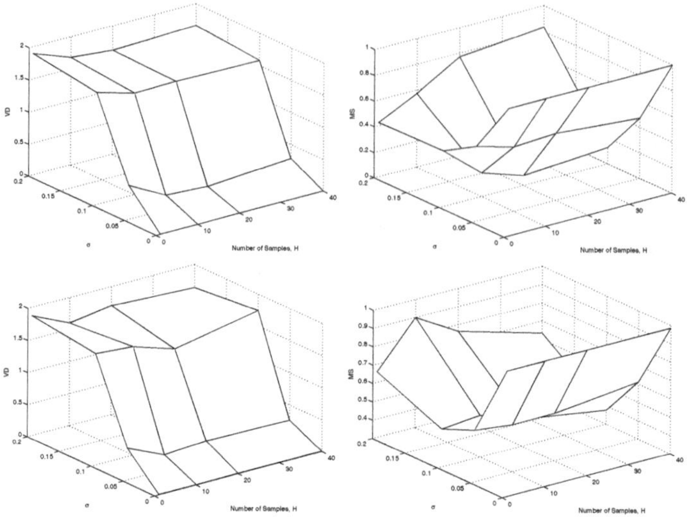

(a) (b)

Fig. 8.9 GTCO5 Performance trend of NSGAII (first row) and SPEA2 (second row) over H={1, 10, 20, 40} and σ={0.0, 0.05, 0.1, 0.2} for (a) VD and (b) MS

PS* transformation, it is obvious that the change in solution density results in different sampling requirements. This is evident by comparing Fig. 8.4(b)-(c) and Fig. 8.6(b)-(c).

Similar to GTCO2 and GTCO3, GTCO4 shows similar performance trends over different σ and H values in Fig. 8.7. However, since the change in PS* is closely linked to the change in PF*, the number of samples required to find PS* is also higher. The PF* changes with σ as shown in Fig. 8.8. Once again, it can be observed from Fig. 8.7(b) and Fig. 8.8(d) (along the x_1 axis) that noise poses considerable challenge to the diversity maintenance mechanism of SPEA2 as it is unable to distribute the solutions uniformly along $PF^A_{\text{eff},\sigma}$.

As in the case of GTCO1, NSGAII and SPEA2 have problems finding PS* for GTCO5 as evident from the metric of VD in Fig. 8.9(a). From Fig. 8.9(b), it appears that algorithmic performances in terms of MS seem to have improved beyond $\sigma = 0.1$. However, it is due to the fact that the span of $PF^A_{\text{eff},\sigma}$ has reduced considerably. Moreover, further investigations reveal that the number of solutions found is very small, even at H=40.

From the simulation results, it is clear that the problems of the GTCO test suite pose different difficulties to NSGAII and SPEA2. Generally speaking, both algorithms exhibit similar performances for the five problems over different H and σ settings. Nonetheless, it is also noted that SPEA2 did not fare as well in terms of the discovery of an uniformly distributed and diverse $PF^A_{\text{eff},\sigma}$. Together with the observations that NSGAII and SPEA2 are unable to handle noise-induced features of multi-modality and deception, this suggests that the state-of-the-art MOEAs may not be able to evolve robust solutions effectively through simplistic extensions.

8.2.3 Evaluating VRPSD Test Problems

In this section, simulation studies are conducted to analyze the behavior of the VRPSD, paying particular attention on the impact that customer topology has on the behavior of multi-objective routing and scheduling algorithms. Fig. 8.10 shows the PF^A_{det} and $\mathrm{PF}^A_{\mathrm{eff},\sigma}$ for three instances of the VRPSD, each having a different customer topology. The Pareto fronts in Fig. 8.10(a) are obtained based on

$$\langle \text{VRPSD1} : [C_d, C_m, C_v],\ 100,\ \text{Type-R},\ (50,50) \\ ,\ N\big(\mu, U(0, \tfrac{1}{3}\mu)\big),\ 10\rangle,$$

and the fronts in Fig. 8.10(b) are obtained based on

$$\langle \text{VRPSD2} : [C_d, C_m, C_v],\ 100,\ \text{Type-C},\ (50,50) \\ ,\ N\big(\mu, U(0, \tfrac{1}{3}\mu)\big),\ 10\rangle,$$

while those in Fig. 8.10(c) are obtained based on

$$\langle \text{VRPSD3} : [C_d, C_m, C_v],\ 100,\ \text{Type-RC},\ (50,50) \\ ,\ N\big(\mu, U(0, \tfrac{1}{3}\mu)\big),\ 10\rangle.$$

The PF^A_{det} for each of the test problems is obtained by treating the mean demand of each customer as its deterministic demand when evaluating the expected costs of solutions. This approach converts the stochastic problem into a deterministic one. On the other hand, the $\mathrm{PF}^A_{\mathrm{eff},\sigma}$ is obtained by evaluating a VRPSD solution multiple times (H=10), each time using a different set of customer demands randomly generated based on their demand distributions. The costs are then averaged and taken as the expected cost of the solution. In Fig. 8.10, the filled points correspond to the respective fronts obtained immediately after the optimization process. These points are what the logistic manager sees when he is deciding which one of the solutions along the front to implement. The hollow points are obtained by implementing the corresponding filled points on 100 sets of randomly generated customer demands and then taking the average to get the implementation cost. This cost would be a good indication of the quality of the solution and the Euclidean distance between corresponding filled and hollow points would give an indication of the robustness of the solution. From Fig. 8.10(a)-(c), it is obvious that the solutions along $\mathrm{PF}^A_{\mathrm{eff},\sigma}$ are more robust than those along PF^A_{det} since their expected cost values do not differ much from their implementation costs and would provide the logistic manager with more accurate information about the quality of the solutions based on which decision will be made.

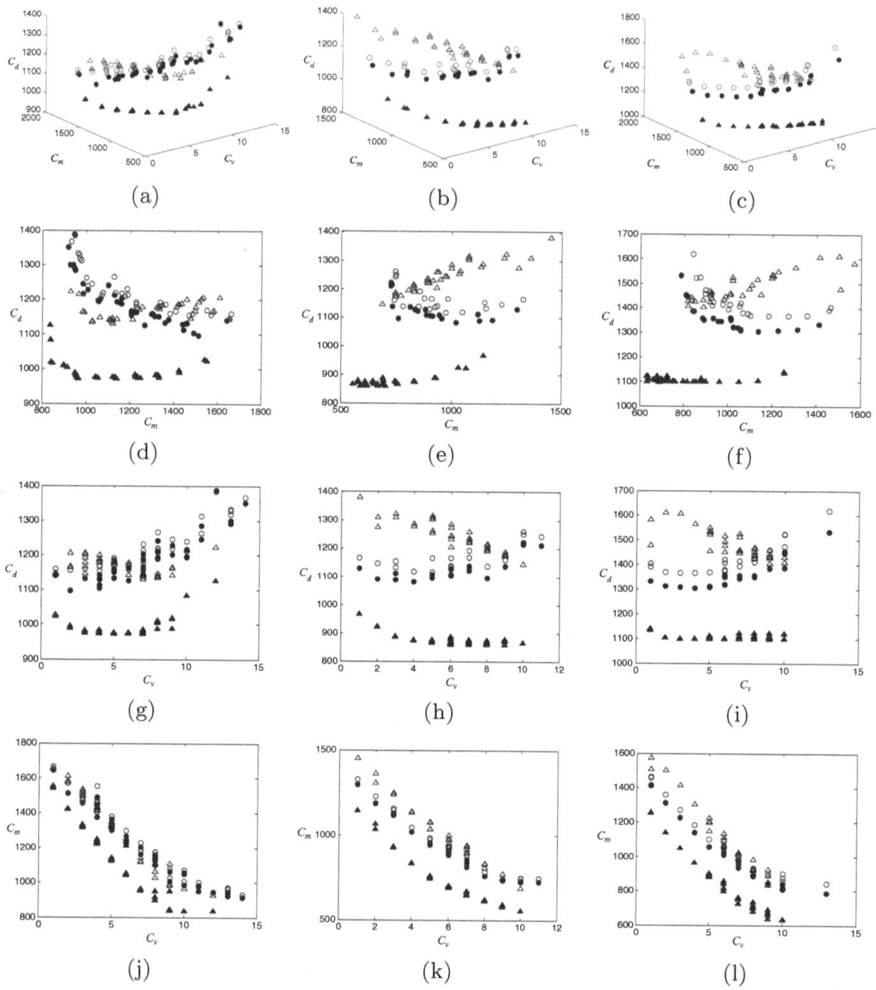

Fig. 8.10 Pareto fronts for (a) VRPSD1, (b) VRPSD2, (c) VRPSD3 test problems. The first row shows the three-dimensional Pareto fronts, the second row shows the same fronts along C_d and C_m, the third row shows the same fronts along C_d and C_v and the fourth row shows the same front along C_m and C_v. • denote solutions evolved using averaging while ▲ denote solution evolved deterministically. ○ and △ represent the corresponding solutions after averaging over 5000 samples.

In order to further examine the behaviors of the PF^A_{det} and $\text{PF}^A_{\text{eff},\sigma}$ for each of the VRPSD test problems, separate two-dimensional graphs, each time considering only two of the objectives, are plotted in Fig. 8.10(d)-(l). It can be observed from the figures that the $\text{PF}^A_{\text{eff},\sigma}$ considering the expected and implementation costs have the same shape. The spacing between corresponding solutions is also uniform. This implies that all the solutions along $\mathrm{PF}^{A}_{\mathrm{eff},\sigma}$ have almost equal levels of robustness. This cannot be said of the solutions along PF^{A}_{det}. The front for the expected costs changes shape in most cases when the implementation costs of the solutions along the front are considered. In Fig. 8.10(f), the front for the expected costs is horizontal but curves upwards when the implementation costs are considered. This shows that solutions with higher C_m are less robust. Similarly, it can also be seen from Fig. 8.10(h), 8.10(i), 8.10(k), and 8.10(l) that solutions for the Type-CS and Type-RCS test problems become more robust as C_v increases. It is also observed from Fig. 8.10(g) that the robustness of the solutions for the Type-RS test problem is the worst when there are four or five vehicles. All these observations point to the fact that the deterministic approach is not reliable in solving the VRPSD.

It can also be seen from Fig. 8.10(d)-(i) that the $\mathrm{PF}^{A}_{\mathrm{eff},\sigma}$ and PF^{A}_{det} obtained have very different shapes. In Fig. 8.10(d)-(f), $\mathrm{PF}^{A}_{\mathrm{eff},\sigma}$ shows that C_d decreases with increasing C_m but PF^{A}_{det} shows the exact opposite relationship, i.e. C_d increases with increasing C_m. Similarly, in Fig. 8.10(g)-(i), $\mathrm{PF}^{A}_{\mathrm{eff},\sigma}$ shows that C_d increases with C_v, while PF^{A}_{det} shows the reverse relationship. This finding demonstrates that the stochastic nature of the problem is not trivial and that inadequate handling of the problem will give the logistic manager a false understanding of the situation at hand.

8.3 Conclusion

This chapter describes some design issues that should be considered in the development of robust MOEA. In order to reduce computational requirements incurred by calculating the robust measures, a fitness inheritance scheme is suggested and incorporated into NSGAII and SPEA2. Subsequently, the modified MOEAs are applied to demonstrate the difficulties posed by the GTCO test problems described in Chapter 7. The study suggests that robust multi-objective problems can offer greater challenges to optimization algorithms when noise is introduced, highlighting the necessity to design more effective MOEAs as well as more rigorous simulation studies. Empirical analysis of the VRPSD test problems also reveal the importance of robust optimization for real-world multi-objective problems as the problem characteristics and hence solutions tend to change with noise.

Chapter 9
Evolving Robust Routes*

In the previous chapter, we described the VRPSD as an instance of real-world robust problem. The VRPSD differs from its deterministic counterparts in that when some data are random, it is no longer possible to require that all constraints be satisfied for all realizations of the random variables (170). In addition, the actual cost of a particular solution to the VRPSD cannot be known with certainty before the actual implementation of the solution. Optimization of the VRPSD deterministically may yield very good route schedules which we will show in this chapter to be very sensitive to variations in customer demand.

In this chapter, a hybrid MOEA (HMOEA) (249) is applied to solve the VRPSD problem described previously. The algorithm incorporates two heuristics which exploit knowledge of the VRPSD route structures to improve algorithmic performance. In addition, an intuitive route simulation method (RSM), which is an elaborate form of explicit averaging, is proposed to address the issue of evaluating the expected costs of solutions. A procedure based on the RSM is also proposed to assess the quality of solutions on top of comparing their expected transportation costs which has been used as the main performance measure hitherto.

9.1 Overview of Existing Works

Many researchers have studied the VRPSD in two frameworks, namely as a chance constrained program (CCP) (37, 38) or as a stochastic program with recourse (SPR). In CCP, the problem consists of designing a set of vehicle routes for which the probability of route failure is constrained to be below a certain threshold. It is shown by Steward and Golden (246) that, under some

---

* © 2008 Elsevier. Reprinted, with permission, from K. C. Tan, C. Y. Cheong and C. K. Goh, "Solving multiobjective vehicle routing problem with stochastic demand via evolutionary computation", *European Journal of Operational Research*, vol. 177, pp. 813-839, 2007.

C.-K. Goh and K.C. Tan: Evolutionary Multi-objective Optimization, SCI 186, pp. 229–247.
springerlink.com

restrictive assumptions, the problem can be reduced to a deterministic VRP and then solved using existing deterministic algorithms. Although the CCP tries to control the probability of route failure, the cost of such failures is ignored. In contrast, the SPR tries to minimize the expected transportation cost, which includes the travel cost as well as the additional cost generated by recourse policies. Gendreau *et al* (88) commented that SPR are typically more difficult to solve than CCP but their objective functions are more meaningful. As such, most of the recent researches revolve around SPR and results obtained are compared and assessed based on the expected transportation costs of solutions. In using SPR, various recourse policies have been explored and there are three common recourse policies (88, 170). In the first approach, also known as the simple recourse policy (89, 90, 170, 171, 260, 261), a vehicle returns to the depot to restock when its capacity becomes attained or exceeded. The vehicle will then resume service at the customer on the planned route where route failure had occurred. In the second approach (19, 28, 29, 283), preventive restocking is planned at strategic points, preferably when the vehicle is near to the depot and its capacity is almost empty, along the scheduled route instead of waiting for route failures to occur. The third approach sought to optimize the remaining portion of a route after each failure or knowledge of the actual demand of each customer (234, 235).

Yang *et al* (283) proposed a dynamic programming recursive objective function for the VRPSD. The Or-opt operator is adapted to the stochastic case using a fast approximation computation for the change in the objective function when performing a local search move where the objective function needs to be repeatedly computed. Yang *et al* also showed that the optimal route, in terms of travel distance, is always a single route, if only capacity constraints are considered.

Bianchi *et al* (28, 29) also employed the recursive objective function and its approximation in Or-opt operations in the analysis of various meta-heuristics such as iterated local search, tabu search, simulated annealing, ant colony optimization, and evolutionary algorithm. However, it should be noted that the dynamic programming recursive objective function ((28, 29, 283) is applicable only if demands take on integer values, i.e. the stochastic demands follow discrete distributions.

Dror and Trudeau (65) showed that given that the customers' demands are independent random variables with non-negative means, route failures are more likely to occur at the end of a route. They also showed that the expected transportation cost of a route is dependent on the direction in which the route is traversed.

9.2 Hybrid Evolutionary Multi-Objective Optimization

Section 9.2.1 describes the structure of the variable-length chromosome (253, 254) used to encode solutions in the HMOEA. Section 9.2.2 presents two local search heuristics that exploit the intrinsic structures of a VRPSD solution. The HMOEA has previously been applied to solve the vehicle routing problem with time windows (VRPTW) (254) and the truck and trailer vehicle routing problem (TTVRP) (253). However, due to the absence of hard constraint in the VRPSD, the search space of the problem is considerably larger than those of the VRPTW and the TTVRP. As such, the genetic operators of the HMOEA are enhanced here to allow the exploration of the larger search space and are presented in Section 9.2.3 and 9.2.4. Furthermore, a route simulation method (RSM) that allows effective evaluation of VRPSD solutions is presented in Section 9.2.5. The computing budget and the algorithmic flow of the HMOEA are presented in Section 9.2.6 and Section 9.2.7 respectively.

9.2.1 Variable-Length Chromosome

In the HMOEA, a variable-length chromosome representation, shown in Fig. 9.1, is applied such that each chromosome encodes a complete solution, including the number of vehicles and the customers served by these vehicles. A chromosome may consist of several routes and each route or gene is not a constant but a sequence of customers to be served. Such a representation is efficient and allows the number of vehicles to be manipulated and minimized directly for multi-objective optimization in the VRPSD.

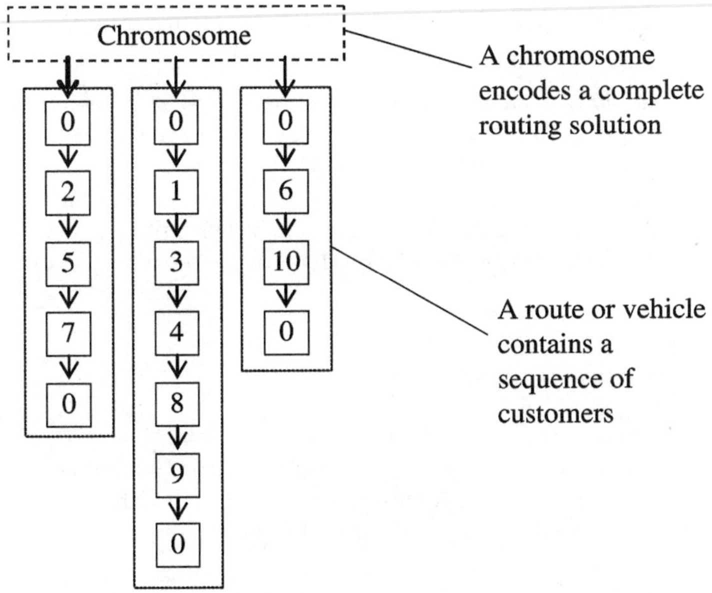

Fig. 9.1 Variable-length chromosome representation

9.2.2 Local Search Exploitation

The role of local search is vital in multi-objective evolutionary optimization in order to encourage better convergence and to discover any gap in the Pareto front. The local search approach can contribute to the intensification of the optimization results, which is usually regarded as a complement to the evolutionary operators that mainly focus on global exploration. The two local search methods presented here are inspired by the underlying structures of a VRPSD solution identified by Dror and Trudeau (65).

1) Shortest Path Search: Shortest Path Search (SPS) is designed to exploit the fact that route failures are more likely to occur at the end of a route. The SPS attempts to rearrange the order of customers in a particular route. For example, given a route that contains five customers, a new route is built by choosing the customer that is furthest from the depot as the first customer in the route, while the customer that is nearest to the depot is chosen as the last customer of the route. Next, the customer that is nearest to the first customer is chosen as the second customer, while the customer that is nearest to the last customer is chosen as the second last customer of the new route. This step continues until all the customers in the original route are re-routed. The new route will be compared against the original one and the better route will be retained. By re-routing customers in such a manner, customers that are further from the depot will be at the beginning of the route whereas those that are nearer to the depot will be at the end of the route. The rationale is to reduce the additional transportation cost that will be incurred by the recourse policy.

2) Which Directional Search: Which Directional Search (WDS) is designed to exploit the fact that the expected transportation cost of a route is dependent on the traversed direction. In contrast, for the deterministic VRP, the transportation cost of a route is the same regardless of the direction in which the route is traversed. To be specific, given a route, the WDS builds a new route that runs in the opposite direction. Similarly, the new route will replace the original one if it is better.

9.2.3 Route-Exchange Crossover

In contrast to classical one-point crossover which may produce infeasible route sequences, we adopt a simple route-exchange crossover operator that allows good sequences of routes or genes in a chromosome to be shared with other chromosomes in the evolving population. The operation of this crossover is shown in Fig. 9.2.

In the route-exchange crossover, only the best routes of the selected chromosomes are eligible for exchange. In the case where one of the selected chromosomes has only one route, a segment of the route is randomly selected to exchange with the other chromosomes best route which will be inserted as a new

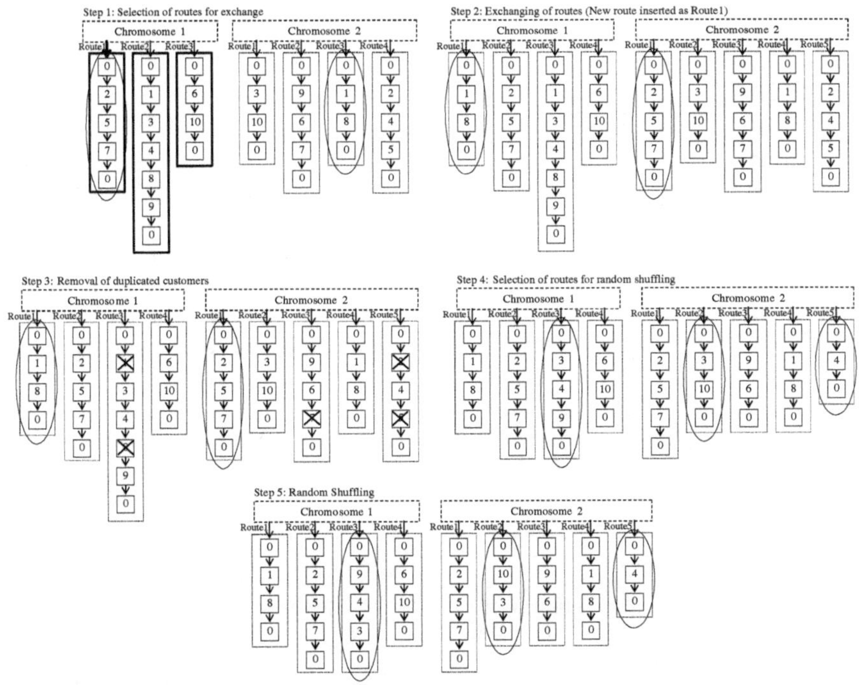

Fig. 9.2 Illustration of route-exchange crossover

route in the first chromosome. To ensure the feasibility of chromosomes after the crossover, duplicated customers are deleted. These customers are deleted from the original routes while the newly inserted route is left intact. The route-exchange crossover is enhanced with a random shuffling operator to increase the diversity of chromosomes to explore the larger VRPSD search space. In the random shuffling operation, with the exception of the newly inserted route, the order of customers in each of the remaining routes will be shuffled with a probability equal to the shuffle rate.

9.2.4 Multi-mode Mutation

The algorithm implements a multi-mode mutation operator (253, 254) to complement the crossover operator in allowing a larger search space to be explored. There are four parameters associated with the multi-mode mutation, namely mutation rate, elastic rate, squeeze rate, and shuffle rate. Fig. 9.3 shows the operation of the multi-mode mutation.

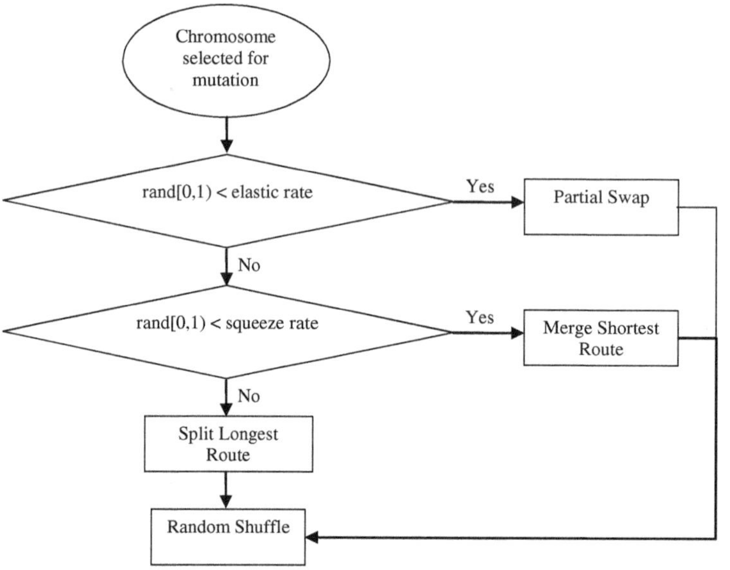

Fig. 9.3 Operation of multi-mode mutation

1) Partial Swap: The partial swap operator of Tan *et al* (254), which involves only a single swap move that swaps two routes at a random point, is modified in (249) to create a more diverse pool of chromosomes. The operation involves a number of swap moves and for each move, two routes will be randomly chosen. A segment is then randomly selected from each route and swapped to the other route. This new segment takes the place of the previous segment that has been swapped out. In the situation where either one of the routes has only a customer in it, a random segment is still selected from the route with more than one customer. This segment is then swapped with the solitary customer in the other route. In addition, a mechanism is in place such that the same two routes will not be selected twice in a particular partial swap operation.

2) Merge Shortest Route: This operation searches for the two routes of the chromosome with the smallest sum of travel distance and driver remuneration, and appends one route to the other. The merge shortest route will not operate on any chromosome with only one route.

3) Split Longest Route: This operation searches for the route with the largest sum of travel distance and driver remuneration, and breaks the route into two at a random point.

Like the route-exchange crossover, at the end of the multi-mode mutation, the random shuffling operation is applied on every route of each chromosome with a probability equal to the shuffle rate.

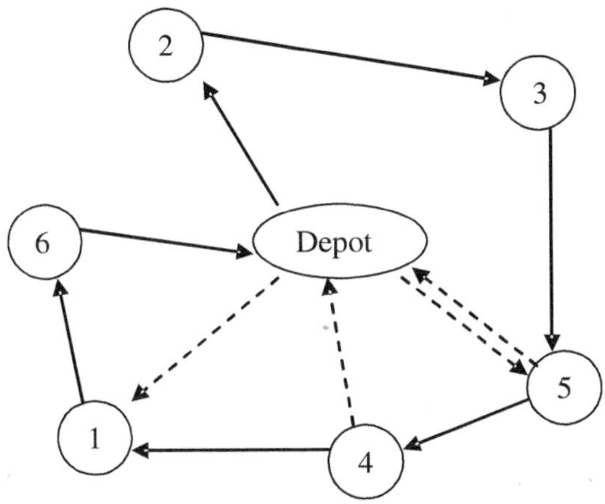

Customer	Real demand
1	5
2	6
3	2
4	13
5	9
6	5

Fig. 9.4 Example illustrating the operation of the RSM

9.2.5 Route Simulation Method

As mentioned earlier, one of the main difficulties of solving the VRPSD is in finding an objective function that is able to define properly the expected transportation cost of a solution. In this section, the route simulation method (RSM) is proposed to evaluate the expected costs of solutions. The fundamental idea behind the RSM is based on the sampling method of Lee and Chew (179) who applied the method for numerical optimization. Fig. 9.4 will be used to illustrate the operation of the RSM.

Fig. 9.4 shows a route sequence, Depot→2→3→5→4→1→6→Depot. The solid arrows indicate the route that the vehicle will take if this were a deterministic VRP. In the VRPSD, due to the recourse policies in the event of a route failure, the actual route taken by the vehicle cannot be known with certainty before the route is actually implemented. However, the implementation of the route can be simulated by generating a set of demands of all the customers based on their demand distributions and treating these demands as if they were the real demands revealed when a vehicle first arrives at the customer. The set of demands generated is tabulated in Fig. 9.4. For this particular example, it is assumed that the vehicle capacity is 15 and each arrow indicates a unit of distance.

The vehicle first leaves the depot and arrives at customer 2. It is able to satisfy its demand with a remaining capacity of 9. The vehicle then travels to customer 3 and satisfies its demand. The capacity of the vehicle is 7 when it reaches customer 5. The vehicle then finds that it is unable to satisfy the demand of customer 5, so it unloads all remaining goods and makes a return trip to the depot to restock. This recourse is indicated by the dashed arrows between the depot and customer 5. The vehicle then unloads two units of goods and leaves customer 5 for customer 4 with a capacity of 13.

After serving customer 4, the vehicle is empty and returns to the depot to restock. Since the demand of customer 4 has been satisfied, the vehicle travels to customer 1 from the depot. The vehicle then satisfies the demands of customers 1 and 6 and returns to the depot. From this simulation, the total distance traveled by the vehicle (10 units for this example) and the remuneration for the driver for a particular realization of the set of customer demands can be obtained.

Due to the stochastic nature of the cost considered, there is a need to sample or repeat the above operation H times for every route of a particular solution, using a different set of demands randomly generated based on the demand distributions of the customers each time and then taking the average to obtain the expected transportation cost of the solution. Three different RSM settings are studied. Generate Every Generation (GEG) refers to the setting where the H demand sets, which are used by the RSM, are refreshed every generation. Generate Every M (GEM) refers to the setting where the H demand sets are only refreshed at the end of every M generations. Lastly, Alternate Every M (AEM) refers to the setting where for M generations, the RSM uses the H randomly generated demand sets and for the next M generations, the RSM uses the mean values of the customers demand distributions to simulate the implementations of the routes. The H demand sets are then refreshed for use over the next M generations and the process repeats.

9.2.6 Computing Budget

The computing budget (39, 179) represents a fixed amount of computational work. As can be observed from the previous section, the RSM requires intensive computations and can be regarded as the bottleneck of the whole algorithm. As such, it makes sense to define one unit of the computing budget as one run of the RSM on a particular solution using a particular demand set. Thus, the computing budget puts a cap on the total number of times the RSM is run on solutions in the HMOEA and can be regarded as the stopping criterion of the evolutionary algorithm. For GEG and GEM, the computing budget set is actually approximately equal to the product of the size of the search population, the number of generations used in the HMOEA, and H, the number of samples of RSM is performed to obtain the expected transportation cost of a solution. This product gives only an approximate value due to the additional runs of the RSM during local search. As for AEM, the following equation applies

$$\text{Computing Budget} \approx 0.5 \cdot (\mathrm{T} \cdot |\mathrm{P}_t|) \cdot (\mathrm{H}+1) \tag{9.1}$$

where T is the number of generations.

This equation takes into account the fact that AEM alternates between using H randomly generated demand sets and the mean demand set for the RSM every M generations. As such, the RSM needs to run H times when using the H randomly generated demand sets but only runs one time when using the mean demand set. The equation only gives an approximation, again due to the additional runs of the RSM during local search.

9.2.7 Algorithmic Flow of HMOEA

The major steps of HMOEA are shown below:

Hybrid Multi-Objective Evolutionary Algorithm

Step 1: Build customer database.
Step 2: Initialize population.
Step 3: Update archive.
Step 4: Perform binary tournament selection to form mating pool.
Step 5: Apply route-exchange crossover and multi-mode mutation.
Step 6: Perform local search if criterion is satisfied.
Step 7: Route simulation method and Pareto ranking.
Step 8: Elitism.
Step 9: If stopping criterion is not satisified, repeat process from Step 3. Otherwise output archive.

At the start of the algorithm, a database of customers' information is built. The information consists of the coordinates, the mean and variance of the demand distribution, and the service time of each customer. Other information includes the capacity and time window of a vehicle, and the coordinates and service time of the depot.

Initialization: The first chromosome is built such that the sum of the mean values of the customer demands on each route does not exceed the vehicle capacity. Furthermore, the sum of the travel and service times on each route must not exceed the vehicle time window. The number of vehicles required in this first chromosome is then taken as the maximum number of vehicles that each of the remaining chromosomes can use. For each chromosome, the number of vehicles is randomly picked from the feasible range. The routes are then built such that each route has approximately the same number of customers. This procedure is done so that the initial population has a wide range of chromosomes with different number of vehicles to start with.

Evaluation: After the initial evolving population is formed, all the chromosomes are evaluated based on the RSM and ranked according to their Paretos dominance in the population. Following the ranking process, an archive population is updated. The archive population has the same size as the evolving population and is used to store all the best solutions found during the search. The archive population updating process consists of a few steps. The evolving population is first appended to the archive population. All repeated chromosomes, in terms of the objective domain, are deleted. Pareto ranking is then performed on the remaining chromosomes in the population. The higher ranked (weaker) chromosomes are then deleted such that the size of the archive population remains the same as before the updating process. The evolving population remains intact during the updating process.

Genetic operations: The binary tournament selection scheme is then performed. All the chromosomes in the evolving population are randomly grouped into pairs and from each pair, the chromosome with the lower rank is selected for reproduction. This procedure is performed twice to preserve the original population size. The genetic operators consist of the route-exchange crossover and the multi-mode mutation. To further improve the internal routings of customers, the SPS and the WDS are applied to the evolving and archive populations every 50 generations for better local exploitation in the evolutionary search.

Elitism: A simple elitism mechanism is employed in the HMOEA for faster convergence and better routing solutions. The elitism strategy involves randomly picking a number of good chromosomes (3% of the population size) from the pool of chromosomes in the archive population belonging to the best three Pareto ranks. The chosen chromosomes then replace the worst ranked chromosomes in the evolving population.

This is one complete generation of the HMOEA and the evolution process repeats until the computing budget is exhausted.

9.3 Simulation Results and Analysis

The HMOEA is programmed in C++ and simulations were performed on an Intel Pentium 4 2.8 GHz computer. Table 9.1 shows the parameter settings chosen after some preliminary experiments.

The test problem applied in this section can be described by

$$\begin{aligned}\langle \text{VRPSD1} : [C_d, C_m, C_v],\ 75,\ \text{Type-R},\ (40, 40),\\ N(\mu, U(0, \tfrac{1}{3}\mu)),\ \max\{1, \mu - 10\}\rangle,\end{aligned} \tag{9.2}$$

The actual location and μ of each customer are based on the test function in (65). It is to be noted that all the customers' demands in VRPSD1 are normally distributed. The service time of each customer is filled in by subtracting 10 from the mean demand or one unit, whichever is larger. This is done so that a customer with a higher mean demand would require a longer service time. A fixed service time of 10 is also set for the depot. Since VRPSD1 uses a 70x80 map, the vehicle time window is calculated to be 212 units. This time window is equivalent to 8 hours. As such, each hour corresponds to 26.5 units, which is used to compute the remuneration for drivers according to the rates given in the previous chapter.

The subsequent sections present the extensive simulation results and analysis. Section 9.3.1 demonstrates the effectiveness of the proposed hybrid local search, as well as analyzes how the various settings in which the local search

Table 9.1 Parameter settings for HMOEA

Parameter	Settings
Population size	500
Crossover rate	0.7
Mutation rate	0.4
Elastic rate	0.5
Squeeze rate	0.5
Shuffle rate	0.3
Computing budget	2,000,000

heuristics are incorporated with the HMOEA will affect the performance of the algorithm. Section 9.3.2 presents a new way of assessing the quality of solutions to the VRPSD on top of comparing their expected transportation costs and shows that the HMOEA, equipped with the RSM, is able to produce solutions that are robust to the stochastic nature of the problem. Section 9.3.3 shows how the value of H affects the performance of GEG, whereas section 9.3.4 attempts to study how the value of M affects the performance of GEM.

9.3.1 Performance of Hybrid Local Search

The HMOEA incorporates the local search heuristics in order to exploit local routing solutions in parallel with global evolutionary optimization. The local search heuristics are specially designed to exploit the route structures of a solution to the VRPSD. This section demonstrates the effectiveness of local exploitation in the HMOEA and also analyzes the effectiveness of various settings in which the local search heuristics are incorporated with the HMOEA.

Simulations were conducted using six different settings. Three of the settings include the HMOEA with no local search (NLS), with only WDS (WD), and with only SPS (SP). WD/SP is a setting which involves the application of WDS for the first two local exploitations in the HMOEA, i.e. on the 50th and the 100th generation, and alternates between the two local search heuristics every 100 generations, while applying local search every 50 generations. On the other hand, SP/WD starts with SPS on the 50th and the 100th generation, and alternates between the two local search heuristics every 100 generations. The final setting is RAN, where during the application of local search every 50 generations of the HMOEA, each chromosome will have equal chance of being applied by either SPS or WDS on all of its routes. Each of the six settings underwent 10 simulation runs. The simulations were conducted using the GEG setting with H set to 10.

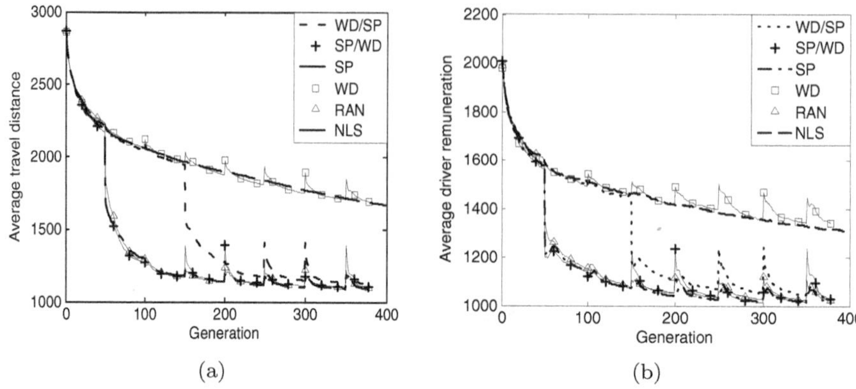

Fig. 9.5 (a) Average travel distance and (b) average driver remuneration of archive populations for different local search settings

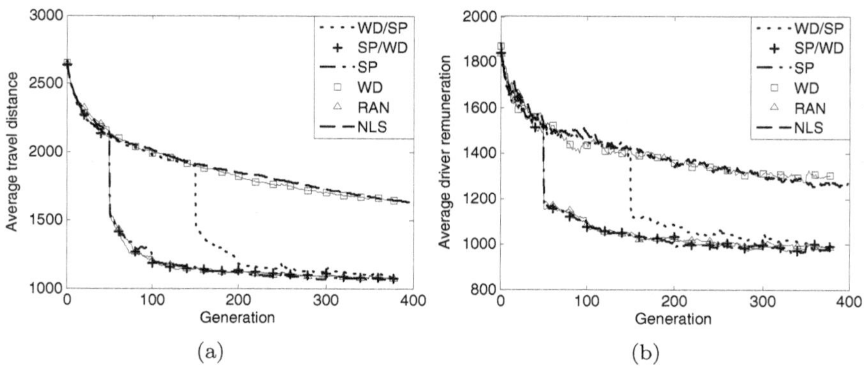

Fig. 9.6 (a) Average travel distance and (b) average driver remuneration of non-dominated solutions for different local search settings

The convergence traces of the travel distance and the driver remuneration for the six settings are plotted in Fig. 9.5(a)-(b) and Fig. 9.6(a)-(b). Fig. 9.5 shows the convergence of the respective costs, averaged over all the solutions in the archive population, over the generations. Fig. 9.6(a)-(b) shows the same costs averaged over all the non-dominated solutions in the archive population. The costs are further averaged over the 10 simulation runs performed. The plots show the effectiveness of local exploitation in the HMOEA as the five settings which use local search perform better than NLS. The effectiveness of the SPS is evident since the four settings, namely WD/SP, SP/WD, SP, and RAN, which make use of the local search operator, are able to find solutions with travel distance and driver remuneration significantly lower than those found by WD and NLS. The SPS is able to speed up convergence as it causes sharp dips in the respective costs of the solutions found whenever it is performed. The performances of WD/SP, SP/WD, SP, and RAN are comparable and the setting WD/SP is selected as the default setting for any further analysis unless otherwise stated.

In Fig. 9.5, it is observed that there are distinctive spikes in the convergence traces which coincide with the occurrences of local search. This is despite the fact that during local search, a new route is constructed and compared with the original one and the better route is retained. This happens because in comparing the new and original routes during local search, the solutions in the archive population are re-evaluated by the RSM. This re-evaluation acts to complement the RSM and is important in the stochastic problem where the costs of solutions are sensitive to the demand sets that are used by the RSM. For a particular solution, the fitness evaluated using the RSM can be very different depending on the demand sets used. As such, it is essential that a solution to the VRPSD be robust to the stochastic nature of the problem and its fitness should not differ too much with each evaluation by the RSM. The re-evaluation of all the solutions in the archive population during local search ensures that only robust solutions stay non-dominated. The effect of this can be seen in Fig. 9.6 which considers only non-dominated solutions in the archive population. The spikes in these convergence traces during local search are significantly smaller, if not negligible.

9.3.2 Comparison with a Deterministic Approach

In the absence of a stochastic procedure to deal with stochastic demands, one can generate the routes using a deterministic vehicle routing algorithm by treating the expected demand at each customer as its deterministic demand. The attraction of this deterministic approach is its relative simplicity and familiarity to practitioners. The HMOEA can in fact be modified into a deterministic vehicle routing algorithm by solely using the mean demand set in the RSM. However, what makes the HMOEA different from a deterministic vehicle routing algorithm is the RSM's ability to operate on demand sets which are randomly generated based on the demand distributions. This section will show that the RSM's ability to operate on randomly generated demand sets can lead to solutions which are more robust to the stochastic nature of the problem compared to the deterministic approach and that the expected transportation costs of such solutions are good estimates of the true performance of the solutions. In addition, a RSM-based procedure is proposed to assess the quality of solutions on top of comparing their expected costs. For comparison purposes, simulations were conducted on the three RSM settings, GEG, GEM, and AEM, with H and M both set to 10. The results of these settings are compared with the deterministic approach (DET). These four settings provide a spectrum, from pure stochastic to pure deterministic, of approaches to the VRPSD. Ten simulation runs of each of the four settings were performed.

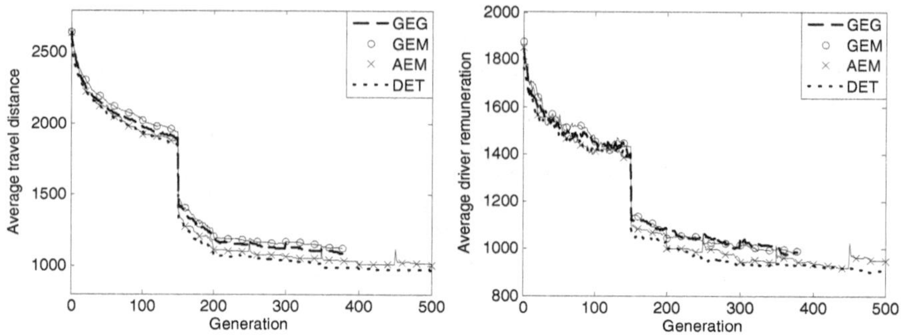

(a) (b)

Fig. 9.7 (a) Average travel distance and (b) average driver remuneration of non-dominated solutions of GEG, GEM, AEM, and DET

The convergence traces of the travel distance and the driver remuneration, averaged over the non-dominated solutions and over the 10 simulation runs, for the four settings are plotted in Fig. 9.7. Due to the nature with which the RSM is run in AEM and DET, i.e. the RSM is run only once, instead of H times, when evaluating solutions using the mean demand set, the HMOEA for these two settings took more than 500 generations to complete. However, by the 500th generation, these two settings have converged and the plots in Fig. 9.7 show only the convergence traces up to the 500th generation. By comparing the convergence traces of the four settings, it appears that DET is able to churn out the best solutions since both the average travel distance and the average driver remuneration are the lowest among the four settings at the termination of the algorithm.

It is highlighted in the introduction that due to the stochastic nature of the problem, the actual cost of a particular solution to the VRPSD cannot be known with certainty before the actual implementation of the solution. During the decision making process, the logistic manager will look at the expected transportation costs of all the candidate solutions and choose the solution that best suits the companys logistic condition, in terms of the available vehicle fleet size, and the companys priorities of whether to take the solution with a shorter travel distance but is likely to incur greater cost in the form of the remuneration for the drivers. In view of this, for the logistic manager to make correct decisions, it is important for the expected cost of each solution to give a good estimate of the true performance of the solution, i.e. the actual cost of implementing the solution should not deviate too much from the expected cost. As such, it is necessary to compare the results to the VRPSD based on this aspect on top of comparing their expected costs.

To perform such a comparison, a test demand set is randomly generated based on the customers demand distributions. This test demand set will represent the real demands that the vehicles of a particular solution would

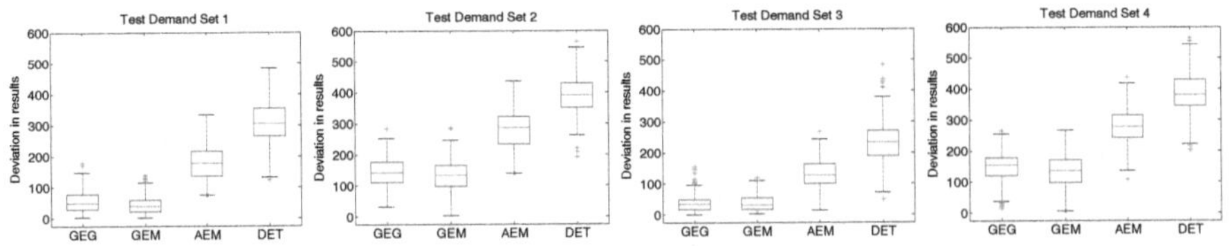

Fig. 9.8 Deviation between actual and expected costs of non-dominated solutions of GEG, GEM, AEM, and DET for four test demand sets

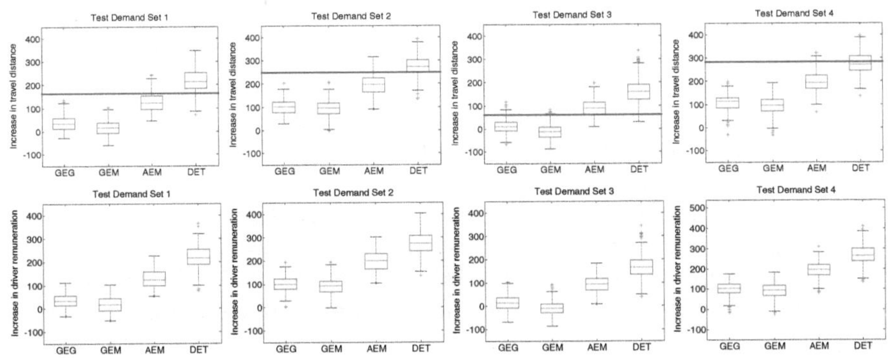

Fig. 9.9 Increase in travel distance (first row) and driver remuneration (second row) after implementing non-dominated solutions of GEG, GEM, AEM, and DET

experience when the solution is implemented. The RSM is then operated, using that test demand set, on all the non-dominated solutions found at the termination of each of the four settings, GEG, GEM, AEM, and DET, to simulate the actual costs of implementing the solutions. The deviation between the actual and expected costs of each solution is then calculated using the following equation:

$$Dev = \sqrt{(C_{d,exp} - C_d)^2 + (C_{m,exp} - C_m)^2} \tag{9.3}$$

This deviation is essentially the Euclidean distance in the objective domain between the actual and expected costs of each solution. To ensure that the results are not biased towards any test demand set, the same process is repeated for three other randomly generated test demand sets. The above procedures are repeated for the non-dominated solutions found by the 10 simulation runs of each setting that were performed. The results of these comparisons are represented in box plots and are shown in Fig. 9.8.

It can be seen from Fig. 9.8 that the expected costs of solutions obtained by GEG and GEM deviate less from the corresponding actual costs for all the test demand sets compared to the other two settings. AEM and DET produce solutions that have expected costs that are poor estimates of the actual costs. The spreads of their deviations are also larger compared to those of GEG and GEM which will result in poorer predictability in the deviations. It is noted that test demand sets 2 and 4 resulted in greater deviations for all the four settings but the order of performances of the four settings remains the same. Although the above results show the magnitudes of deviations between the actual and expected costs of solutions, they do not show the direction of these deviations. The actual cost of a particular solution can be better than the expected cost of the solution even though the deviation between the two costs is large. To compare the performances of the four settings based on this aspect, two separate comparisons were made. The first involves comparing the increase in travel distance, from the expected value, after implementing a particular solution, whereas the other compares the increase in driver remuneration. The results of these two comparisons are shown in Fig. 9.9. The figures again show the same pattern where GEG and GEM produced the most robust solutions which have expected costs that are good approximations of the actual costs.

A test is also conducted to compare the robustness of the solution found by (65) with those found by the four settings. The solution of Dror and Trudeau (65) is implemented using the simple recourse policy. The increases in travel distances for the four test demand sets are obtained and plotted as four horizontal lines in the respective box plots in Fig. 9.9 since (65) only considered the single objective of travel distance. From Fig. 9.9, it can be seen that the solution of Dror and Trudeau (1986) is not as robust as those found by GEG and GEM. For test demand sets 2 and 4, the increases in distances after implementing the solution are comparable with those found by DET. This is despite the fact that Dror and Trudeau (65) used a worst case recourse policy where in case of a route failure, all the remaining customers in the route are served through individual deliveries.

9.3.3 Effects of Sample Size, H

This section analyzes the effect of H on the performance of the HMOEA using the GEG setting in terms of the expected costs of solutions found and how well these expected costs approximate the actual costs of implementation. Ten simulation runs of seven settings with H=$\{1, 3, 5, 10, 30, 50, 70\}$ are performed. The RAN local search setting is used in all the simulations here to allow a fair comparison since the H settings of 30, 50, and 70 run for less than 150 generations and would not be operated by the SPS if the default local search setting of WD/SP were used.

The convergence traces of the travel distance and the driver remuneration, averaged over the non-dominated solutions in the archive population and over the 10 simulation runs, for the seven settings are plotted in Fig. 9.10. Due to space constraints, the convergence traces for the H settings of 1, 3, and 5 are shown only up to the 400th generation. From Fig. 9.10, it can be observed that as the value of H increases, the number of generations used in

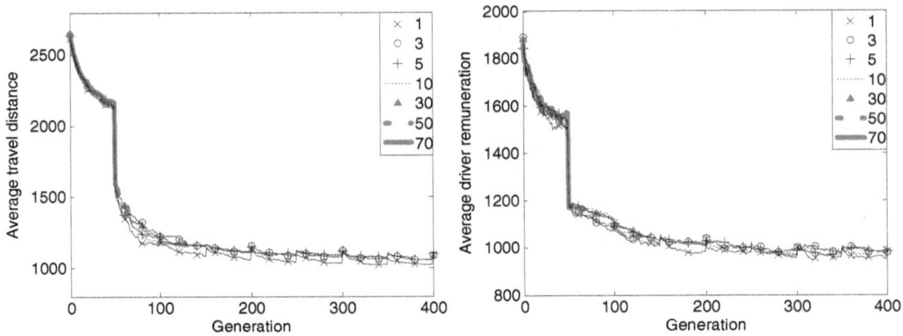

(a) (b)

Fig. 9.10 (a) Average travel distance and (b) average driver remuneration of non-dominated solutions of GEG using different H values

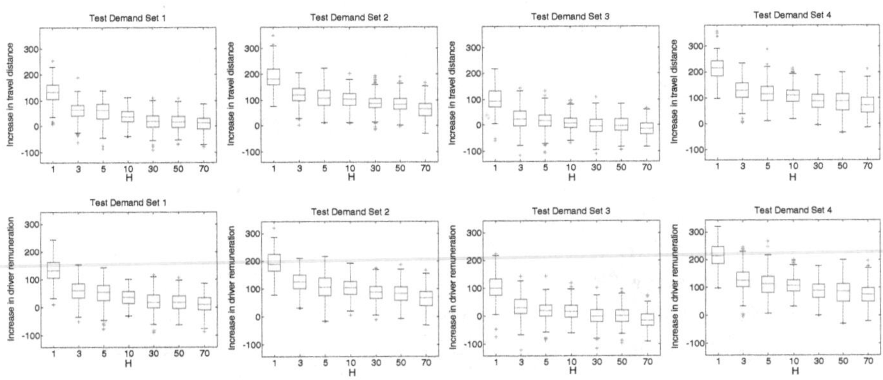

Fig. 9.11 Increase in travel distance (first row) and driver remuneration (second row) after implementing non-dominated solutions of GEG using different H values

the HMOEA decreases (A portion of the convergence traces has been enlarged to highlight this point). This is because as H increases, more runs of the RSM is applied each time the fitness of a chromosome is evaluated and since the total number of times the RSM is applied throughout the algorithm for the seven settings is fixed at the computing budget, the number of generations used in the HMOEA is reduced accordingly. This reduction in the number of generations used in the HMOEA results in poorer routing solutions as the HMOEA does not have sufficient time to explore the search space.

The deviation between expected and actual cost for the different settings are shown in Fig. 9.11. It can be observed that as the value of H increases, the solutions found are more robust to the stochastic nature of the problem in that the expected costs of the solutions are better estimates of the actual costs. From the above results, it can be seen that while setting a larger value of H for GEG will produce more robust solutions whose expected costs are better approximations of the actual costs of implementation, due to the fixed computing budget, the corresponding smaller number of generations used in the HMOEA will result in poorer routing solutions as there is insufficient time to explore the search space. As such, there is a trade-off between the number of generations used in the HMOEA and H, the number of repetitions of the RSM to obtain the expected cost of a chromosome.

9.3.4 Effects of M

Having analyzed how H affects the performance of GEG, this section attempts to study if M, the number of generations of the HMOEA before the H demand sets used by the RSM are refreshed, has any effect on the performance of GEM. It is to be noted that if M = 1, GEM becomes GEG since the H demand sets used by the RSM will be refreshed every generation of the HMOEA, and if M is equal to the maximum number of generations of the HMOEA, i.e. the H demand sets remain unchanged throughout the algorithm, and H = 1, then GEM becomes DET except that the mean demand set is replaced by a demand set that is randomly generated from the demand distributions of the customers.

Ten simulation runs of three settings with M=$\{100, 200, 300\}$ are performed. H is set to 10 for all the three settings. The convergence traces and box plots for the three settings are plotted in Fig. 9.12 and Fig. 9.13 respectively. The plots for GEG (M = 1), M = 10, and DET are also plotted for comparison. From the figures, it can be seen that the performance of GEM is the same regardless of the value of M. The average travel distance and the average driver remuneration of the non-dominated solutions found are almost the same. The robustness of the solutions is also comparable. The large disparity between M = 300 and DET in terms of how well the expected

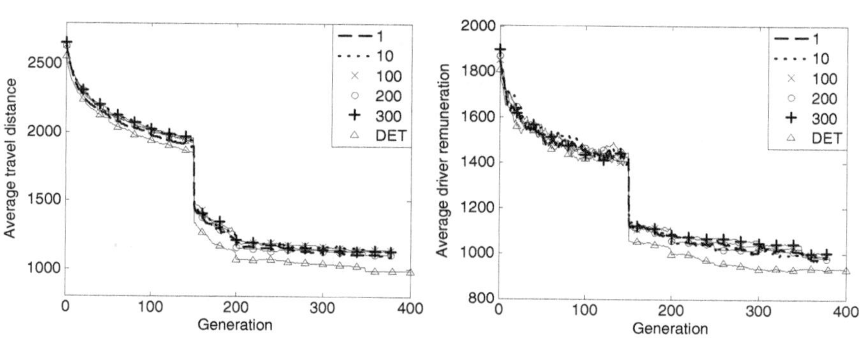

Fig. 9.12 (a) Average travel distance and (b) average driver remuneration of non-dominated solutions of GEM using different M values

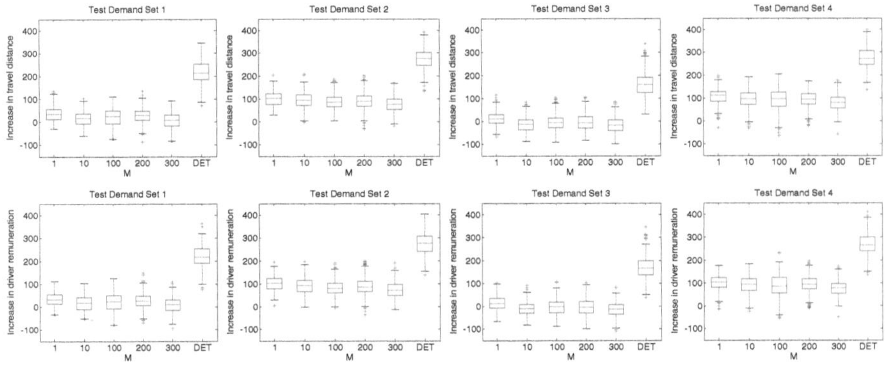

Fig. 9.13 Increase in travel distance (first row) and driver remuneration (second row) after implementing non-dominated solutions of GEM using different M values

costs of solutions estimate the actual costs also highlights the contribution of the RSM to finding robust solutions to the VRPSD since the major difference between the two settings is the RSMs ability to use H randomly generated demand sets to evaluate the fitness of chromosomes.

9.4 Conclusion

A HMOEA featured with enhanced genetic operators and two VRPSD-specific local search heuristics has been presented in this chapter. To evaluate the cost of a VRPSD solution, which is stochastic, a route simulation method (RSM) has also been implemented and incorporated with the HMOEA. The effectiveness of the two VRPSD-specific local search heuristics and the various settings in which local exploitation is incorporated with the HMOEA have been studied. A new way of assessing the quality of solutions to the VRPSD on top of comparing their expected costs has also been proposed. Extensive simulations have been performed to show that the solutions obtained by the HMOEA, equipped with the RSM, are robust to the stochastic nature of the problem. The expected costs of such solutions are good approximations of the actual costs of implementing the solutions, thus providing the logistic manager with accurate information based on which decision will be made.

Chapter 10
Final Thoughts

Our primary objective is to provide readers with a comprehensive introduction on the design of evolutionary algorithms for multi-objective optimization with uncertainties. This goal has been accomplished through a relatively easy walk-through on the basic ideas found in the single-objective literature that can be applied to handle multi-objective noisy, dynamic and robust fitness landscapes. These ideas have been introduced in the context of uncertain multi-objective optimization, together with the different multi-objective evolutionary algorithms designed specifically for handling uncertainties in the preceeding chapters. In doing so, we hope to expose the readers to a wide range of design optimization issues and concepts, ranging from the use of possibilistic theory to dampen the effects of noise to the use of predictive models to forecast future solutions in dynamic optimization. With this exposure, readers will have a better appreciation of the generic nature of evolutionary computation techniques, which allows one to extend and explore new ideas that can better handle uncertainties.

When we embarked on this research of evolutionary multi-objective optimization with uncertainties, there are only a few publications on this topic. However, the number of publications has since increased steadily though there remains many issues to consider in this area. An in-depth discussion for each of the three forms of uncertainties covered here would warrant a book on its own each. In squeezing all three topics within a single cover, we have inevitably neglected some related issues. This very last chapter aims to identify some directions for future development of better evolutionary algorithms for multi-objective optimization in uncertain and dynamic environments.

In Chapter 2 and 3, we assumed that noise in the fitness functions can be described by a Gaussian distribution. Noise is commonly modeled as such but this assumption is not valid for all cases. The fact that other noise models such as Cauchy and χ^2 have different impacts on the optimization process implies that techniques developed on the grounds of any particular noise model may not extend well for others. Examples of such techniques include probabilistic selection schemes of ESPEA and MOPSEA, and the heuristics of ELDP and GASS.

C.-K. Goh and K.C. Tan: Evolutionary Multi-objective Optimization, SCI 186, pp. 249–251.
springerlink.com

In Chapter 1, we pointed out that multi-objective test functions must pose sufficient difficulties to impede MOEAs' search for Pareto optimal solutions. While it is not possible to ascertain with absolute certainty any correlation between real-world problem characteristics and test function features, our categorization of dynamic multi-objective problems in Chapter 5 has, at least, revealed some possible types of test functions that have yet been explored. Therefore, it is important to examine the algorithms against various difficult dynamic test functions before applying them in a practical context. In addition, one common assumption in evolutionary dynamic optimization is that changes in the landscape can be detected easily by checking for discrepancies between the old and the re-evaluated objective values. However, this may not be the case in the event where new peaks are introduced without affecting the fitness values of existing non-dominated solutions.

In Chapter 7, we showed that MOEAs have a tendency of finding solutions with broad basins of attraction. This does not show the algorithm's ability in finding robust solutions with narrow basins. Addressing the problem, we have researched in this direction and presented a new robust multi-objective test suite. Unlike most existing research, sensitivities are introduced in the environmental variables instead of the decision variables. Such a feature has enabled us to generate problems with noise-induced difficulties. This necessitates the design of robust MOEAs that are capable of handling such parametric sensitivities.

For real-world problems, we realized that there is sufficient overlap between the different types of uncertainties, which can be examined at the same time. For instance, in adapting the controller for a dynamic process, the sensors may return noisy performance readings from the process plant. Also, it may not be practical to implement a new solution everytime a change in the landscape is discovered due to the implementation cost and the time required to effect the change. In such a situation, it may be better to employ a robust solution that can work well despite the environmental changes. This effectively transforms the problem into one that requires us to find a new robust solution whenever performance deteriorates beyond certain criterion.

The importance of reducing the number of evaluations is becoming more apparent in the face of increasing problem complexity. In presenting the heuristics of ELDP and GASS to guide the evolutionary process through the noisy multi-objective fitness landscape in Chapter 3, we have sought to reduce the number of re-evaluations that is required to improve the reliability of the selection process. Likewise, in speeding up the convergence process by means of the competitive-cooperative co-evolutionary framework to track the dynamic Pareto solution set in Chapter 6, we are essentially reducing the total number of function evaluations. Obviously, a straightforward approach to improve the run-time is to utilize grid or parallel computing approaches. However, such an improvement in run-time is achieved through sheer brute computational force rather than the actual reduction of computational requirements and such computing resources may also not be available to all. A promising approach is to replace computationally the expensive high fidelity fitness functions with computationally cheap approximate models. These approximate models, also known as surrogate or meta-models, have been applied to robust single-objective optimization. However, there exist few works in the context of multi-objective optimization for both dynamic and noisy optimization problems. One application of surrogate models is thus in the optimization of noisy fitness functions, where noise may be filtered out through the approximation process, which is an advantage that may have been overlooked thus far.

A closely related issue, which often arises in real-time systems, is the need to discover implementable solutions quickly within a limited time. Addressing this issue goes beyond improving the soft aspects of evolutionary algorithms, e.g., it encompasses algorithm intelligence, coding efficiency, and approximate fitness functions. Given the iterative nature of evolutionary algorithms and the demand of handling uncertainties, executing the population based algorithms on desktops or workstations may not necessarily meet the requirements of real-time applications. To this end, it would be worth investigating evolutionary algorithms in dedicated hardware such as VLSI or FPGA.

In our approaches, it is assumed that the decision-maker is capable of reliably selecting a solution from the diverse Pareto front for implementation. For multi-objective problems with low-dimensional objective space or Pareto front, the tradeoffs can be easily analyzed by the decision-maker for selecting the final solution. On the other hand, the amount of information present in the final solution set can be overwhelming for high-dimensional problems. Since only one solution will usually be implemented as the final design eventually, it would be useful to incorporate some form of preferences in the optimization process to present a smaller set of "short-listed" solutions. Adding to the fact that considerable computational expense is often needed to reduce noise present in the final evolved Pareto front, having such a sub-set of solutions also reduces the computational effort required in the simulation. In Chapter 5, we have mentioned as a design consideration that the decision-maker may not be available at all time in the case of dynamic optimization. Thus the incorporation of a mechanism that encapsulates the essence of decision-makers is important in this aspect.

References

1. Abido, M.A.: Multiobjective Evolutionary Algorithms for Electric Power Dispatch Problem. IEEE Transactions on Evolutionary Computation 10(3), 315–329 (2006)
2. Abbass, H., Sarker, R., Newton, C.: PDE: A Pareto-frontier Differential Evolution Approach for Multi-objective Optimization Problems. In: Proceedings of the 2001 IEEE Congress on Evolutionary Computation, pp. 27–30 (2001)
3. Abbass, H.: A Memetic Pareto Evolutionary Approach to Artificial Neural Networks. In: Proceedings of the Australian Joint Conference on Artificial Intelligence, pp. 1–12 (2001)
4. Abbass, H.A.: Speeding up backpropagation using multiobjective evolutionary algorithms. Neural Computation 15, 2705–2726 (2003)
5. Angeline, P.J., Pollack, J.B.: Competitive environments evolve better solutions for complex tasks. In: Proceedings of the Fifth International Conference on Genetic Algorithms, pp. 264–270 (1993)
6. Antonie, M.-L., Zaiane, O.R., Holte, R.C.: Learning to Use a Learned Model: A Two-Stage Approach to Classification. In: Proceedings of the Sixth International Conference on Data Mining, pp. 33–42 (2006)
7. Ascia, G., Catania, V., Palesi, M.: A GA-Based Design Space Exploration Framework for Parameterized System-On-A-Chip Platforms. IEEE Transactions on Evolutionary Computation 8(4), 329–346 (2004)
8. Atkinson-Abutridy, J., Mellish, C., Aitken, S.: A Semantically Guided and Domain-Independent Evolutionary Model for Knowledge Discovery From Texts. IEEE Transactions on Evolutionary Computation 7(6), 546–560 (2003)
9. Barbosa, H.J.C., Barreto, A.M.S.: An Interactive Genetic Algorithm with Coevolution of Weights for Multiobjective Problems. In: Proceedings of the 2001 Genetic and Evolutionary Computation Congress, pp. 203–210 (2001)
10. Arnold, D.V., Beyer, H.G.: Local performance of the (1+1)-ES in a noisy environment. IEEE Transactions on Evolutionary Computation 6(1), 30–41 (2002)
11. Arnold, D.V., Beyer, H.G.: A General Noise Model and Its Effects on Evolution Strategy Performance. IEEE Transactions on Evolutionary Computation 10(4), 380–391 (2006)
12. Babbar, M., Lakshmikantha, A., Goldberg, D.E.: Modified NSGA-II to solve Noisy Multi-objective Problems. In: Proceedings of the 2003 Genetic and Evolutionary Computation Conference, Late-Breaking Papers, pp. 21–27 (2003)

13. Back, T., Hammel, U.: Evolution strategies applied to perturbed objective functions. In: Proceedings of the First IEEE Conference on Evolutionary Computation, vol. 1, pp. 40–45 (1994)
14. Bambha, N.K., Bhattacharyya, S.S., Teich, J., Zitzler, E.: Systematic Integration of Parameterized Local Search Into Evolutionary Algorithms. IEEE Transactions on Evolutionary Computation 8(2), 137–154 (2004)
15. Barr, R.S., Golden, B.L., Kelly, J.P., Resendex, M.G.C., Stewart, W.R.: Designing and Reporting on Computational Experiments with Heuristic Methods. Journal of Heuristics 1(1), 9–32 (1995)
16. Basseur, M., Zitzler, E.: Handling Uncertainty in Indicator-Based Multiobjective Optimization. International Journal of Computational Intelligence Research 2(3), 255–272 (2006)
17. Benedetti, A., Farina, M., Gobbi, M.: Evolutionary Multiobjective Industrial Design: The Case of a Racing Car Tire-Suspension System. IEEE Transactions on Evolutionary Computation 10(3), 230–244 (2006)
18. Beielstein, T., Markon, S.: Threshold selection, hypothesis tests, and DOC methods. In: Proceedings of the 2002 IEEE Congress on Evolutionary Computation, vol. 1, pp. 777–782 (2002)
19. Bertsimas, D.J., Chervi, P., Peterson, M.: Computational approaches to stochastic vehicle routing problems. Transportation Science 29(4), 342–352 (1995)
20. Bertsimas, D.J., Jaillet, P., Odoni, A.R.: A priori optimization. Operations Research 38, 1019–1033 (1990)
21. Beyer, H.G.: Evolutionary algorithms in noisy environments: Theoretical issues and guidelines for practice. Computer Methods in Applied Mechanics and Engineering 186, 239–267 (2000)
22. Beyer, H.-G., Sendhoff, B.: Evolution Strategies for Robust Optimization. In: Proceedings of the 2006 IEEE Congress on Evolutionary Computation (2006)
23. Bleuler, S., Laumanns, M., Thiele, L., Zitzler, E.: PISA - A Platform and Programming Language Independent Interface for Search Algorithms. In: Fonseca, C.M., Fleming, P.J., Zitzler, E., Deb, K., Thiele, L. (eds.) EMO 2003. LNCS, vol. 2632, pp. 494–508. Springer, Heidelberg (2003)
24. Bosman, P., Thierens, D.: The balance between proximity and diversity in multiobjective evolutionary algorithms. IEEE Transactions on Evolutionary Computation 7(2), 174–188 (2003)
25. Bosman, P., Thierens, D.: The naive MIDEA: A baseline multi-objective EA. In: Coello Coello, C.A., Hernández Aguirre, A., Zitzler, E. (eds.) EMO 2005. LNCS, vol. 3410, pp. 428–442. Springer, Heidelberg (2005)
26. Branke, J.: Evolutionary Optimization in Dynamic Environments. Kluwer, Norwell (2001)
27. Branke, J., Schmidt, C., Schmeck, H.: Efficient fitness estimation in noisy environments. In: Proceedings of Genetic and Evolutionary Computation, pp. 243–250. Morgan Kaufmann, San Francisco (2001)
28. Bianchi, L., Birattari, M., Manfrin, M., Mastrolilli, M., Paquete, L., Rossi-Doria, O., Schiavinotto, T.: Hybrid metaheuristics for the vehicle routing problem with stochastic demands. Journal of Mathematical Modelling and Algorithms 5(1), 91–110 (2006)

29. Bianchi, L., Birattari, M., Chiarandini, M., Manfrin, M., Mastrolilli, M., Paquete, L., Rossi-Doria, O., Schiavinotto, T.: Metaheuristics for the vehicle routing problem with stochastic demands. In: Yao, X., Burke, E.K., Lozano, J.A., Smith, J., Merelo-Guervós, J.J., Bullinaria, J.A., Rowe, J.E., Tiňo, P., Kabán, A., Schwefel, H.-P. (eds.) PPSN 2004. LNCS, vol. 3242, pp. 450–460. Springer, Heidelberg (2004)
30. Branke, J.: Reducing the sampling variance when searching for robust solution. In: Proceedings of the Genetic and Evolutionary Computation Conference, pp. 235–424 (2001)
31. Branke, J.: Creating robust solutions by means of evolutionary algorithms. In: Proceedings of the Fifth International Conference on Parallel Problem Solving from Nature, pp. 119–128 (1998)
32. Buche, D., Stoll, P., Dornberger, R., Koumoutsakos, P.: Multiobjective Evolutionary Algorithm for the Optimization of Noisy Combustion Processes. IEEE Transactions on Systems, Man, and Cybernetics-Part C: Applications and Reviews 32(4) (2002)
33. Bui, L.T., Abbass, H.A., Essam, D.: Fitness Inheritance For Noisy Evolutionary Multi-Objective Optimization. In: Proceedings of the 2005 Genetic and Evolutionary Computation Congress, pp. 779–785 (2005)
34. Branke, J., Schmeck, H.: Designing evolutionary algorithms for dynamic optimization problems. In: Tsutsui, S., Ghosh, A. (eds.) Theory and Application of Evolutionary Computation: Recent Trends, pp. 239–262. Springer, Heidelberg (2002)
35. Cantú-Paz, E.: A survey of parallel genetic algorithms. Calculateurs Paralleles, Reseaux et Systems Repartis 10(2), 141–171 (1998)
36. Cantú-Paz, E., Kamath, C.: An empirical comparison of combinations of evolutionary algorithms and neural networks for classification problems. IEEE Transactions on Systems, Man, and Cybernetics-Part B: Cybernetics, 915–927 (2005)
37. Charnes, A., Cooper, W.: Deterministic equivalents for optimizing and satisfying under chance constraints. Operations Research 11, 18–39 (1963)
38. Charnes, A., Cooper, W.: Chance-constrained programming. Management Science 6, 73–79 (1959)
39. Chen, C.H., Wu, S.D., Dai, L.: Algorithm comparison for manufacturing scheduling problems. In: Proceedings of the 36th IEEE Conference on Decision and Control, vol. 2, pp. 1214–1215 (1997)
40. Cobb, H.G.: An investigation into the use of hypermutation as an adaptive operator in genetic algorithms having continuous, time-dependent nonstationary environments, Technical Report AIC-90-001, Naval Research Laboratory, Washington, DC (1990)
41. Coello Coello, C.A.: Evolutionary Multiobjective Optimization: A Historical View of the Field. IEEE Computational Intelligence Magazine 1(1), 28–36 (2006)
42. Coello Coello, C.A.: An empirical study of evolutionary techniques for multiobjective optimization in engineering design, Ph.D. dissertation, Department of Computer Science, Tulane University, New Orleans, LA (1996)
43. Coello Coello, C.A., Cruz Cortés, N.: Solving Multiobjective Optimization Problems using an Artificial Immune System. Genetic Programming and Evolvable Machines 6(2), 163–190 (2005)

44. Coello Coello, C.A., Pulido, G.T., Lechuga, M.S.: Handling Multiple Objectives With Particle Swarm Optimization. IEEE Transactions on Evolutionary Computation 8(3), 256–279 (2004)
45. Coello Coello, C.A., Sierra, M.R.: A Coevolutionary Multi-Objective Evolutionary Algorithm. In: Proceedings of the 2003 IEEE Congress on Evolutionary Computation, vol. 1, pp. 482–489 (2003)
46. Corne, D.W., Knowles, J.D., Oates, M.J.: The Pareto Envelope-based Selection Algorithm for Multiobjective Optimization. In: Proceedings of the Sixth International Conference on Parallel Problem Solving from Nature, pp. 839–848 (2000)
47. Cover, T.M.: Geometrical and statistical properties of systems of linear inequalities with applications in pattern recognition. IEEE Transactions on Electronic Computation 14, 326–334 (1965)
48. Chipperfield, A.J., Fleming, P.J.: Multiobjective Gas Turbine Engine Controller Design Using Genetic Algorithms. IEEE Transactions on Industrial Electronics 43(5), 583–587 (1996)
49. Crowder, H., Dembo, R., Mulvey, J.: On reporting computational experiments with mathematical software. ACM Transactions on Mathematical software 5(2), 193–203 (1979)
50. Cui, X., Li, M., Fang, T.: Study of Population Diversity of Multiobjective Evolutionary Algorithm Based on Immune and Entropy Principles. In: Proceedings of the 2001 IEEE Congress on Evolutionary Computation, vol. 2, pp. 1316–1321 (2001)
51. Cunha, A.G., Oliviera, P., Covas, J.: Use genetic algorithms in multicriteria optimization to solve industrial problems. In: Proceedings of the Seventh International Conference on Genetic Algorithms, pp. 682–688 (1997)
52. Cvetkovic, D., Parmee, I.C.: Preferences and their Application in Evolutionary Multiobjective Optimisation. IEEE Transactions on Evolutionary Computation 6(1), 42–57 (2002)
53. Darwen, P.J., Yao, X.: Speciation as automatic categorical modularization. IEEE Transactions on Evolutionary Computation 1(2), 100–108 (1997)
54. de Castro, L.N., Timmis, J.: Artificial Immune Systems: A New Computational Paradigm. Springer, Heidelberg (2002)
55. Deb, K.: Multi-objective genetic algorithms: problem difficulties and construction of test problem. Evolutionary Computation 7(3), 205–230 (1999)
56. Deb, K., Thiele, L., Laumanns, M., Zitzler, E.: Scalable multi-objective optimization test problems. In: Proceedings of the 2002 IEEE Congress on Evolutionary Computation, vol. 1, pp. 825–830 (2002)
57. Deb, K., Gupta, H.: Introducing robustness in multiobjective optimization, Kanpur Genetic Algorithms Lab (KanGAL), Indian Institue of Technology, Kanpur, India, Technical Report 2004016 (2004)
58. Deb, K., Goel, T.: Controlled Elitist Non-dominated Sorting Genetic Algorithms for Better Convergence. In: Proceedings of the First International Conference on Evolutionary Multi-Criterion Optimization, pp. 67–81 (2001)
59. Deb, K., Udaya Bhaskara Rao, N., Karthik, S.: Dynamic Multi-Objective Optimization and Decision-Making Using Modified NSGA-II: A Case Study on Hydro-Thermal Power Scheduling, Kanpur Genetic Algorithms Lab. (KanGAL), Indian Institue of Technology, Kanpur, India, Technical Report 2006008 (2006)

60. Deb, K., Padmanabhan, D., Gupta, S., Kumar Mall, A.: Reliability-based multi-objective optimization using evolutionary algorithms. In: Proceedings of the Fourth International Conference on Evolutionary Multi-Criterion Optimization, pp. 66–80 (2007)
61. Deb, K., Zope, P., Jain, A.: Distributed computing of Pareto-optimal solutions with evolutionary algorithms. In: Proceedings of the Second International Conference on Evolutionary Multi-Criterion Optimization, pp. 534–549 (2003)
62. Deb, K., Agrawal, S., Pratap, A., Meyarivan, T.: A fast and elitist multi-objective genetic algorithm: NSGA-II. IEEE Transactions on Evolutionary Computation 6(2), 182–197 (2002)
63. Deb, K., Goldberg, D.E.: An investigation on niche and species formation in genetic function optimization. In: Proceedings of Third International Conference on Genetic Algorithms, pp. 42–50 (1989)
64. De Jong, K.A.: An analysis of the behaviour of a class genetic adaptive systems, Ph.D thesis, University of Michigan (1975)
65. Devireddy, V., Reed, P.: Efficient and Reliable Evolutionary Multiobjective Optimization Using ϵ-Dominance Archiving and Adaptive Population Sizing. In: Proceedings of the 2004 Genetic and Evolutionary Computation Conference, pp. 130–131 (2004)
66. Di Barba, P., Farina, M., Savini, A.: An improved technique for enhancing diversity in Pareto evolutionary optimization of electromagnetic devices. The International Journal for Computation and Mathematics in Electrical and Electronic Engineering 20(2), 482–496 (2001)
67. Dror, M., Trudeau, P.: Stochastic vehicle routing with modified savings algorithm. European Journal of Operational Research 23(2), 228–235 (1986)
68. Dror, M., Ball, M.O., Golden, B.L.: Computational comparison of algorithms for inventory routing. Annals of Operations Research 4, 3–23 (1985)
69. Dubois, D., Prade, H.: Possibility Theory: An Approach to Computerized Processing and Uncertainty. Plenum Press, New York (1988)
70. Eiben, A.E., Jelasiy, M.: A Critical Note on Experimental Research Methodology in EC. In: Proceedings of the 2002 IEEE Congress on Evolutionary Computation, pp. 582–587 (2002)
71. Emmerich, M., Beume, N., Naujoks, B.: An EMO Algorithm Using the Hypervolume Measure as Selection Criterion. In: Proceedings of the Third Conference on Evolutionary Multi-Criterion Optimization, pp. 62–76 (2005)
72. Emmerich, M., Giannakoglou, K.C., Naujoks, B.: Single- and Multiobjective Evolutionary Optimization Assisted by Gaussian Random Field Metamodels. IEEE Transactions on Evolutionary Computation 10(4), 421–439 (2006)
73. Erbas, C., Cerac-Erbas, S., Pimentel, A.D.: Multiobjective Optimization and Evolutionary Algorithms for the Application Mapping Problem in Multiprocessor System-on-Chip Design. IEEE Transactions on Evolutionary Computation 10(3), 358–374 (2006)
74. Everson, R.M., Fieldsend, J.E.: Multiobjective Optimization of Safety Related Systems: An Application to Short-Term Conflict Alert. IEEE Transactions on Evolutionary Computation 10(2), 187–198 (2006)
75. Farina, M., Deb, K., Amato, P.: Dynamic Multiobjective Optimization Problems: Test Cases, Aproximations, and Applications. IEEE Transactions on Evolutionary Computation 8(5), 425–442 (2004)

76. Farina, M., Amato, P.: A fuzzy definition of "optimality" for many-criteria optimization problems. IEEE Transactions on Systems, Man, and Cybernetics-Part A: Systems and Humans 34(3), 315–326 (2003)
77. Farhang-Mehr, A., Azarm, S.: Diversity Assessment of Pareto Optimal Solution Sets: An Entropy Approach. In: Proceedings of the 2002 Congress on Evolutionary Computation, pp. 723–728 (2002)
78. Farina, M.: A Minimal Cost Hybrid Strategy for Pareto Optimal Front Approximation. Evolutionary Optimization 3(1), 41–52 (2001)
79. Fieldsend, J.E., Everson, R.M., Singh, S.: Using Unconstrained Elite Archives for Multiobjective Optimization. IEEE Transactions on Evolutionary Computation 7(3), 305–323 (2003)
80. Fieldsend, J.E., Singh, S.: Pareto evolutionary neural networks. IEEE Transactions on Neural Networks 16(2), 338–354 (2005)
81. Fieldsend, J.E., Everson, R.M.: Multi-objective Optimisation in the Presence of Uncertainty. In: Proceedings of the 2005 IEEE Congress on Evolutionary Computation, vol. 1, pp. 243–250 (2005)
82. Fleischer, M.: The Measure of Pareto Optima. Applications to Multi-objective Metaheuristics. In: Fonseca, C.M., Fleming, P.J., Zitzler, E., Deb, K., Thiele, L. (eds.) EMO 2003. LNCS, vol. 2632, pp. 519–533. Springer, Heidelberg (2003)
83. Fogel, L.J., Owens, A.J., Walsh, M.J.: Artificial Intelligence through Simulated Evolution. John Wiley, Chichester (1966)
84. Fogel, D.B., Wasson, E.C., Boughton, E.M.: Evolving neural networks for detecting breast cancer. Cancer Letters 96(1), 49–53 (1995)
85. Fonseca, C.M., Fleming, P.J.: Multi-objective genetic algorithm made easy: Selection, sharing and mating restriction. In: International Conference on Genetic Algorithm in Engineering Systems: Innovations and Application, pp. 12–14 (1995)
86. Fonseca, C.M., Fleming, P.J.: Multiobjective Optimal Controller Design with Genetic Algorithms. In: Proceedings on IEE Control, pp. 745–749 (1994)
87. Fonseca, C.M., Fleming, P.J.: Genetic algorithm for multiobjective optimization, formulation, discussion and generalization. In: Proceedings of the Fifth International Conference on Genetic Algorithms, pp. 416–423 (1993)
88. Frank, E., Witten, I.H.: Generating accurate rule sets without global optimization. In: Proceedings of the Fifteenth International Conference Machine Learning, vol. 22, pp. 144–151 (1998)
89. Gallagher, M., Yuan, B.: A General-Purpose Tunable Landsscape Generator. IEEE Transactions on Evolutionary Computation 10(5) (2006)
90. Garcia-Pedrajas, N., Hervas-Martinez, C., Ortiz-Boyer, D.: Cooperative Coevolution of Artificial Neural Network Ensembles for Pattern Classification. IEEE Transactions on Evolutionary Computation 9(3), 271–302 (2005)
91. Gendreau, M., Laporte, G., Séguin, R.: Stochastic vehicle routing. European Journal of Operational Research 88(1), 3–12 (1996)
92. Gendreau, M., Laporte, G., Séguin, R.: A tabu search heuristic for the vehicle routing problem with stochastic demands and customers. Operations Research 44(3), 469–477 (1996)
93. Gendreau, M., Laporte, G., Séguin, R.: An exact algorithm for the vehicle routing problem with stochastic demands and customers. Transportation Science 29, 143–155 (1995)

94. Ghosh, A., Tstutsui, S., Tanaka, H.: Function optimization in non-stationary environment using steady state genetic algorithms with aging of individuals. In: Proceedings of 1998 IEEE Congress on Evolutionary Computation (CEC 1998), pp. 666–671 (1998)
95. Ghosh, R., Verma, B.: Finding Optimal Architecture and Weights Using Evolutionary Least Square Based Learning. Proceedings of Neural Information Processing 1, 528–532 (2002)
96. Giustolisi, O., Simeone, V.: Optimal design of artificial neural networks by a multi-objective strategy: groundwater level predictions. Hydrological Sciences Journal 51(3) (2006)
97. Goh, C.K., Tan, K.C.: A Competitive-Cooperation Coevolutionary Paradigm for Dynamic Multi-objective Optimization. IEEE Transactions on Evolutionary Computation (accepted)
98. Goh, C.K., Tan, K.C.: An investigation on noisy environments in evolutionary multiobjective optimization. IEEE Transactions on Evolutionary Computation 11(3), 354–381 (2007)
99. Goh, C.K., Tan, K.C.: Evolving the tradeoffs between Pareto-optimality and Robustness in Multi-Objective Evolutionary Algorithms. In: Yang, S., Ong, Y.S., Jin, Y. (eds.) Evolutionary Computation in Dynamic and Uncertain Environments, pp. 457–478. Springer, Heidelberg (2007)
100. Goh, C.K., Tan, K.C., Cheong, C.Y., Ong, Y.S.: Noise Induced Features in Robust Multiobjective Optimization Problems. In: Proceedings of 2007 IEEE Congress on Evolutionary Computation, pp. 568–575 (2007)
101. Goh, C.K., Teoh, E.J., Tan, K.C.: Hybrid multiobjective evolutionary design for artificial neural networks. IEEE Transactions on Neural Networks 19(9) (2008)
102. Goldberg, D.E.: The Design of Innovation: Lessons from and for Competent Genetic Algorithms. Kluwer Academic Publishers, Dordrecht (2002)
103. Goldberg, D.E.: Genetic Algorithms for Search, Optimization, and Machine Learning. Addison-Wesley, Reading (1989)
104. Goldberg, D.E.: Sizing populations for serial and parallel genetic algorithms. In: Proceedings of the Third International Conference on Genetic Algorithms, pp. 70–79 (1989)
105. Goldberg, D.E., Richardson, J.: Genetic algorithms with sharing for multimodal function optimization. In: Proceedings of the Second International Conference on Genetic Algorithms, pp. 41–49 (1987)
106. Golden, B., Steward, W.: Empirical Analysis of heuristics. In: Lawler, E., Lenstra, J., Kan, A.R., Shmoys, D. (eds.) The Travelling Salesman Problem, A Guilded Tour of Combinatorial Optimization, pp. 207–249. John Wiley & Sons, Chichester (1985)
107. Goodenough, J.B., Gerhart, S.L.: Towards a theory of test data selection. IEEE Transactions on Software Engineering 1(2), 156–173 (1975)
108. Goulermas, J.Y., Liatsis, P.: A Collective-Based Adaptive Symbiotic Model for Surface Reconstruction in Area-Based Stereo. IEEE Transactions on Evolutionary Computation 7(5), 482–502 (2003)
109. Greenberg, H.: Computational testing: Why, how and how much. ORSA Journal on Computing 2, 7–11 (1990)
110. Grefenstette, J.J.: Genetic algorithms for changing environments. In: Proceedings of the Second International Conference on Parallel Problem Solving from Nature, pp. 137–144 (1992)

111. Grefenstette, J.J., Ramsey, C.L.: An approach to anytime learning. In: Proceedings of the Ninth International Conference on Machine Learning, pp. 41–49 (1987)
112. Guan, S.-U., Zhang, S.: An Evolutionary Approach to the Design of Controllable Cellular Automata Structure for Random Number Generation. IEEE Transactions on Evolutionary Computation 7(1), 23–36 (2003)
113. Gunawan, S., Azarm, S.: Multi-objective robust optimization using a sensitivity region concept. Structural and Multidisciplinary Optimization 29, 50–60 (2005)
114. Gupta, H., Deb, K.: Handling constraints in robust multi-objective optimization. In: Proceedings of the 2005 IEEE Congress on Evolutionary Computation, pp. 25–32 (2005)
115. Hajela, P., Lin, C.Y.: Genetic Search Strategies in Multicriterion Optimal Design. Structural Optimization 4, 99–107 (1992)
116. Hallam, N., Blanchfield, P., Kendall, G.: Handling Diversity in Evolutionary Multiobjective Optimisation. In: Proceedings of the 2005 IEEE Congress on Evolutionary Computation, pp. 2233–2240 (2005)
117. Hansen, N., Ostermeier, A.: Completely derandomized self-adaptation in evolution strategies. Evolutionary Computation 9(2), 159–195 (2001)
118. Hansen, P.C.: Rank-deficient and discrete ill-posed problems: Numerical aspects of linear inversion. Society for Industrial and Applied Mathematics (1998)
119. Hansen, M.P., Jaszkiewicz, A.: Evaluating the quality of approximations to the non-dominated set, Technical Report IMM-REP1998-7, Technical University of Denmark (1998)
120. Hatzakis, I., Wallace, D.: Dynamic Multi-Objective Optimization with Evolutionary Algorithms: A Foward-Looking Approach. In: Proceedings of the 2006 Genetic and Evolutionary Computation Congress, pp. 1201–1208 (2006)
121. Hillis, D.W.: Coevolving parasites improve simulated evolution as an optimization procedure. In: Langton, C., Taylor, C., Farmer, J.D., Rasmussen, S. (eds.) Artificial Life 2, pp. 313–324 (1991)
122. Hiroyasu, T., Nakayama, S., Miki, M.: Comparison Study of SPEA2+, SPEA2, and NSGA-II in Diesel Engine Emissions and Fuel Economy Problem. In: Proceedings of the 2005 IEEE Congress on Evolutionary Computation, pp. 236–242 (2005)
123. Ho, S.Y., Shu, L.S., Chen, J.H.: Intelligent Evolutionary Algorithms for Large Parameter Optimization Problems. IEEE Transactions on Evolutionary Computation 8(6), 522–541 (2004)
124. Holland, J.H.: Adaptation in Natural Artificial Systems: An Introductory Analysis with Applocations to Biology, Control, and Artificial Intelligence. MIT press, Cambridge (1992)
125. Homberger, J., Gehring, H.: Two evolutionary meta-heuristics for the vehicle routing problem with time windows. INFORMS Journal on Computing 37(3), 297–318 (1999)
126. Horn, J., Nafpliotis, N.: Multiobjective optimization using the niched Pareto genetic algorithm, Technical Report No. 930005, Illinois Genetic Algorithms Laboratory (IlliGAL), University of Illinois at Urbana-Champaign (1993)
127. Huang, S.C., Huang, Y.F.: Bounds on number of hidden neurons of multilayer percep-trons in classification and recognition. In: Proceedings of IEEE International Symposium on Circuits and Systems, vol. 4, pp. 2500–2503 (1990)

128. Hughes, E.J.: Evolutionary Many-Objective Optimisation: Many Once or One Many? In: Proceedings of 2005 IEEE Congress on Evolutionary Computation, vol. 1, pp. 222–227 (2005)
129. Hughes, E.J.: Multiple single objective pareto sampling. In: Proceedings of 2003 IEEE Congress on Evolutionary Computation, pp. 2678–2684 (2003)
130. Hughes, E.J.: Evolutionary multi-objective ranking with uncertainty and noise. In: Proceedings of the First Conference on Evolutionary Multi-Criterion Optimization, pp. 329–343 (2001)
131. Hughes, E.J.: Constraint handling with uncertain and noisy multi-objective evolution. In: Proceedings of the 2001 Congress on Evolutionary Computation, vol. 2, pp. 963–970 (2001)
132. Hughes, E.J.: Multi-objective Probabilistic Selection Evolutionary Algorithm (MOPSEA), Technical Report No. DAPS/EJH/56/2000, Department of Aerospace, POwer & Sensors, Cranfield University (2000)
133. Hughes, E.J., Leyland, M.: Using Multiple Genetic Algorithms to Generate Radar Point-Scatterer Models. IEEE Transactions on Evolutionary Computation 4(2), 147–163 (2000)
134. Huband, S., Barone, L., White, L., Hingston, P.: A Scalable Multi-objective Test Problem Toolkit. In: Proceedings of the Third International Conference on Evolutionary Multi-Criterion Optimization, pp. 280–294 (2005)
135. Ikeda, K., Kita, H., Kobayashi, S.: Does Non-dominated Really Mean Near to Optimal? In: Proceedings of the 2001 IEEE Conference on Evolutionary Computation, vol. 2, pp. 957–962 (2001)
136. Inoue, H., Narihisa, H.: Self-Organizing Neural Grove and Its Applications. In: Proceedings of International Joint Conference on Neural Networks, pp. 1205–1210 (2005)
137. Iorio, A.W., Li, X.: A Cooperative Coevolutionary Multiobjective Algorithm Using Non-dominated Sorting. In: Proceedings of the 2004 Genetic and Evolutionary Computation Congress, pp. 537–548 (2004)
138. Igel, C., Hansen, N., Roth, S.: Covariance matrix adaptation for multi-objective optimization. Evolutionary Computation 15(1), 1–28 (2007)
139. Ishibuchi, H., Shibata, Y.: An Empirical Study on the Effect of Mating Restriction on the Search Ability of EMO Algorithms. In: Proceedings of the Second International Conference on Evolutionary Multi-Criterion Optimization, pp. 433–447 (2003)
140. Ishibuchi, H., Shibata, Y.: A Similarity-Based Mating Scheme for Evolutionary Multiobjective Optimization. In: Proceedings of the Second International Conference on Evolutionary Multi-Criterion Optimization, pp. 1065–1076 (2003)
141. Ishibuchi, H., Murata, T.: A Multi-Objective Genetic Local Search Algorithm and Its Application to Flowshop Scheduling. IEEE Transactions on Systems, Man, and Cybernetics - Part C 28(3), 392–403 (1998)
142. Ishibuchi, H., Yoshida, T., Murata, T.: Balance between Genetic Search and Local Search in Memetic Algorithms for Multiobjective Permutation Flowshop. IEEE Transactions on Evolutionary Computation 7(2), 204–223 (2003)
143. Jaszkiewicz, A.: On the Performance of Multiple-Objective Genetic Local Search on the 0/1 Knapsack Problem-A Comparative Experiment. IEEE Transactions on Evolutionary Computation 6(4), 402–412 (2002)
144. Jaszkiewicz, A.: Do multi-objective metaheuristics deliver on their promises? A computational experiment on the set-covering problem. IEEE Transactions on Evolutionary Computation 7(2), 133–143 (2003)

145. Jensen, M.T.: Reducing the run-time complexity of multiobjective EAs: The NSGA-II and other algorithms. IEEE Transactions on Evolutionary Computation 7(5), 503–515 (2003)
146. Jiao, L., Gong, M., Shang, R., Du, H., Lu, B.: Clonal Selection with Immune Dominance and Anergy Based Multiobjective Optimization. In: Proceedings of the Third International Conference on Evolutionary Multi-Criterion Optimization, pp. 474–489 (2005)
147. Jin, Y., Branke, J.: Evolutionary Optimization in Uncertain Environments-A Survey. IEEE Transactions on Evolutionary Computation 9(3), 303–317 (2005)
148. Jin, Y., Sendhoff, B.: Constructing Dynamic Optimization Test Problems Using the Multi-objective Optimization Concept. In: Proceedings of the 2004 EvoWorkshops, pp. 525–536 (2004)
149. Jin, Y., Sendhoff, B.: Tradeoff between performance and robustness: An evolutionary multiobjective approach. In: Proceedings of the Second Conference on Evolutionary Multi-Criterion Optimization, pp. 237–251 (2003)
150. Jin, Y., Okabe, T., Sendhoff, B.: Adapting Weighted Aggregation for Multiobjective Evolution Strategies. In: Proceedings of the First Conference on Evolutionary Multi-Criterion Optimization, pp. 96–110 (2001)
151. Jin, Y., Olhofer, M., Sendhoff, B.: Dynamic Weighted Aggregation for Evolutionary Multi-Objective Optimization: Why Does It Work and How? In: Proceedings of the 2001 Genetic and Evolutionary Computation Conference, pp. 1042–1049 (2001)
152. John, G.H., Langley, P.: Estimating continuous distributions in Bayesian classifiers. In: Proceedings of the Eleventh Conference on Uncertainty in Artificial Intelligence, pp. 338–345 (1995)
153. Karzrlis, S.A., Papadakis, S.E., Theocharis, J.B., Petridis, V.: Microgenetic Algorithms as Generalized Hill-Climbing Operators for GA Optimization. IEEE Transactions On Evolutionary Computation 5(3), 204–217 (2001)
154. Keerativuttiumrong, N., Chaiyaratana, N., Varavithya, V.: Multiobjective cooperative coevolutionary genetic algorithm. In: Proceedings of the Seventh International Conference on Parallel Problem Solving from Nature, pp. 288–297 (2002)
155. Khare, V., Yao, X., Sendhoff, B.: Credit assignment among neurons in coevolving populations. In: Proceedings of the Eighth International Conference on Parallel Problem Solving from Nature, pp. 882–891 (2004)
156. Khare, V., Yao, X., Deb, K.: Performance scaling of multi-objective evolutionary algorithms. In: Proceedings of the Second International Conference on Evolutionary Multi-Criterion Optimization, pp. 376–390 (2003)
157. Khor, E.F., Tan, K.C., Lee, T.H., Goh, C.K.: A study on distribution preservation mechanism in evolutionary multi-objective optimization. Artificial Intelligence Review 23(1), 31–56 (2005)
158. Khor, K.F., Tan, K.C., Lee, T.H.: Tabu-based exploratory evolutionary algorithm for effective multi-objective optimization. In: Proceedings of the First Conference on Evolutionary Multi-Criterion Optimization, pp. 344–358 (2001)
159. Khoshgoftaar, T., Liu, Y., Seliya, N.: A Multiobjective Module-Order Model for Software Quality Enhancement. IEEE Transactions on Evolutionary Computation 8(6), 593–608 (2004)
160. Kinnebrock, W.: Accelerating the standard backpropagation method using a genetic approach. Neurocomputing 6(5-6), 583–588 (1994)

161. Kita, H., Yabumoto, Y., Mori, N., Nishikawa, Y.: Multi- Objective Optimization by Means of the Thermodynamical Genetic Algorithm. In: Proceedings of the Fourth Parallel Problem Solving from Nature, pp. 504–512 (1996)
162. Klema, V., Laub, A.: The Singular Value Decomposition: Its Computation and Some Applications. IEEE Transanction on Automatic Control 2, 164–176 (1980)
163. Knowles, J.D.: ParEGO: A hybrid algorithm with on-line landscape approximation for expensive multiobjective optimization problems. IEEE Transactions on Evolutionary Computation 10(1), 50–66 (2006)
164. Knowles, J.D., Corne, D.W., Fleischer, M.: Bounded archiving using the Lebesgue measure. In: Proceedings of the 2003 IEEE Congress on Evolutionary Computation, vol. 4, pp. 2490–2497 (2003)
165. Knowles, J.D., Corne, D.W.: Properties of an adaptive archiving algorithm for storing nondominated vectors. IEEE Transactions on Evolutionary Computation 7(2), 100–116 (2003)
166. Knowles, J.D., Corne, D.W.: On Metrics for Comparing Nondominated Sets. In: Proceedings of the 2002 Congress on Evolutionary Computation, vol. 1, pp. 711–716 (2002)
167. Konstantinides, K., Yao, K.: Statistical analysis of effective singular values in matrix rank determination. IEEE Transactions on Acoustics, Speech, and Signal Processing 36(5), 757–763 (1988)
168. Koppen, M., Vicente-Garcia, R., Nickolay, B.: Fuzzy-Pareto-Dominance and Its Application in Evolutionary Multi-objective Optimizationin. In: Proceedings of the Third International Conference on Evolutionary Multi-Criterion Optimization, pp. 399–412 (2005)
169. Kursawe, F.: A Variant of Evolution Strategies for Vector Optimization. In: Schwefel, H.-P., Männer, R. (eds.) PPSN 1990. LNCS, vol. 496, pp. 193–197. Springer, Heidelberg (1991)
170. Knowles, J.D., Corne, D.W.: Approximating the non-dominated front using the Pareto archived evolution strategy. Evolutionary Computation 8(2), 149–172 (2000)
171. He, J., Yao, X.: Towards An Analytic Framework For Analysing The Computation Time Of Evolutionary Algorithms. Artificial Intelligence 145(1-2), 59–97 (2003)
172. Huband, S., Hingston, P., Barone, L., While, L.: A review of Multiobjective Test Problems and a Scable Test Problem Toolkit. IEEE Transactions on Evolutionary Computation 10(5), 477–506 (2006)
173. Lambert, V., Laporte, G., Louveaux, F.V.: Designing Collection Routes through Bank Branches. Computers and Operations Research 20, 783–791 (1993)
174. Laporte, G., Louveaux, F.V.: Solving stochastic routing Problems with the Integer L-shaped Method. In: Crainic, T.G., Laporte, G. (eds.) Fleet Management and Logistics, pp. 159–167. Kluwer Academic Publishers, Boston (1998)
175. Laporte, G., Louveaux, F.V.: The integer L-shape method for stochastic integer problems with complete recourse. Operations Research Letters 13, 133–142 (1993)
176. Larson, R.C.: Transportation of sludge to the 106-mile site: An inventory routing algorithm for fleet sizing and logistic system design. Transportation Science 22, 186–198 (1988)

177. Laumanns, M., Thiele, L., Zitzler, E., Deb, K.: Archiving with Guaranteed Convergence and Diversity in Multi-Objective Optimization. In: Proceedings of the Genetic and Evolutionary Computation Conference, pp. 439–447 (2002)
178. Laumanns, M., Thiele, L., Zitzler, E., Welzl, E., Deb, K.: Running time analysis of multi-objective evolutionary algorithms in a simple discrete optimization problem. In: Proceedings of the Seventh International Conference on Parallel Problem Solving from Nature, pp. 44–53 (2002)
179. Laumanns, M., Zitzler, E., Thiele, L.: On the effects of archiving, elitism, and density based selection in evolutionary multi-objective optimization. In: Proceedings of the First International Conference on Evolutionary Multi-Criterion Optimization, pp. 181–196 (2001)
180. Laumanns, M., Zitzler, E., Thiele, L.: A unified model for multi-objective evolutionary algorithms with elitism. In: Proceedings of the 2000 IEEE Congress on Evolutionary Computation, vol. 1, pp. 46–53 (2000)
181. Laumanns, M., Rudolph, G., Schwefel, H.-P.: A Spatial Predator-Prey Approach to Multi-Objective Optimization: A Preliminary Study. In: Proceedings of the Second International Conference on Parallel Problem Solving from Nature, vol. 1, pp. 46–53 (1998)
182. LeCun, Y., Bottou, L., Orr, G., Muller, K.: Efficient BackProp. In: Orr, G.B., Müller, K.-R. (eds.) Neural Networks: Tricks of the trade. Springer, Heidelberg (1998)
183. Lee, L.H., Chew, E.P.: A simulation study on sampling and selecting under fixed computing budget. In: Proceedings of the 2003 Winter Simulation Conference, vol. 1, pp. 535–542 (2003)
184. Leyland, G.: Multi-objective optimization applied to Industrial Energy Problems, Ph.D thesis, EPFL, Lausanne (2002)
185. Li, M., Azarm, S., Aute, V.: A multi-objective genetic algorithm for robust design optimization. In: Proceedings of the 2005 Genetic and Evolutionary Computation Conference, pp. 771–778 (2005)
186. Lim, D., Ong, Y.-S., Lim, M.-H., Jin, Y.: Single/Multi-objective Inverse Robust Evolutionary Design Methodology in the Presence of Uncertainty. In: Yang, S., Ong, Y.S., Jin, Y. (eds.) Evolutionary Computation in Dynamic and Uncertain Environments, pp. 437–456. Springer, Heidelberg (2007)
187. Limbourg, P.: Multi-objective optimization of Problems with Epistemic Uncertainty. In: Proceedings of the Third International Conference on Evolutionary Multi-Criterion Optimization, pp. 413–427 (2005)
188. Liu, Y., Yao, X., Higuchi, T.: Evolutionary Ensembles with Negative Correlation Learning. IEEE Transactions On Evolutionary Computation 4(4), 380–387 (2000)
189. Liu, Y., Yao, X., Zhao, Q., Higuchi, T.: Scaling up fast evolutionary programming with cooperative coevolution. In: Proceedings of the 2001 IEEE Congress on Evolutionary Computation, pp. 1101–1108 (2001)
190. Liu, T.H., Mills, K.J.: Robotic Trajectory Control System Design for Multiple Simultaneous Specifications: Theory and Experimentation. In: Transactions of ASME, vol. 120, pp. 520–523 (1998)
191. Lohn, J.D., Kraus, W.F., Haith, G.L.: Comparing a coevolutionary genetic algorithm for multiobjective optimization. In: Proceedings of the 2002 IEEE Congress on Evolutionary Computation, pp. 1157–1162 (2002)

192. Lu, H., Yen, G.G.: Rank-based multiobjective genetic algorithm and benchmark test function study. IEEE Transactions on Evolutionary Computation 7(4), 325–343 (2003)
193. Luh, G.C., Chueh, C.H., Liu, W.W.: MOIA: Multi-Objective Immune Algorithm. Engineering Optimization 35(2), 143–164 (2003)
194. Maneeratana, K., Boonlong, K., Chaiyaratana, N.: Multi-objective Optimisation by Co-operative Co-evolution. In: Proceedings of the Eighth International Conference on Parallel Problem Solving from Nature, pp. 772–781 (2004)
195. Maniezzo, V.: Genetic evolution of the topology and weight distribution of neural networks. IEEE Transactions on Neural Networks 5(1), 39–53 (1994)
196. Mehnen, J., Wagner, T., Rudolph, G.: Evolutionary Optimization of Dynamic Multi-objective Test Functions. In: Proceedings of the Second Italian Workshop on Evolutionary Computation (2006)
197. Merz, P., Freisleben, B.: A comparison of memetic algorithms, Tabu search, and ant colonies for the quadratic assignment problem. In: Proceedings of the 1999 IEEE Congress on Evolutionary Computation, vol. 1, pp. 2063–2070 (1999)
198. Michie, D., Spiegelhalter, D.J., Taylor, C.C.: Machine Learning, Neural and Statistical Classification. Ellis Horwood, London (1994)
199. Mori, N., Kita, H., Nishikawa, Y.: Adaptation to a changing environment by means of the thermodynamical genetic algorithm. In: Proceedings of the Fourth International Conference on Parallel Problem Solving from Nature, pp. 513–522 (1996)
200. Morrison, R.: Designing Evolutionary Algorithms for Dynamic Environments. Springer, Berlin (2004)
201. Moriarty, D.E.: Symbiotic Evolution of Neural Networks in Sequential Decision Tasks. Ph.D. Thesis, The University of Texas at Austin (1997)
202. Mumford, C.L.: A Hierarchical Solve-and-Merge Framework for Multi-Objective Optimization. In: Proceedings of the 2005 IEEE Congress on Evolutionary Computation, pp. 2241–2247 (2005)
203. Murata, T., Ishibuchi, H.: MOGA: Multi-objective genetic algorithms. In: Proceedings of the 1995 IEEE Congress on Evolutionary Computation, pp. 289–294 (1995)
204. Nissen, V., Propach, J.: On the robustness of population based versus point-based optimization in the presence of noise. IEEE Transactions on Evolutionary Computation 2(3), 107–119 (1998)
205. Obayashi, S., Tsukahara, T., Nakamura, T.: Multiobjective Evolutionary Computation for Supersonic Wing-Shape Optimization. IEEE Transactions on Evolutionary Computation 4(2), 182–187 (2000)
206. Okabe, T., Jin, Y., Sendhoff, B., Olhofer, M.: Voronoi-based estimation of distribution algorithm for multi-objective optimization. In: Proceedings of the 2004 IEEE Congress on Evolutionary Computation, pp. 1594–1601 (2004)
207. Okuda, T., Hiroyasu, T., Miki, M., Watanabe, S.: DCMOGA: Distributed Cooperation model of Multi-Objective Genetic Algorithm. In: Proceedings of the Seventh International Conference on Parallel Problem Solving from Nature, pp. 155–160 (2002)
208. Ong, Y.S., Keane, A.J.: Meta-Lamarckian Learning in Memetic Algorithms. IEEE Transactions on Evolutionary Computation 8(2), 99–110 (2004)
209. Ong, Y.S., Nair, P.B., Lum, K.Y.: Min-Max Surrogate Assisted Evolutionary Algorithm for Robust Aerodynamic Design. IEEE Transactions on Evolutionary Computation 10(4), 392–404 (2006)

210. Osyczka, A., Krenich, S.: Evolutionary Algorithms for Multicriteria Optimization with Selecting a Representative Subset of Pareto Optimal Solutions. In: Proceedings of the First International Conference on Evolutionary Multi-Criterion Optimization, pp. 141–153 (2001)
211. Paenke, I., Branke, J., Jin, Y.: Efficient Search for Robust Solutions by Means of Evolutionary Algorithms and Fitness Approximation. IEEE Transactions on Evolutionary Computation 10(4), 405–420 (2006)
212. Palmes, P.P., Hayasaka, T., Usui, S.: Mutation-Based Genetic Neural Network. IEEE Transactions on Neural Networks 16(3), 587–600 (2005)
213. Parks, G., Li, J., Balazs, M., Miller, I.: An empirical investigation of elitism in multiobjective genetic algorithms. Foundations of Computing and Decision Sciences 26(1), 51–74 (2001)
214. Parzen, E.: On the estimation of a probability density function and mode. Annals of Mathematical Statistics 33, 1065–1076 (1962)
215. Potter, M.A., De Jong, K.A.: A cooperative coevolutionary approach to function optimization. In: Proceedings of the Third International Conference on Parallel Problem Solving from Nature, Berlin, Germany, pp. 249–257 (1994)
216. Potter, M.A.: The Design and Analysis of a Computational Model of Cooperative Coevolution, Ph.D Thesis, George Mason University (1997)
217. Potter, M.A., De Jong, K.A.: Cooperative coevolution: An architecture for evolving coadapted subcomponents. Evolutionary Computation 8(1), 1–29 (2000)
218. Paredis, J.: Coevolutionary constraint satisfaction. In: Proceedings of the Third International Conference on Parallel Problem Solving from Nature, pp. 46–55 (1994)
219. Quinlan, J.R.: C4.5: Programs for Machine Learning. Morgan Kaufmann, San Mateo (1993)
220. Rana, S., Whitney, D., Cogswell, R.: Searching in the presence of noise. In: Proceedings of the Fourth International Conference on Parallel Problem Solving from Nature, pp. 198–207. Springer, Heidelberg (1996)
221. Rattray, M., Shapiro, J.: Noisy fitness evaluations in genetic algorithms and the dynamics of learning. In: Belew, R.K., Vose, M.D. (eds.) Foundations of Genetic Algorithms 4, pp. 117–139. Morgan Kaufmann, San Francisco (1997)
222. Ray, T.: Constrained robust optimal design using a multiobjective evolutionary algorithm. In: Proceedings of the 2002 IEEE Congress on Evolutionary Computation, pp. 419–424 (2002)
223. Rechenberg, I.: Evolutionsstrategie, Frommann-Holzboog (1994)
224. Reeves, C.R.: Modern Heuristic Techniques for Combinatorial Problems. Blackwell Scientific Publication, Malden (1993)
225. Rivera, W.: Scalable parallel genetic algorithms. Artificial Intelligence Review 16, 153–168 (2001)
226. Rosin, C.D., Belew, R.K.: New methods for competitive coevolution. Evolutionary Computation 5(1), 1–29 (1997)
227. Rowe, J., Vinsen, K., Marvin, N.: Parallel GAs for Multiobjective Functions. In: Second Nordic Workshop on Genetic Algorithms and Their Applications, pp. 61–70 (1996)
228. Rudolph, G.: A partial order approach to noisy fitness functions. In: Proceedings of the 2001 IEEE Congress on Evolutionary Computation, vol. 1, pp. 318–325 (2001)

229. Rudolph, G., Agapie, A.: Convergence Properties of Some Multi-Objective Evolutionary Algorithms. In: Proceedings of the 2000 Conference on Evolutionary Computation, pp. 1010–1016 (2000)
230. Rudolph, G.: On a Multi-Objective Evolutionary Algorithm and Its Convergence to the Pareto Set. In: Proceedings of the 1998 IEEE Conference on Evolutionary Computation, pp. 511–516 (1998)
231. Sano, Y., Kita, H.: Optimization of noisy fitness functions by means of genetic algorithms using history of search with test of estimation. In: Proceedings of the 2002 IEEE Congress on Evolutionary Computation, vol. 1, pp. 360–365 (2002)
232. Sarker, R., Liang, K., Newton, C.: A New Evolutionary Algorithm for Multiobjective Optimization. European Journal of Operational Research 140(1), 12–23 (2002)
233. Sartori, M.A., Antsaklis, P.J.: A simple method to derive bounds on the size and to train multi-layer neural networks. IEEE Transactions on Neural Networks 2(4), 467–471 (1991)
234. Sato, H., Aguirre, H.E., Tanaka, K.: Enhanced Multi-objective Evolutionary Algorithms Using Local Dominance. In: Proceedings of the 2004 RISP International Workshop on Nonlinear Circuits and Signal Processing, pp. 319–322 (2004)
235. Schaffer, J.D.: Multi-Objective Optimization with Vector Evaluated Genetic Algorithms. In: Proceedings of the First International Conference on Genetic Algorithms, pp. 93–100 (1985)
236. Schaffer, W.M., Zeh, D.W., Buchmann, S.L., Kleinhaus, S., Schaffer, M.V., Antrim, J.: Competition for nectar between introduced honeybees and native North American bees and ants. Ecology 64, 564–577 (1983)
237. Scott, J.R.: Fault Tolerant Design Using Single and Multi-criteria Genetic Algorithms, Master's Thesis, Department of Aeronautics and Astronautics, Massachusetts Institute of Technology (1995)
238. Secomandi, N.: A rollout policy for the vehicle routing problem with stochastic demands. Operations Research 49(5), 796–802 (2001)
239. Secomandi, N.: Comparing neuro-dynamic programming algorithms for the vehicle routing problem with stochastic demands. Computers and Operations Research 27(11-12), 1201–1225 (2000)
240. Sendhoff, B., Beyer, H.-G., Olhofer, M.: On Noise Induced Multi-modality in evolutionary algorithms. In: Wang, L., Tan, K., Furuhashi, T., Kim, K.-H., Sattar, F. (eds.) Recent Advances in Simulated Evolution and Learning, pp. 219–224 (2002)
241. Serre, D.: Matrices: Theory and Applications. Springer, New York (2002)
242. Shannon, C.E.: A mathematical theory of communications. Bell System Technical Journal 27, 379–423 (1948)
243. Shin, S.-Y., Lee, I.-H., Kim, D., Zhang, B.-T.: Multiobjective Evolutionary Optimization of DNA Sequences for Reliable DNA Computing. IEEE Transactions on Evolutionary Computation 9(2), 143–158 (2005)
244. Sim, K.B., Kim, J.Y., Lee, D.W.: Game Model Based Co-evolutionary Solution for Multiobjective Optimization Problems. Internation Journal of Contol, Automation, and Systems 2(2) (2004)
245. Singh, A.: Uncertainty based Multi-objective Optimization of Groundwater Remediation Design, Master's Thesis, University of Illinois at Urbana-Champaign (2003)

246. Solomon, M.M.: Algorithms for the vehicle routing and scheduling problems with time window constraints. Operations Research 35(2), 254–265 (1987)
247. Srinivas, N., Deb, K.: Multiobjective optimization using non-dominated sorting in genetic algorithms. Evolutionary Computation 2(3), 221–248 (1994)
248. Stanley, K., Miikkulainen, R.: Evolving neural networks through augmenting topologies. Evolutionary Computation 10(2), 99–127 (2002)
249. Stewart, G.: Determining Rank in the Presence of Error, Technical Report (TR-92-108) Institute for Advanced Computer Studies (TR-2972) Department of Computer Science, University of Maryland, College Park (October 1992)
250. Stewart, W.R., Golden, B.L.: Stochastic vehicle routing: A comprehensive approach. European Journal of Operational Research 14(4), 371–385 (1983)
251. Stroud, P.D.: Kalman-extended genetic algorithms for search in nonstationary environments with noisy fitness evaluations. IEEE Transactions on Evolutionary Computation 5(1), 66–77 (2001)
252. Tan, K.C., Goh, C.K., Mamun, A.A., Zin, E.E.: An Evolutionary Artificial Immune System for Multi-Objective Optimization. European Journal of Operational Research 187(2), 371–392 (2008)
253. Tan, K.C., Cheong, C.Y., Goh, C.K.: Solving multiobjective vehicle routing problem with stochastic demand via evolutionary computation. European Journal of Operational Research 177, 813–839 (2007)
254. Tan, K.C., Yu, Q., Ang, J.H.: A coevolutionary algorithm for rules discovery in data mining. International Journal of Systems Science 37(12), 835–864 (2006)
255. Tan, K.C., Yang, Y.J., Goh, C.K.: A distributed cooperative coevolutionary algorithm for multiobjective optimization. IEEE Transactions on Evolutionary Computation 10(5), 527–549 (2006)
256. Tan, K.C., Goh, C.K., Yang, Y.J., Lee, T.H.: Evolving better population distribution and exploration in evolutionary multi-objective optimization. European Journal of Operational Research 171(2), 463–495 (2006)
257. Tan, K.C., Lee, T.H., Chew, Y.H., Lee, L.H.: A hybrid multiobjective evolutionary algorithm for solving truck and trailer vehicle routing problems. In: Proceedings of the 2003 IEEE Congress on Evolutionary Computation, vol. 3, pp. 2134–2141 (2003)
258. Tan, K.C., Lee, T.H., Chew, Y.H., Lee, L.H.: A multiobjective evolutionary algorithm for solving vehicle routing problem with time windows. In: Proceedings of the IEEE International Conference on Systems, Man and Cybernetics, vol. 1, pp. 361–366 (2003)
259. Tan, K.C., Khor, E.F., Lee, T.H., Sathikannan, R.: An evolutionary algorithm with advanced goal and priority specification for multiobjective optimization. Journal of Artificial Intelligence Research 18, 183–215 (2003)
260. Tan, K.C., Tay, A., Cai, J.: Design and implementation of a distributed evolutionary computing software. IEEE Transactions on Systems, Man and Cybernetics: Part C 33(3), 325–338 (2003)
261. Tan, K.C., Lee, T.H., Khor, E.F.: Evolutionary algorithms for multi-objective optimization: performance assessments and comparisons. Artificial Intelligence Review 17(4), 251–290 (2002)
262. Tan, K.C., Lee, T.H., Khor, E.F.: Evolutionary algorithms with dynamic population size and local exploration for multiobjective optimization. IEEE Transactions on Evolutionary Computation 5(6), 565–588 (2001)
263. Teo, J., Abbass, H.A.: Multiobjectivity and Complexity in Embodied Cognition. IEEE Transactions on Evolutionary Computation 9(4), 337–360 (2005)

264. Teodorovic, D., Lucic, P.: Intelligent vehicle routing system. In: Proceedings of the IEEE International Conference on Intelligent Transportation Systems, pp. 482–487 (2000)
265. Teodorovic, D., Pavkovic, G.: The fuzzy set theory approach to the vehicle routing problem when demand at nodes is uncertain. Fuzzy Sets and Systems 82(3), 307–317 (1996)
266. Teoh, E.J., Tan, K.C., Xiang, C.: Estimating the number of hidden neurons in a feedforward network using the Singular Value Decomposition. IEEE Transactions on Neural Networks 17(6), 1623–1629 (2006)
267. Teich, J.: Pareto-front exploration with uncertain objectives. In: Proceedings of the First Conference on Evolutionary Multi-Criterion Optimization, pp. 314–328. Springer, Heidelberg (2001)
268. Thompson, A.: Evolutionary techniques for fault tolerance. In: Proceedings of the UKACC International Conference on Control, pp. 693–698 (1996)
269. Thompson, H.A., Fleming, P.J.: An Integrated Multi-Disciplinary Optimisation Environment for Distributed Aero-engine Control System Arhitectures. In: Proceedings of the Fourteenth World Congress of International Federation of Automatic Control, pp. 407–412 (1999)
270. Toffolo, A., Benini, E.: Genetic Diversity as an Objective in Multi-Objective Evolutionary Algorithms. Evolutionary Computation 11(2), 151–167 (2003)
271. Tsutsui, S., Ghosh, A.: A comparative study on the effects of adding perturbations to phenotypic parameters in genetic algorithms with a robust solution searching scheme. In: Proceedings of the 1999 IEEE International Conference on Systems, Man, and Cybernetics, pp. 585–591 (1999)
272. Tsutsui, S., Ghosh, A.: Genetic algorithms with a robust solution searching scheme. IEEE Transactions on Evolutionary Computation 1(3), 201–208
273. Turkcan, A., Akturk, M.S.: A problem space genetic algorithm in multiobjective optimization. Journal of Intelligent Manufacturing 14, 363–378 (2003)
274. Ursem, R.K.: Mutinational GA optimization techniques in dynamic environments. In: Proceedings of the 2000 Genetic and Evolutionary Computation Congress, pp. 19–26 (2000)
275. Van den Bergh, F., Engelbrecht, A.P.: A cooperative approach to particle swarm optimization. IEEE Transactions on Evolutionary Computation 8(3), 225–239 (2004)
276. Venter, G., Haftka, R.T.: A Two Species Genetic Algorithm for Designing Composite Laminates Subjected to Uncertainty. In: Proceedings of the 37th AIAA/ASME/ASCE/AHS/ASC Structures, Structural Dynamics, and Materials Conference, AIAA Paper 96-1535 (1996)
277. Vavak, F., Jukes, K., Fogarty, T.C.: Adaptive combustion balancing in multiple burner boiler using a genetic algorithm with variable range of local search. In: Proceedings of the Seventh International Conference on Genetic Algorithms, pp. 719–726 (1997)
278. Van Veldhuizen, D.A., Zydallis, J.B., Lamont, G.B.: Considerations in engineering parallel multiobjective evolutionary algorithms. IEEE Transactions on Evolutionary Computation 7(2), 144–173 (2003)
279. Van Veldhuizen, D.A., Lamont, G.B.: On measuring multiobjective evolutionary algorithm performance. In: Proceedings of the 2000 IEEE Congress on Evolutionary Computation, vol. 1, pp. 204–211 (2000)
280. Veldhuizen, D.A.V., Lamont, G.B.: Multiobjective Evolutionary Algorithm Test Suites. In: ACM Symposium on Applied Computing, pp. 351–357 (1999)

281. Van Veldhuizen, D.A., Lamont, G.B.: Multiobjective Evolutionary Algorithm Research: A History and Analysis, Technical Report TR-98-03, Department of Electrical and Computer Engineering, Air Force Institute of Technology, Ohio (1998)
282. Verma, B., Ghosh, R.: A novel evolutionary Neural Learning Algorithm. In: Proceedings of the 2002 IEEE Congress on Evolutionary Computation, vol. 2, pp. 1884–1889 (2002)
283. Weyuker, E.J.: Axiomatizing software test data adequacy. IEEE Transactions on Software Engineering 12(12), 1128–1138 (1986)
284. Weicker, N., Szabo, G., Weicker, K., Widmayer, P.: Evolutionary Multiobjective Optimization for Base Station Transmitter Placement with Frequency Assignment. IEEE Transactions on Evolutionary Computation 7(2), 189–203 (2003)
285. Weicker, K.: Performance measures in dynamic environments. In: Guervós, J.J.M., Adamidis, P.A., Beyer, H.-G., Fernández-Villacañas, J.-L., Schwefel, H.-P. (eds.) PPSN 2002. LNCS, vol. 2439, pp. 64–73. Springer, Heidelberg (2002)
286. Wineberg, M., Oppacher, F.: Enhancing the GA's ability to cope with dynamic environments. In: Proceedings of the 2000 Genetic and Evolutionary Computation Congress, pp. 3–10 (2000)
287. Yang, W.H., Mathur, K., Ballou, R.H.: Stochastic vehicle routing problem with restocking. Transportation Science 34, 99–112 (2000)
288. Yao, X.: Evolving Artificial Neural Networks. Proceedings of the IEEE 87(9), 1423–1447 (1999)
289. Yao, X., Liu, Y.: A new evolutionary system for evolving artificial neural networks. IEEE Transactions on Neural Networks 8(3), 694–713 (1997)
290. Yao, X., Liu, Y.: Making use of population information in evolutionary artificial neural networks. IEEE Transaction on Systems, Man, and Cybernetics-Part B: Cybernetics 28, 417–425 (1998)
291. Yen, G.G., Lu, H.: Dynamic Multiobjective Evolutionary Algorithm: Adaptive Cell-Based Rank and Density Estimation. IEEE Transactions on Evolutionary Computation 7(3), 253–274 (2003)
292. Zeng, S.Y., de Garis, H., Kang, L., Liu, Y.: An efficient multi-objective evolutionary algorithm: OMOEA-II. In: Proceedings of the Third International Conference on Evolutionary Multi-Criterion Optimization, pp. 108–119 (2005)
293. Zeng, S.Y., Kang, L., Ding, L.: An Orthogonal Multi-objective Evolutionary Algorithm for multi-objective optimization problems with constraints. Evolutionary Computation 12(1), 77–98 (2006)
294. Zhang, Z.: Multiobjective optimization immune algorithm in dynamic environments and its application to greenhouse control. In: Proceedings of the 2005 IEEE Congress on Evolutionary Computation, pp. 714–719 (2005)
295. Zhang, Q., Zhou, A., Jin, Y.: Modelling the regularity in an estimation of distribution algorithm for continuous multiobjective optimization with variable linkages. IEEE Transactions on Evolutionary Computation (accepted)
296. Zhang, X.-H., Meng, H.-Y., Jiao, L.-C.: Intelligent Particle Swarm Optimization in Multiobjective Optimization. Applied Soft Computing 8, 959–971 (2008)
297. Zhao, Q., Higuchi, T.: Evolutionary learning of nearest-neighbor MLP. IEEE Transactions on Neural Networks 7, 762–767 (1996)

298. Zhao, Q.: Stable on-line evolutionary learning of NN-MLP. IEEE Transactions on Neural Networks 8, 1371–1378 (1997)
299. Zhu, H., Hall, P.A.V., May, J.: Software Unit Test Coverage and Adequacy. ACM Computing Surveys 29(4), 366–427 (1997)
300. Zitzler, E., Kunzli, S.: Indicator-Based Selection in Multiobjective Search. In: Proceedings of the Eighth International Conference on Parallel Problem Solving from Nature, pp. 832–842 (2004)
301. Zitzler, E., Thiele, L., Laumanns, M., Fonseca, C.M., Fonseca, V.G.: Performance assessment of multiobjective optimizers: An analysis and review. IEEE Transactions on Evolutionary Computation 7(2), 117–132 (2003)
302. Zitzler, E., Laumanns, M., Thiele, L.: SPEA2: Improving the Strength Pareto Evolutionary Algorithm, Technical Report 103, Computer Engineering and Networks Laboratory (TIK), Swiss Federal Institute of Technology (ETH) Zurich, Switzerland (2001)
303. Zitzler, E., Deb, K., Thiele, L.: Comparison of multiobjective evolutionary algorithms: empirical results. Evolutionary Computation 8(2), 173–195 (2000)
304. Zitzler, E., Thiele, L.: Multiobjective evolutionary algorithms: a comparative case study and the strength Pareto approach. IEEE Transactions on Evolutionary Computation 3(4), 257–271 (1999)
305. Zitzler: Evolutionary Algorithms for Multiobjective Optimization: Methods and Applications, Ph.D Thesis, Swiss Federal Institute of Technology, Zurich (1999)

FSC
www.fsc.org
MIX
Papier | Fördert gute Waldnutzung
FSC® C083411

Kolibri 360 GmbH
Ferdinand-Jühlke-Straße 7
99095 Erfurt, Deutschland
produktsicherheit@kolibri360.de